T0358965

ADVANCES IN GAS-PHASE ION CHEMISTRY

Volume 3 • 1998

ADVANCES IN GAS-PHASE ION CHEMISTRY

Editors: NIGEL G. ADAMS
LUCIA M. BABCOCK
Department of Chemistry
University of Georgia

VOLUME 3 • 1998

JAI PRESS INC.

Greenwich, Connecticut   *London, England*

Copyright © 1998 by JAI PRESS INC.
55 Old Post Road, No. 2
Greenwich, Connecticut 06836

JAI PRESS LTD.
38 Tavistock Street
Covent Garden
London WC2E 7PB
England

All rights reserved. No part of this publication may be reproduced, stored on a retrieval system, or transmitted in any form, or by any means, electronic, mechanical, photocopying, filming, recording, or otherwise without prior permission in writing from the publisher.

ISBN: 0-7623-0204-6

Printed and bound in the United Kingdom

Transferred to Digital Print 2011

CONTENTS

LIST OF CONTRIBUTORS

Nigel G. Adams

Department of Chemistry
University of Georgia
Athens, Georgia

Arthur T. Blades

Department of Chemistry
University of Alberta
Edmonton, Alberta, Canada

A.W. Castleman, Jr.

Department of Chemistry
The Pennsylvania State University
University Park, Pennsylvania

Neyoka D. Fisher

Department of Chemistry
University of Georgia
Athens, Georgia

Peter W. Harland

Department of Chemistry
University of Canterbury
Christchurch, New Zealand

William L. Hase

Department of Chemistry
Wayne State University
Detroit, Michigan

Eric Herbst

Departments of Physics and Astronomy
The Ohio State University
Columbus, Ohio

Yeunghaw Ho

Department of Chemistry
University of Alberta
Edmonton, Alberta, Canada

R. Johnsen

Department of Physics and Astronomy
University of Pittsburgh
Pittsburgh, Pennsylvania

Paul Kebarle

Department of Chemistry
University of Alberta
Edmonton, Alberta Canada

John S. Klassen

Department of Chemistry
University of Alberta
Edmonton, Alberta, Canada

Kueih-Tzu Lu

Synchrotron Radiation Research Center
Hsinchu, Taiwan, Republic of China

J.B.A. Mitchell

Department of Physics and Astronomy
University of Western Ontario
London, Ontario, Canada; and
PALMS
Université de Rennes I
Rennes, France

Gilles H. Peslherbe

Department of Chemistry
Wayne State University
Detroit, Michigan

Erik C. Richard

National Oceanic and Atmospheric
Administration
Boulder, Colorado

Claire Vallance

Department of Chemistry
University of Canterbury
Christchurch, New Zealand

Robert A. Walker

Department of Chemistry
University of Oregon
Eugene, Oregon

Haobin Wang

Department of Chemistry
Wayne State University
Detroit, Michigan

James C. Weisshaar

Department of Chemistry
University of Wisconsin-Madison
Madison, Wisconsin

PREFACE

Gas-phase ion chemistry is a broad field which has many applications and which encompasses various branches of chemistry and physics. An application that draws together many of these branches is the synthesis of molecules in interstellar clouds (Herbst). This was part of the motivation for studies on the neutralization of ions by electrons (Johnsen and Mitchell) and on isomerization in ion-neutral associations (Adams and Fisher). The results of investigations of particular aspects of ion dynamics are presented in these association studies, in studies of the intermediates of binary ion–molecule S_N2 reactions (Hase, Wang, and Peslherbe), and in those of excited states of ions and their associated neutrals (Richard, Lu, Walker, and Weisshaar). Solvation in ion–molecule reactions is discussed (Castleman) and extended to include multiply charged ions by the application of electrospray techniques (Klassen, Ho, Blades, and Kebarle). These studies also provide a wealth of information on reaction thermodynamics which is critical in determining reaction spontaneity and availability of reaction channels. More focused studies relating to the ionization process and its nature are presented in the final chapter (Harland and Vallance).

The articles collected here represent only a subset of the advances that are being made in this rapidly developing area. Successive volumes will emphasize the progress being made in other areas of the subject. The editors are making a specific effort to include contributions from those relatively new to the field, who bring in new ideas and perspectives, as well as those more established in the field who also bring their wealth of experience. Throughout each volume, a blend will be sought among experiment, theory, and applications so that the reader, in addition to having up-to-date information on developments in forefront areas, will have a broad base from which to view the subject as a whole. To edit such a series gives the opportunity and privilege to look into the laboratories and the minds of the scientists who are advancing our understanding in the research area of gas-phase ion chemistry, a privilege which, through these volumes, is passed on to our fellow scientists.

Nigel G. Adams
Lucia M. Babcock
Editors

ION–MOLECULE CHEMISTRY IN INTERSTELLAR CLOUDS:

SUCCESSES AND PROBLEMS

Eric Herbst

Advances in Gas-Phase Ion Chemistry
Volume 3, pages 1–47
Copyright © 1998 by JAI Press Inc.
All rights of reproduction in any form reserved.
ISBN: 0-7623-0204-6

ABSTRACT

High resolution observations have uncovered the existence of more than 110 different molecules, ranging up to 13 atoms in size, in the gas phase of dense interstellar clouds. Ion–molecule reactions play a leading role in the gas-phase chemistry of these clouds, which are giant accumulations of cold gas and dust located in between stars in our galaxy and others. In this review, we first discuss the astronomical observations of dense interstellar clouds. We then discuss the basic ion–molecule chemistry and the role it plays in detailed chemical models. A review of both the successes and limitations of the models follows. Some outstanding chemical problems in the field are highlighted. The possible synthesis of molecules much larger than observed at high resolution is explored, and the changes in interstellar chemistry induced by star formation are considered.

I. INTRODUCTION

Within the last 25 years, it has become apparent that there is a rich chemistry in interstellar space. Starting in the 1960s, radioastronomers turned their telescopes towards rather dark regions of interstellar space known as dense interstellar clouds, or dark nebula, and, to the surprise of most, began to find large numbers of discrete spectral lines, mainly in emission.[1,2] The spectral lines were soon assigned as rotational spectral features of simple gas-phase molecules. From these spectral features, the identities and abundances of the molecules could be deduced as well as the physical conditions necessary to generate the lines in emission.[1,2] Although dense interstellar clouds are certainly not homogeneous, the standard physical conditions determined by the observations are a gas density n of 10^{3-5} cm^{-3} and a temperature T of 10–50 K.

At the current time, more than 110 different molecules have been detected via high resolution spectral techniques in interstellar clouds and in the somewhat similar expanding circumstellar clouds surrounding old stars. Table 1 contains a list of these molecules; they range in complexity from molecular hydrogen to a 13-atom linear nitrile ($HC_{10}CN$) and encompass both normal and highly unusual species in a terrestrial sense. Determining the unusual carriers of some of the spectral transitions has proved a challenging task for spectroscopists.[1] Molecular hydrogen is by far the most abundant species; the second most abundant molecule, carbon

Table 1. Gas Phase Interstellar Molecules in Order of Increasing Complexity

H_2	$AlCl$	HCO	NH_3	CH_4	CH_3OH	HC_4CN
CH	AlF	HCO^+	H_3O^+	SiH_4	CH_3SH	C_7H
CH^+	PN	HOC^+	H_2CO	CH_2NH	C_2H_4	$H_2C_6(lin)$
NH	SiN	HN_2^+	H_2CS	$H_2C_3(lin)$	CH_3CN	$HCOOCH_3$
OH	SiO	HNO	$HCCH$	$c\text{-}C_3H_2$	CH_3NC	CH_3COOH
C_2	SiS	HCS^+	$HCNH^+$	CH_2CN	HC_2CHO	CH_3C_2CN
CN	CO^+	C_3	H_2CN	NH_2CN	NH_2CHO	C_8H
CO	SO^+	C_2O	$C_3H(lin)$	CH_2CO	HC_3NH^+	$(CH_3)_2O$
CSi	H_3^+	C_2S	$c\text{-}C_3H$	$HCOOH$	$H_2C_4(lin)$	C_2H_5OH
CP	CH_2	SiC_2	$HCCN$	C_4H	C_5H	C_2H_5CN
CS	NH_2	SO_2	$HNCO$	HC_2CN	CH_3NH_2	CH_3C_4H
NO	H_2O	OCS	$HOCO^+$	$HCCNC$	CH_3CCH	HC_6CN
NS	H_2S	$MgNC$	$HNCS$	$HNCCC$	CH_3CHO	$(CH_3)_2CO$
SO	C_2H	$NaCN$	C_2CN	H_2COH^+	CH_2CHCN	$CH_3C_4CN?$
HCl	HCN	$MgCN$	C_3O	C_4Si	CH_2OCH_2	$NH_2CH_2COOH?$
$NaCl$	HNC	N_2O	C_3S	C_5	C_6H	HC_8CN
KCl		$+$	ISOTOPOMERS		....	$HC_{10}CN$

Notes: Question marks denote ambiguity. The designation "lin" means a linear molecule, whereas the prefix "c-" stands for a cyclic molecule. The many isotopomers are not listed.

monoxide, has a fractional abundance with respect to H_2 of 10^{-4} by number. Despite its lower abundance, the existence of a permanent dipole on CO makes it the easier molecule to detect and it is the standard species radioastronomers use to map out the details of the dense interstellar medium.[1,2] The more complex molecules have fractional abundances as low as 10^{-10}, while the most abundant molecular ion, HCO^+, has a fractional abundance of 10^{-8}. These concentrations reflect both chemistry and the relative proportions of the elements. Typical elemental abundances, as determined from stellar spectra, are dominated by hydrogen. By number with respect to hydrogen, helium lies at the 10% level while the elements C, N, and O lie at the 10^{-3}–10^{-4} level, with oxygen exceeding carbon in most stars.[3] Cloud-to-cloud and intracloud variations in molecular concentrations do exist, with dramatic changes especially in the areas around newly forming stars, known as "protostars," born as part of the material in a cloud collapses and heats up, or in the areas around the hot and ionized gas surrounding bright young stars that have already formed.[4] Although most of the molecules listed in Table 1 have been detected by radioastronomers working in the microwave and millimeter-wave regions of the electromagnetic spectrum, some molecules have been detected in the visible and infrared regions. The visible detections[1] have been made in absorption in rather diffuse clouds, through which background stellar radiation can mainly penetrate, while the infrared detections have been made in absorption against both background stars, and stars and protostars embedded in dense clouds.[5]

The darkness associated with dense interstellar clouds is caused by dust particles of size ≈0.1 microns, which are a common ingredient in interstellar and circumstellar space, taking up perhaps 1% of the mass of interstellar clouds with a fractional number density of 10^{-12}. These particles both scatter and absorb external visible and ultraviolet radiation from stars, protecting molecules in dense clouds from direct photodissociation via external starlight. They are rather less protective in the infrared, and are quite transparent in the microwave.[6] The chemical nature of the dust particles is not easy to ascertain compared with the chemical nature of the interstellar gas; broad spectral features in the infrared have been interpreted in terms of core-mantle particles, with the cores consisting of two populations, one of silicates and one of carbonaceous, possibly graphitic material. The mantles, which appear to be restricted to dense clouds, are probably a mixture of ices such as water, carbon monoxide, and methanol.[7]

Broad spectral features in both the infrared and the visible have been assigned to molecules far more complex than those listed in Table 1. In the infrared, a series of previously unidentified emission lines at 3050, 1610, 1300, 1150, and 885 cm^{-1} has been assigned to a class of molecules known as polycyclic aromatic hydrocarbons (PAHs) based on a functional group analysis rather than a detailed comparison between interstellar and laboratory spectra for specific species.[8] The lines are found near bright young stars, such as exist in the well-known bar region of the Orion nebula, which are thought to pump the infrared emission via internal conversion from excited electronic states to highly excited vibrational levels of the ground electronic state. Another class of broad interstellar spectral lines is found in the visible, in diffuse matter along the line of sight to background stars. These lines, seen mainly in absorption, are labeled the diffuse interstellar bands (DIBs) for their spectral width rather than their environment. They have been known for more than 60 years, and now number more than 150. Although no definitive assignments have been made, speculation now centers on complex molecules such as long carbon chains, fullerenes, and fulleranes, which would have to exist in a harsh environment pervaded by ultraviolet radiation.[9]

Detailed radioastronomical studies indicate that, even for regions far removed from areas of star formation and young stars, dense interstellar clouds are far from homogeneous. Perhaps the simplest physical picture is obtained by imagining clouds to consist of spherical shells of material, with the material getting denser and colder as one traverses towards the center. The outer shells are less rich in molecules and are not able to extinguish the flux of external stellar radiation, whereas the inner shells, known as the interior or core, are richer in molecules and exclude most external radiation. A small degree of ionization occurs throughout the cloud as a result of a small flux of cosmic rays.[10] This "onion" model of dense clouds is simplistic, however. Not only are clouds not spherical, the material in them tends to be clumpy and even possibly fractal in nature.[11] Small condensations exist on many distance scales, ranging down to amounts of material smaller than 1 solar mass. For example, the well-studied TMC-1 (Taurus Molecular Cloud 1) condensation in the Taurus molecular clouds probably has 4–6 cloudlets within it. In addition to this clumpiness, some condensations collapse to form single stars or groups of stars, although other condensations show more stability against gravitation. Certain interstellar clouds, such as the Taurus region, show evidence only for the inefficient formation of low mass stars, whereas other clouds, such as the Orion nebula, show evidence for efficient formation of high mass stars.

II. BASIC ION–MOLECULE CHEMISTRY

A. Formation of Interstellar Molecules

The first question to ask about the formation of interstellar molecules is where the formation occurs. There are two possibilities: the molecules are formed within the clouds themselves or they are formed elsewhere. As an alternative to local formation, one possibility is that the molecules are synthesized in the expanding envelopes of old stars, previously referred to as circumstellar clouds. Both molecules and dust particles are known to form in such objects, and molecular development is especially efficient in those objects that are carbon-rich (elemental C > elemental O) such as the well-studied source IRC+10216.[12] Chemical models of carbon-rich envelopes show that acetylene is produced under high-temperature thermodynamic equilibrium conditions and that as the material cools and flows out of the star, a chemistry somewhat akin to an acetylene discharge takes place, perhaps even forming molecules as complex as PAHs.[13,14] As to the contribution of such chemistry to the interstellar medium, however, all but the very large species will be photodissociated rapidly by the radiation field present in interstellar space once the molecules are blown out of the protective cocoon of the stellar envelope in which they are formed. Consequently, the material flowing out into space will consist mainly of atoms, dust particles, and possibly PAHs that are relatively immune to radiation because of their size and stability. It is therefore necessary for the observed interstellar molecules to be produced locally.

Starting with a cloud formed by the influence of gravity on stellar atomic and particulate ejecta, how can molecules be produced? The chemistry is constrained by the low temperatures and gas densities. The low densities eliminate the importance of three-body processes while the low temperatures eliminate endothermic reactions and even exothermic reactions with significant activation energy barriers. The vast majority of ion–molecule reactions are known to have no activation energy barriers and to proceed rapidly at low temperatures.[15] Indeed, reactions involving a polar neutral occur more rapidly at lower temperatures than at room temperature, in agreement with capture theory predictions.[16] Activation energy barriers are the norm in neutral–neutral reactions, the major exceptions being reactions between atoms or radicals.[17] Until recently, it was therefore thought that ion–molecule reactions dominate interstellar chemistry. Recent work from several laboratories on neutral reactions at room temperature and below involving one atom or radical and one stable molecule show that at least some such reactions can be quite rapid.[18] The extent of the contribution of neutral–neutral reactions to interstellar chemistry is currently uncertain.

In addition to gas-phase processes, interstellar chemistry can occur on the surface of dust grains. Although both chemisorption and physisorption can be envisaged on bare grains, once an ice mantle builds up, gas-phase molecules will bind only weakly to the surface. Most discussions of grain chemistry invoke the classical Langmuir–Hinshelwood mechanism, in which reaction occurs when two weakly bound migrating species collide with one another.[19] The alternative Eley–Rideal mechanism, in which a gas-phase species collides with an adsorbed molecule, has been relatively neglected.[20] In the absence of strong catalytic surfaces, low temperature reactions on grains will involve mainly those pairs of reactants that do not require large amounts of activation energy.[21]

The first step in interstellar chemistry is the production of diatomic molecules, notably molecular hydrogen. Observations of atomic hydrogen in dense clouds show that this species cannot be detected except in a diffuse halo surrounding the cloud, so that an efficient conversion of H into H_2 is necessary. In the gas phase this might be accomplished by the radiative association reaction,

$$H + H \rightarrow H_2 + h\nu \tag{1}$$

except that it is calculated to be far too inefficient. A competitive ionic process,

$$H + e^- \rightarrow H^- + h\nu \tag{2a}$$

$$H^- + H \rightarrow H_2 + e^- \tag{2b}$$

is also too inefficient.[19] The alternative of H_2 association on grains is generally believed to be the dominant mechanism, with the excess energy going into the phonon modes of the grain.[19] The molecular hydrogen can then evaporate rather readily into the gas phase even at the low prevailing temperatures. Other chemical processes on grain surfaces also occur, especially involving rapidly migrating and reactive hydrogen atoms.[21,22] For example, a rapidly moving hydrogen atom can associate with a more slowly moving oxygen atom to form an OH radical. But the heavy species produced will not evaporate into the gas and the extent of nonthermal desorption is unclear.[23]

Once a significant amount of molecular hydrogen is produced, a rich gas-phase chemistry ensues.[24] Ion–molecule processes are initiated in the interiors of dense clouds mainly via cosmic ray ionization, the most important reaction being,

$$H_2 + \text{cosmic ray} \rightarrow H_2^+ + e^- + \text{cosmic ray} \tag{3}$$

which occurs with a first order rate $\zeta \approx 10^{-17}$ s^{-1} determined via integration of the cross section for ionization by high energy particles with the known cosmic ray flux in the local neighborhood.[10] The value of ζ may be different in external galaxies since cosmic rays are formed in supernovae, and the supernova rate is not thought to be the same in all galaxies. Other product channels ($H + H$; $H + H^+ + e^-$) are less important. As to direct ionization of atoms via cosmic rays, the only major process is the ionization of helium; other atoms are ionized more efficiently by chemical reactions discussed below.

The hydrogen molecular ion is rapidly (within a day at a standard gas density n of 10^4 cm^{-3}) converted to H_3^+ via the well-studied reaction:

$$H_2^+ + H_2 \rightarrow H_3^+ + H \tag{4}$$

The H_3^+ ion, recently detected in the interstellar medium via infrared transitions,[25] can subsequently react with a variety of neutral atoms present in the gas. The reaction with oxygen leads to a chain of reactions that rapidly produce the hydronium ion H_3O^+ via well-studied H atom-transfer reactions:

$$O + H_3^+ \rightarrow OH^+ + H_2 \tag{5}$$

$$OH^+ + H_2 \rightarrow H_2O^+ + H \tag{6}$$

$$H_2O^+ + H_2 \rightarrow H_3O^+ + H \tag{7}$$

Since the hydronium ion does not react with molecular hydrogen, its abundance can be sizeable and, indeed, it has been detected in dense clouds (see Table 1). The destruction of H_3O^+ occurs mainly via dissociative recombination with electrons, a process known to be rapid ($k \approx 10^{-6} - 10^{-7}$ cm^3 s^{-1}) for most if not all polyatomic ions but for which little information concerning the neutral fragments is generally available (the recombination process is discussed by Johnsen and Mitchell, this volume). In recent years, two experimental techniques have been brought to bear on the problem—flowing afterflow[26] and storage ring methods,[27] which are discussed later. For $H_3O^+ + e^-$, the two methods have both been utilized but are only in partial agreement; both yield the branching fraction leading to OH to be ≈ 0.65, with the flowing afterglow results showing the OH channels equally subdivided into atomic (2H) and molecular hydrogen co-products, and the storage ring results showing the OH + 2H channel to be dominant.[28,29] A far more serious discrepancy concerns the other channels; the flowing afterglow result is that the remaining 0.35 is overwhelmingly $O + H + H_2$, while the storage ring result is that it is overwhelmingly $H_2O + H$.[28,29] Most interstellar models contain the latter assumption.

If the H_3^+ ion reacts with atomic carbon, an analogous series of reactions leads to the methyl ion, CH_3^+, although the initial reaction to form CH^+ has not been studied in the laboratory. The methyl ion does not react rapidly with H_2 but does undergo a relatively slow radiative association reaction,

$$CH_3^+ + H_2 \rightarrow CH_5^+ + h\nu \tag{8}$$

which has been studied in the laboratory at low density and found to have a rate coefficient of $10^{-13} - 10^{-14}$ cm^3 s^{-1} at temperatures under 100 K,[30] in good agreement with theoretical estimates using statistical approximations.[31] A rate coefficient of this size, implying that one in 10^{4-5} strong collisions leads to sticking, is still critical given that H$_2$ is 4 orders of magnitude more abundant than any heavy species. Dissociative recombination of CH$_3^+$ into CH$_2$ and CH is also competitive.[29] Another, but slower, radiative association of some importance is the reaction,

$$C^+ + H_2 \rightarrow CH_2^+ + h\nu \tag{9}$$

since the initial form of carbon, an easily ionizable atom, may well be C$^+$, and since the H-atom transfer reaction between C$^+$ and H$_2$ is quite endothermic. Reaction 9 has also been studied in the laboratory at low temperature and an upper limit to the rate coefficient found to be in reasonable agreement with theory.[30,31]

The dissociative recombination of CH$_5^+$ leads to an uncertain mixture of products such as methane and the methyl and methylene radicals:

$$CH_5^+ + e^- \rightarrow CH_4 + H \tag{10a}$$

$$\rightarrow CH_3 + 2H(H_2) \tag{10b}$$

$$\rightarrow CH_2 + H_2 + H \tag{10c}$$

Although theoretical estimates for neutral branching fractions of reactions such as 10 have been attempted, their reliability is suspect.[32] In the absence of measurement, modelers have typically assumed equal branching fractions between channels in which one and two hydrogen atoms are separated from the molecular skeleton (e.g. 0.50 for reactions 10a and 10b). New assumptions based on recent experimental work have also been made.[33]

The initial nitrogen chemistry has until recently been thought to be somewhat different because the reaction,

$$N + H_3^+ \rightarrow NH^+ + H_2 \tag{11}$$

is endothermic and the exothermic reaction leading to the alternative products NH$_2^+$ + H calculated not to occur.[34] Experimental work by Scott et al.[35] has found, in contradiction with the theoretical result, that these latter products are produced with a standard ion–atom rate coefficient k at room temperature of 4.5×10^{-10} cm^3 s^{-1}. Two H-atom transfer reactions with H$_2$ then convert NH$_2^+$ into NH$_4^+$, a precursor to ammonia as well as NH$_2$, a detected interstellar radical. The reaction,

$$NH_3^+ + H_2 \rightarrow NH_4^+ + H \qquad (12)$$

has a small activation energy barrier and a measured rate coefficient that decreases as the temperature is reduced below 300 K. At still lower temperatures, the rate begins to increase, however, due, it would appear, to tunneling under the activation energy barrier, according to a phase–space statistical calculation with a one-dimensional tunneling algorithm added.[36]

The older approach to the formation of nitrogen-containing compounds starts with the neutral–neutral chemistry leading to molecular nitrogen:

$$N + OH \rightarrow NO + H \qquad (13)$$

$$N + NO \rightarrow N_2 + O \qquad (14)$$

Molecular oxygen is formed analogously via,

$$O + OH \rightarrow O_2 + H \qquad (15)$$

whereas CO can be formed via an assortment of neutral–neutral reactions:

$$C + OH \rightarrow CO + H \qquad (16)$$

$$C + O_2 \rightarrow CO + O \qquad (17)$$

$$O + CH_2 \rightarrow CO + 2H \qquad (18)$$

Atom–radical chemistry also leads to the formation of the radical CN via processes such as:

$$C + NO \rightarrow CN + O \qquad (19)$$

Only some of the atom–radical reactions invoked here have been studied in the laboratory.

After CO is formed, the abundant formyl ion can be produced via the reaction:

$$CO + H_3^+ \rightarrow HCO^+ + H_2 \qquad (20)$$

The reactants in 20 can also produce the higher energy isomer HOC^+, which is detected in dense interstellar clouds,[37] albeit with a lower abundance than HCO^+ due mainly to the "catalytic" reaction,

$$HOC^+ + H_2 \rightarrow HCO^+ + H_2 \qquad (21)$$

which occurs at a non-negligible rate at low temperature according to a phase–space calculation due to tunneling through an activation energy barrier,[38] in analogy with reaction 12. A variety of other protonated ions can be generated via analogous reactions.

Once formed, molecular nitrogen can be broken apart by reaction with He^+ (generated from He via cosmic ray ionization),

$$He^+ + N_2 \rightarrow N^+ + N + He \qquad (22)$$

and the N^+ can undergo two H-atom transfer reactions with H_2 leading to NH_2^+, which is also produced by the reaction between N and H_3^+. The first of the two reactions,

$$N^+ + H_2 \rightarrow NH^+ + H \qquad (23)$$

is very slightly endothermic but occurs rapidly anyway due to residual nonthermal excitations in both reactants.[39]

The ion He^+ reacts with a variety of neutral species other than N_2 to break them up. One notable example is,

$$He^+ + CO \rightarrow C^+ + O + He \qquad (24)$$

which produces the carbon ion.

Another variation in the initial chemistry occurs for the case of sulfur. The reaction between S and H_3^+ to form SH^+ and H_2 is exothermic although it has not been studied in the laboratory. The SH^+ product does not react exothermically with H_2 and only a rather slow radiative association, in which an electronic spin flip must occur, can produce H_3S^+, the precursor to HS and H_2S.[40]

The above examples should suffice to show how ion–molecule, dissociative recombination, and neutral–neutral reactions combine to form a variety of small species. Once neutral species are produced, they are destroyed by ion–molecule and neutral–neutral reactions. Stable species such as water and ammonia are depleted only via ion–molecule reactions. The dominant reactive ions in model calculations are the species HCO^+, H_3^+, H_3O^+, He^+, C^+, and H^+; many of their reactions have been studied in the laboratory.[41] Radicals such as OH can also be depleted via neutral–neutral reactions with atoms (see reactions 13, 15, 16) and, according to recent measurements, by selected reactions with stable species as well.[18] Another loss mechanism in interstellar clouds is adsorption onto dust particles. Still another is photodestruction caused by ultraviolet photons produced when secondary electrons from cosmic ray-induced ionization excite H_2, which subsequently fluoresces.[42]

B. Formation of Larger Species

The list of molecules in Table 1 contains species with as many as 13 atoms. Ion–molecule dominated syntheses for many of these species have been considered although many of the critical reactions have not been measured in the laboratory.

Perhaps the synthesis of hydrocarbons is best understood. There are three main classes of reactions leading to complex hydrocarbons: carbon insertion, condensation, and radiative association. Carbon insertion reactions are between C^+ ions and smaller hydrocarbon neutrals; viz.,

$$C^+ + CH_4 \rightarrow C_2H_3^+ + H \tag{25a}$$

$$\rightarrow C_2H_2^+ + H_2 \tag{25b}$$

and form larger hydrocarbon ions with one or more hydrogen atoms removed. A large number of these reactions have been studied in the laboratory.[41] An analogous class of reactions, included in interstellar models but not yet studied in the laboratory, is thought to occur between neutral carbon atoms and hydrocarbon ions. Although the dominant mechanism for producing complex species through perhaps 10 atoms in size, carbon insertion reactions require hydrogenation to keep the synthesis going; otherwise only cluster ions will form eventually and these will recombine to form smaller neutral clusters that cannot grow via carbon insertion. Hydrogenation proceeds mainly in the gas phase via H-atom transfer reactions involving H_2; such reactions with hydrocarbon ions containing more than one carbon atom do not generally occur rapidly to form saturated ions, at least at room temperature.[43] Consider the case of two- and three-carbon hydrocarbon ions. The acetylene ion may or may not react with H_2 at low temperatures to form $C_2H_3^+ + H$ since the reaction may be slightly endothermic.[44] Association to form $C_2H_4^+$ does occur with a measured rate coefficient at 80 K of 7×10^{-14} cm^3 s^{-1}.[30] The $C_2H_3^+$ and $C_2H_4^+$ ions definitely do not react with H_2; rather, they combine with electrons to form species such as C_2H_3, acetylene, and CCH. The carbon insertion reaction involving acetylene,[41]

$$C^+ + C_2H_2 \rightarrow C_3H^+ + H \tag{26}$$

leads to the production of an ion which can react with H_2 to form two isomeric forms of the $C_3H_3^+$ ion by association (see, Adams and Fisher, this volume), and possibly to form $C_3H_2^+ + H$, although once again the exothermicity of the H-atom transfer process is in dispute.[30,44] Hydrogenation beyond $C_3H_3^+$ does not occur so that dissociative recombination will only produce species as saturated as C_3H_2.

More saturated species can be formed via well-studied condensation reactions between hydrocarbon ions and neutrals; viz.,

$$CH_3^+ + CH_4 \rightarrow C_2H_5^+ + H_2 \tag{27}$$

followed by recombination reactions, although such processes are generally not as efficient as carbon insertion due to the relatively high abundances of neutral and ionized carbon. The final mechanism involves radiative association between heavy species, which is expected to become more efficient as the reactants become larger in size, an expectation borne out by a variety of recent experiments.[30] In particular, a synthesis of bare carbon cluster ions via the process,

$$C^+ + C_n \rightarrow C_{n+1}^+ + h\nu \tag{28}$$

may well operate in the interstellar medium,[45] although competitive product channels exist.

Neutral–neutral reactions are also involved in the synthesis of hydrocarbons, but here the evidence is less clear since, even for those systems studied in the laboratory, reaction products are rarely available. Unlike reactions involving O atoms, those involving C atoms and unsaturated hydrocarbons appear to be rapid, at least at room temperature and above.[46,47] If the products of these reactions are analogous to ion–molecule insertion reactions, they can lead to molecular synthesis; for example:

$$C + C_2H_2 \rightarrow C_3H + H \tag{29}$$

Recent studies with a crossed-beam apparatus not only show that the products shown above are the correct ones, but that both the linear and cyclic isomers, each of which is a detected interstellar molecule, are formed.[47] Crossed-beam studies also show that other reactions between C atoms and unsaturated hydrocarbons proceed to form similar products.[48]

Radical hydrocarbon-stable hydrocarbon reactions also play a role in synthesis; the reaction,

$$C_2H + C_2H_2 \rightarrow C_4H_2 + H \tag{30}$$

is known to occur without activation energy at room temperature.[17] As discussed in Section IV.D, the role of neutral–neutral reactions may on balance be inimical to synthesis under oxygen-rich conditions, due mainly to reactions between oxygen atoms and radicals. At present, too few reactions at low temperature have been studied for modelers to generalize about which systems are likely to be rapid at low temperatures and what the likely products will be. In addition, the unusual temperature dependence observed for selected reactions between room temperature and very low temperature, in which the rate coefficient actually increases with decreasing temperature, is not in apparent agreement with sophisticated capture model calculations,[49] suggesting that these systems cannot be understood in the same way as ion–molecule reactions except at very low temperatures.[50]

The formation of the cyanopolyynes ($HC_{2n}CN$) is closely tied to the synthesis of hydrocarbons. Both hydrocarbon ion-atomic nitrogen reactions;[41] viz.,

$$N + C_3H_3^+ \rightarrow H_2C_3N^+ + H \tag{31}$$

and CN–hydrocarbon reactions[51]; viz.,

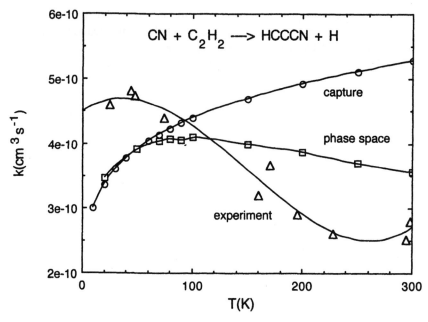

Figure 1. The measured and theoretical rate coefficients for the neutral–neutral reaction between CN and C$_2$H$_2$ are shown vs. temperature. The theoretical values include results obtained via the phase-space approach[52] and via a simple capture theory[49] while the experimental points are from Sims et al.[51] Reproduced with permission from *The Astrophysical Journal*, by the American Astronomical Society.

$$CN + C_2H_2 \rightarrow HC_2CN + H \tag{32}$$

appear to be involved. Reaction 32 has been investigated both in the laboratory[51] and theoretically, with results shown in Figure 1.[52] The theoretical investigation of Woon and Herbst uses a phase–space approach to the dynamics. The unusual temperature dependence—in which the rate coefficient first increases with decreasing temperature and then begins to decrease as the temperature goes below 50 K—is adequately reproduced by the theory, although the phase–space technique, designed for ion–molecule reactions, may not be appropriate for weak long-range forces.[50] In the theoretical approach, the temperature dependence above 50 K is influenced by a small barrier in the exit channel of the reaction. A simple capture theory[49] cannot reproduce this dependence, as shown in Figure 1. Interestingly, only ion–molecule reactions can be invoked in the synthesis of the detected interstellar isomers of HC$_2$CN–HNCCC and HCCNC;[52] these species are of considerably lower abundance than is HC$_2$CN.

The production of oxygen-containing organic molecules is not as well understood. Although methanol appears to be produced at least partially by the radiative association between $CH_3^+ + H_2O$ followed by dissociative recombination, analogous radiative association-based syntheses of more complex, rather saturated species, such as ethanol, acetaldehyde, and methyl formate, are now calculated to be too inefficient to explain observations.[53] The observational evidence for some of these species suggests that they are not produced in the gas of quiescent interstellar clouds, but rather only in so-called hot cores (see Section VI.A). In these warm regions found in the vicinity of star formation activity, the oxygen-containing organic molecules are synthesized by reactions following the desorption of precursor species such as methanol formed previously on grain mantles.[54]

III. GAS-PHASE CHEMICAL MODELS

The construction of gas-phase chemical models of dense interstellar clouds first involves the incorporation of the gas-phase chemical reactions thought to be important in both the formation and destruction of interstellar molecules. The only heterogeneous process considered is the formation of H_2. Despite the great effort of ion–molecule groups such as those headed by Smith, Adams, Bohme, McEwan, Gerlich, Anicich, and others to study reactions of importance to astrochemists, current models are largely composed of reactions that have not been studied in the laboratory, especially not at low temperatures. One reason is the relatively large abundances of atoms and radicals in the interstellar medium. Rate coefficients for unstudied ion–molecule systems are normally estimated based on the Langevin model or, for ion–polar neutral systems, on parameterized results of capture models.[55] Product branching fractions are generally estimated from analogous studied reactions. Rate coefficients for dissociative recombination reactions are known to lie within a very flat range,[56] but product branching fractions must for the most part be estimated despite some measurements.[26-29] For the most part, the results are not particularly sensitive to differing assumptions.[33]

The molecules considered in typical models range in size through 10–12 atoms, which is near the limit for molecules observed with high resolution techniques (see Table 1). The most critical elements in the models are H, He, C, N, and O, although the additional elements S, Si, P, Na, Mg, and Fe are often considered. The elemental abundances utilized vary because it is still uncertain what percentage of elements heavier than helium are found in the solid state in the cores of and on the mantles of dust particles. A standard approach is to use so-called "low metal" abundances in which the stellar abundances of C, N, and O are reduced by modest amounts to the levels of 7.3×10^{-5}, 1.76×10^{-4}, and 2.14×10^{-5}, respectively, with respect to the total hydrogen abundance, while the stellar abundances of the elements S, Si, Na, Mg, and Fe are depleted by more than 2 orders of magnitude.[57] This approach can only be partially justified by observations; it does, however, result in the best agreement between model results and observational results for molecular abundances. Note that astronomers do not really understand what the word "metal" means; most of these authors view the term to mean any element heavier than helium!

In the simplest models, the differential rate equations are solved for a homogeneous source representing the cloud interior in the steady-state limit as a function of gas density and temperature. Although astronomers often use the terms "thermodynamic equilibrium" and "steady state" interchangeably, the two are very different in interstellar clouds. Steady-state abundances are reached when kinetic rates of formation and depletion are equal; these can differ by many orders of magnitude from abundances determined by minimizing the free energy. Essentially, the activation energy barriers which preclude most neutral–neutral reactions from occurring make a kinetic solution necessary. At higher temperatures, such as those that pertain in stellar atmospheres, chemical thermodynamic equilibrium does indeed pertain.

Steady-state models are still regarded as appropriate for so-called diffuse clouds, where low densities and high radiation fields combine to reduce the chemistry to that of atoms and selected diatomic molecules.[58] For dense interstellar clouds, the steady-state approach used originally[59] has been superseded by the so-called pseudo-time-dependent approach[24,60] in which physical conditions (gas density, temperature) are fixed but chemical abundances vary as a function of time from a given set of initial abundances. In the choice of initial abundances, it is customarily assumed that hydrogen is in molecular form and the heavy elements are all in atomic form. Molecular hydrogen can reach a high relative abundance before a typical cloud has had much of a chance to contract to form large densities in the interior. The major reason is that the photodissociation of H_2 occurs for stellar photons ($E < 13.6$ eV) only through line absorption rather than continuous absorption; once the radiation field is depleted in the line frequencies, the molecular hydrogen is self-shielded.[61] It therefore makes sense physically to use H_2 rather than H as the initial material. With given initial abundances, the differential equations representing the formation and loss rates for each species are integrated until steady-state conditions are reached. This takes an astronomical (but fortunately not computational!) time of $\approx 10^7$ years.

The efficiency of the Gear algorithm for solving coupled and "stiff" differential equations allows large numbers of species to be included in pseudo-time-dependent models; our models currently encompass $\approx$4000 reactions involving $\approx$400 species.[62] A model of similar size has been published by the UMIST group.[63] Although not discussed here, more advanced models exist in which the change in physical conditions and/or the inhomogeneity of the material are taken into account with some degree of success.[64,65] Slab models, in which the inhomogeneities are treated as static, have become fairly common.[65] The change in physical conditions is especially important in regions where stars are in the act of forming; chemical models pertaining to such regions are considered in Section VI.

Before reviewing the results of pseudo-time-dependent models of homogeneous cloud interiors in detail, it is worthwhile to mention five successful predictions/explanations in general terms. One successful prediction lies in the calculation of significant abundances of molecular ions, especially protonated ones. The observation of protonated ions such as HCO^+, HN_2^+, HCS^+, H_3O^+, $HCNH^+$, and HC_2CNH^+ bears out this prediction, in most cases quantitatively. Model results show the overall fractional ionization to be in the vicinity of 10^{-7}–10^{-8}, so that the source of ionization—cosmic ray bombardment—does not cause a high degree of ionization. The fractional ionization is higher if metallic elements such as Na are not assumed to be depleted from the gas.

A second successful prediction is that many so-called "metastable" species (i.e. isomers) are abundant even if they are quite reactive in the laboratory.[66] Perhaps the simplest interstellar molecule in this class is HNC, but large numbers of others can be seen in Table 1. It is assumed that most metastable species are formed in dissociative recombination reactions along with their stable counterparts at approximately equal rates, and that both are destroyed by ion–molecule reactions so that the laboratory reactivity, which is normally determined by reactions with neutral species, is irrelevant. Both HCN and HNC, for example, are thought to derive from the dissociative recombination reaction involving a linear precursor ion:

$$HCNH^+ + e^- \rightarrow HCN + H; \; HNC + H \tag{33}$$

Not all metastables derive from dissociative recombination reactions; the metastable ion HOC^+ is produced in a variety of normal ion–molecule reactions along with the lower energy formyl ion.

A third success lies in the area of isotopic fractionation. In addition to the molecules listed in Table 1, a large number of isotopomers have been detected with isotopes such as D, ^{13}C, $^{17,18}O$, and ^{34}S. The most dramatic isotopic fractionation occurs in the case of deuterium; although the deuterium-to-hydrogen elemental abundance ratio in our galaxy is on the order of a few times 10^{-5}, the abundance ratio of selected singly deuterated isotopomers to the normal species in cold clouds can range up to 0.10: a 4 order-of-magnitude effect! The explanation for this very large effect is not kinetic but thermodynamic in origin. At very low temperatures, selected isotopic exchange reactions can occur rapidly in the exothermic direction but not at all in the endothermic direction.

Consider the critical interstellar reaction system,[67]

$$H_3^+ + HD \Leftrightarrow H_2D^+ + H_2 \tag{34}$$

which is exothermic in the left-to-right direction by approximately 230 K when one considers both vibrational and rotational zero-point energies. (Note: astronomers use K as an energy unit; the exothermicity in this case is approximately 0.5 kcal mol^{-1}.) The isotopomer HD is most probably the source of most deuterium in dense clouds, possessing a fractional abundance with respect to H_2 equal to twice the deuterium-to-hydrogen ratio. Both forward and backward reactions have been measured at room temperature and below and there appears to be no activation energy barrier.[68] (In general, the same cannot be said for other reactions in this class; they represent an instance in which ion–molecule processes can possess barriers.[69]) At temperatures considerably below 230 K, the right-to-left (endothermic) reaction cannot occur rapidly and, if there were no other reactions to consider, the "equilibrium" would lie far to the right despite the large amount of H_2. Even with side reactions such as dissociative recombination, the abundance ratio of H_2D^+ to H_3^+ can lie as high as 0.10 in the coldest clouds. The high relative abundance of H_2D^+ can then transmit itself into high abundances of other isotopomers via reactions such as:

$$H_2D^+ + CO \rightarrow DCO^+ + H_2 \tag{35}$$

DCO^+ is detected in space with an abundance with respect to HCO^+ of $\approx 2\%$ in the 10 K source TMC-1, in good agreement with model results that consider a wider variety of exchange reactions than discussed here.[70]

A fourth success concerns the high degree of unsaturation found in the observed list of molecules. Very few highly saturated molecules are detected, and those that are saturated tend to be found in highly localized sources known as hot cores, where they are probably formed via H-atom hydrogenation on grain surfaces.[54] The reason that ion–molecule reactions do not produce more saturated polyatomic species is, as discussed above, the small number of reactions between hydrocarbon ions and H_2 that can occur rapidly.

A fifth success concerns carbon monoxide, the dominant interstellar molecule from an observer's point of view. Despite all the uncertainties and problems with the model calculations, which will be amply brought out in this review, the predicted fractional abundance of CO is large and in the range of 10^{-5} to 10^{-4}, in excellent agreement with observation.

In discussing these general areas of agreement, we have not mentioned time dependence, allowing the reader to think that we have been discussing steady-state abundances. The calculated time dependence of species is somewhat complex. The majority of small molecules, and CO is typical in this respect, are produced at abundances near their steady-state values in approximately 10^5 years at the standard density of 10^4 cm^{-3}. Whether one chooses abundances at so-called "early time" (10^5 years) or at steady state, the degree of agreement with observation is unchanged. There are some species that are produced slowly; for these few, the agreement at steady state is better than at early time. For most complex species, however, the time dependence is quite different from that of smaller species. In particular, the calculated abundances peak at early time and then diminish greatly before reaching steady state. The chemical reason for this is that the major synthetic mechanism—carbon insertion—requires large abundances of C^+ and C, especially the latter. Atomic carbon is calculated to have its largest abundance somewhat before early time, and then is gradually oxidized to form CO. If one makes the unusual assumption that the elemental abundances in the gas are carbon rich (C > O), then a high abundance of C exists at times through steady state, and large abundances of complex species exist likewise at steady state.[57,62] With the normal assumption of oxygen-rich elemental abundances, however, the interstellar sources richest in molecules cannot be accounted for by pseudo-time-dependent homogeneous gas phase models unless one assumes that early-time pertains; i.e, that the source condensed 10^{5-6} years ago and that the chemical abundances reflect this age.

Perhaps the source with the largest fractional abundances of unsaturated complex molecules is the small condensation in the Taurus complex known as TMC-1 (Taurus Molecular Cloud -1) which, although it shows evidence for substructure, can be considered to have average physical conditions ($n = 10^4$ cm^{-3} and T = 10 K). Table 2 contains a comparison of observed[4] abundances in TMC-1 for almost 50

Table 2. Observed Fractional Abundances in TMC-1 (in Relation to H$_2$ Abundance) Compared with Calculated Values ($n = 1 \times 10^4$ cm$^{-3} \approx n(H_2)$; $T = 10$ K) at a Time of 10^5 Years

Species[a]	Observed[b]	Calculated[b]	Species	Observed[b]	Calculated[b]
C	1.0(-04)	4.7(-05)	CH	2.0(-08)	6.6(-09)
OH	3.0(-07)	*1.4(-08)*	CO	8.0(-05)	9.2(-05)
C$_2$	5.0(-08)	2.6(-08)	CN	3.0(-08)	1.1(-07)
NO	3.0(-08)	5.8(-09)	CS	1.0(-08)	7.4(-09)
SO	5.0(-09)	7.6(-10)	H$_2$S	5.0(-10)	*2.4(-11)*
CCH	5.0(-08)	6.6(-08)	HCN	2.0(-08)	2.1(-07)
HNC	2.0(-08)	8.1(-08)	CCO	6.0(-10)	9.5(-11)
CCS	8.0(-09)	3.0(-09)	SO$_2$	1.0(-09)	1.4(-10)
OCS	2.0(-09)	2.7(-09)	HCO$^+$	8.0(-09)	4.3(-09)
HN$_2^+$	5.0(-10)	*3.3(-13)*	HCS$^+$	6.0(-10)	*9.6(-12)*
NH$_3$	2.0(-08)	4.5(-09)	H$_2$CO	2.0(-08)	1.0(-07)
H$_2$CS	3.0(-09)	3.7(-10)	C$_3$H (total)	1.1(-09)	*3.9(-08)*
HCNH$^+$	1.9(-09)	8.7(-10)	C$_3$N	1.0(-09)	*2.1(-08)*
C$_3$O	1.0(-10)	3.0(-11)	C$_3$S	1.0(-09)	7.7(-10)
C$_3$H$_2$ (total)	1.0(-08)	1.9(-08)	H$_2$CN	5.0(-09)	5.1(-08)
CH$_2$CO	1.0(-09)	*4.2(-08)*	HCOOH	2.0(-10)	1.1(-09)
C$_4$H	2.0(-08)	1.1(-07)	HC$_2$CN	6.0(-09)	*1.2(-07)*
CH$_3$OH	2.0(-09)	4.1(-09)	CH$_3$CN	1.0(-09)	1.1(-08)
H$_2$C$_4$	8.0(-10)	<2.3(-08)[c]	C$_5$H	3.0(-10)	4.2(-09)
CH$_3$CCH	6.0(-09)	4.4(-09)	CH$_3$CHO	6.0(-10)	*1.1(-11)*
C$_2$H$_3$CN	2.0(-10)	6.2(-10)	C$_6$H	1.0(-10)	*2.6(-09)*
HC$_4$CN	3.0(-09)	1.1(-08)	CH$_3$CCCN	5.0(-10)	1.1(-09)
CH$_3$C$_4$H	2.0(-10)	*4.6(-09)*	HC$_6$CN	1.0(-09)	1.6(-09)
HC$_8$CN	3.0(-10)	1.9(-10)			

Notes: [a]The word "total" refers to the sum of two isomers—linear and cyclic.

[b]a(-b) signifies a × 10^{-b}. Numbers in italics signify a discrepancy of more than 1 order of magnitude.

[c]Calculated value for the sum of diacetylene and H$_2$CCCC; observed value for the latter isomer only.

different molecules with early-time calculated peak abundances from our "new standard model"[62] using the average physical conditions; the agreement is typically order-of-magnitude, which is generally as good as can be expected given observational uncertainties. Perhaps 10% of the observed molecules (shown in italics in Table 2) show larger discrepancies with theory; one reason for this discrepancy is that some of the molecules also exist in the more diffuse foreground which cannot be separated out in the observations and which is not handled in the homogeneous model, but is treated in recent inhomogeneous models.[65] Although the level of agreement is encouraging, one must remember that both chemical and astronomical uncertainties may make it at least partially fortuitous.

There is a mathematical problem with the gas-phase models: since the differential equations are nonlinear, it is not obvious that there is only one unique solution. After much searching, a second solution to the kinetic equations has finally been found under steady-state conditions, although only for a cosmic ray ionization rate somewhat higher than the rate typically used, and with a rather high elemental abundance of sulfur. This second solution has been dubbed the high ionization phase (HIP) by its discoverers,[71] as opposed to the more normal low ionization phase (LIP). In HIP, the fractional ionization is higher by several orders of magnitude and there is less molecular development. Interestingly, the calculated CO abundance is still large. Another salient feature of HIP is a predicted high abundance of neutral atomic carbon. Such a prediction may partially explain observations of widespread atomic carbon throughout dense interstellar clouds.

Another difficulty is the uncertain role of dust. Even if one assumes dust particles to be purely passive objects except for their ability to make H_2, the time scale for collision between a given species and a dust particle can be estimated to be about 10^{5-6} years in a typical dense cloud core. The sticking probability is known to be high, and only the most weakly bound light species—helium atoms, hydrogen atoms, and hydrogen molecules—can thermally evaporate in reasonable times. Thus, in the absence of nonthermal desorption, the heavy species in the gas phase cannot exist there for more than 10^6 years. Although this time is longer than the early-time age at which chemical models of particularly fecund sources show best agreement, it is shorter than the estimated lifetime of many other sources. The consensus is that nonthermal desorption occurs via a variety of mechanisms, although the laboratory work in support of this view is somewhat equivocal. Two types of desorption mechanisms have been considered: intermittent and continuous ones.[23] The intermittent mechanisms refer to astronomical processes which result in a dramatic rise in temperature so that the grain mantles do indeed evaporate or are lost by sputtering. The most common astronomical processes referred to arise from star formation, which can heat up surrounding areas gradually or through shock waves. Although such processes do indeed result in the removal of grain mantles, there is a need for a continuous mechanism as well in many sources, which can occur under the ambient low temperature conditions. Some possibilities include sputtering and whole grain heating caused by cosmic rays, photodesorption via both visible/ultraviolet and infrared photons, partial utilization of the energy of exothermic reactions to break the physisorption bond, and conversion of some of the energy available when dust particles collide inelastically.[23,72] We have recently used time-dependent quantum calculations to explore the role of infrared photons, since these are not extinguished greatly by dust particles and can cause vibrational excitation of species on grain surfaces.[73] Although vibrational excitation does not lead directly to desorption, we have calculated that it can first produce excited librational (hindered rotational) states of a nearby molecule, and that librational energy can promote rapid desorption.

If one wishes to consider the diffusive chemistry occurring on dust particles, a standard rate treatment analogous to what is performed for the gas-phase chemistry is not necessarily applicable given the small sizes of interstellar dust grains and the small number of reactive species present on one particle at any given time.[74] A Monte Carlo approach can be utilized, in which one follows the adsorption of assorted species onto a particular grain.[21] Another possibility, and one more efficiently incorporated into the gas-phase chemical network, is to use modified rate equations.[75] All treatments show that under most circumstances the dominant species on grains are saturated ices, in agreement with observation.[21,22,74,76]

In addition to these problems, there are specific chemical problems, raised by our uncertain knowledge of the gas-phase chemistry and alluded to in the previous discussion of ion–molecule chemistry, which make the gas-phase model results highly uncertain in many instances. These are now discussed in more detail, in the hope that they can be alleviated by future laboratory and theoretical work.

IV. SOME OUTSTANDING CHEMICAL PROBLEMS

A. Dissociative Recombination Reactions

Dissociative recombination reactions are thought to be the final step in the synthesis of polyatomic interstellar molecules via chains of ion–molecule reactions. Although these reactions are known to be rapid and dissociative, the branching fractions for the various possible neutral fragments have not been studied extensively. Since the recombination of ions and electrons yields an ionization potential worth of energy, many sets of exothermic products can exist, and these exothermic products can be formed in a wide variety of excited electronic, vibrational, and rotational states. The situation is much better than it was a few years ago, as two sets of techniques have recently been used to study the problem of the recombination of fully thermalized ions. In the first approach, which involves a flowing afterglow apparatus,[26,28] experimental methods of product determination consist mainly of the laser-induced fluorescence (LIF) technique for the OH radical, and two techniques for atomic hydrogen: ultraviolet absorption and chemical conversion into OH radicals followed by LIF. These techniques do not normally suffice to determine all of the branching fractions. In the second approach, which involves a storage ring,[27,29] molecular ions are fully thermalized in the ring before undergoing recombination. In principle, the product distribution can be fully measured by placing a grid in front of the particle detector.

As mentioned previously, both techniques have been used to study the dissociative recombination of H_3O^+. This reaction possess four exothermic channels:

$$H_3O^+ + e^- \rightarrow OH + 2H \tag{36a}$$

$$\rightarrow OH + H_2 \tag{36b}$$

$$\rightarrow O + H + H_2 \tag{36c}$$

$$\rightarrow H_2O + H \tag{36d}$$

and, for a complete determination of the branching fractions in the flowing afterglow experiment,[28] it was necessary to convert the O in channel (reaction 36c) into OH by reaction with germane (GeH_4). The product fractions determined in the flowing afterglow are 0.29, 0.36, 0.30, and 0.05, respectively, while those determined in the storage ring are 0.48 ± 0.08, 0.18 ± 0.07, 0.01 ± 0.04, and 0.33 ± 0.08, respectively. There is little agreement for any channel, although the sum of the two OH channels is the same to within the uncertainties. The cause of the discrepancy is still not known; it is certainly true that the products are being sampled at differing times after reaction in the two experiments, but it is difficult to imagine this being the reason for the discrepancy when such small neutral products are sampled.

If experimental measurements of dissociative recombination product branching fractions are difficult to make, what about theoretical estimates? We have previously used a statistical approach called the phase–space technique to attempt to determine the branching fractions.[32] The approach shows a mixed pattern of agreement and disagreement with experimental results[26,28,29] and, in addition, has the drawback that the results are sensitive to parameters which must be estimated. Moreover, the use of a statistical treatment can be disputed since such treatments show best success when long-lived complexes exist, a situation far removed from dissociative recombination. In this process, the electron sticks temporarily to the positive ion by exciting other electrons. Two microscopic mechanisms exist: a direct one, in which a repulsive state of the neutral species crosses the bound potential surface of the ion so that the system can fragment quickly on this repulsive state; and, an indirect one, in which a Rydberg state is first formed which then crosses a repulsive potential curve of the parent neutral species. Detailed theoretical treatments of both processes exist [77] but have not led to product determinations for polyatomic ions. Bates has formulated a variety of simplified treatments which emphasize the direct nature of dissociative recombination but which have shown no greater success than the statistical treatment in estimating product branching fractions.[32,78] His recent approximate treatment of the degree of vibrational excitation of the products seems to be more successful.[79] We are currently in the initial stages of a project to study product fractions using both classical trajectory and time-dependent quantum techniques in collaboration with quantum chemists who will supply repulsive potentials. The first system we are studying, in collaboration with W. Kraemer (Garching), is the dissociative recombination of HCO^+:

$$HCO^+ + e^- \rightarrow CO(v) + H \tag{37}$$

In this system there is only one exothermic set of products and the problem consists of determining the vibrational distribution of the CO product in both its ground and first excited electronic states. The results can be compared with the experimental determination of Adams, Babcock, et al.,[80] as well as the approximate treatment of Bates.[79] Assuming some success in this endeavor, we propose to study reaction 33 to determine the HNC/HCN product ratio with potential surfaces from a French quantum chemistry group.[81]

If one looks at the flowing afterglow results on $H_3O^+ + e^-$, as well as the recent storage ring results on this reaction, $CH_3^+ + e^-$ and $H_2O^+ + e^-$, one detects a trend despite the conflicting results. In particular, one notices that three-body channels are rather prominent. In addition to direct formation of three separate products, there are two possible indirect causes for the prominence of such channels: secondary break-up of primary two-body channels in which one of the product species has sufficient vibrational energy to decompose, and formation of a primary two-body product channel in which a molecular product lies in a repulsive electronic state.[33] If the former explanation is general, then it is expected that three-body channels will often occur when the primary products consist of an atom and a molecule, since much of the energy of reaction can go into vibrational energy of the one molecule.

Herbst and Lee[33] have recently run model calculations for a dense interstellar cloud with our "new standard" model (see Section IV.D) in which the estimated branching fractions for many unstudied dissociative recombination reactions are revised to reflect this newly found prominence for three-body channels in a small number of studied systems. The results are not very sensitive to this revision for the most part because a large fraction of the molecular ions in the model are very unsaturated with sufficiently strong chemical bonds that two of them cannot be broken exothermically. In addition to increasing the prominence of three-body channels, Herbst and Lee[33] generalized the flowing afterglow result that there is little if any of the $H_2O + H$ channel in the dissociative recombination of $H_3O^+ + e^-$. In this second model, the calculated abundances of saturated neutral species are strongly diminished and, in some instances, it is no longer possible to explain observed abundances. A prime case concerns methanol, which is thought to be formed via dissociative recombination from the precursor ion $CH_3OH_2^+$. Since methanol is a fairly abundant interstellar molecule, and since we know of no other plausible gas-phase synthesis, if the product channel $CH_3OH + H$ is not a major one, methanol may well be formed primarily on dust particles and desorbed by some unknown mechanism into the gas phase. Indeed, a sequence of hydrogenation reactions starting from CO and ending in methanol has been studied on cold laboratory surfaces.[82]

As the size of the positive ion becomes very large, statistical considerations indicate that associative recombination via radiative emission should begin to dominate over dissociative channels. Using a simple RRKM technique, Bettens and Herbst[83] have estimated that for ions of the type C_n^+, C_nH^+, and $C_nH_2^+$, the radiative channel begins to dominate for n in the range 15–20. Ions of this size are larger than contained in standard interstellar chemical models, although Bettens and Herbst[84] have extended such models to include molecules as complex as fullerenes. These extensions are discussed in Section V.

It is possible that electrons are not the major carrier of negative charge in dense interstellar clouds. It has been suggested that if a large fractional abundance of polycyclic aromatic hydrocarbons (PAHs) exists throughout dense interstellar clouds (see Section I), then electron sticking reactions of the type,

$$PAH + e^- \rightarrow PAH^- + h\nu \tag{38}$$

might lead to greater concentrations of PAH anions than of electrons.[85] Although such reactions are indeed expected on statistical grounds to be efficient for large neutrals with sizeable electron affinities,[86] it is not clear that the electron affinities of PAHs are large enough.[8] In addition, recent experiments have shown unexpected barriers to reaction between electrons and selected large molecules.[87] If large molecular anions are more abundant than electrons, it will be necessary to consider the branching fractions for products of positive ion–negative ion reactions, another field in which little information exists.[88]

B. Ion Structural Characterization

In most (but certainly not all!) experiments involving ion–molecule reactions, the structure of the product ions is not determined. As the number of atoms in the product ions increases, the multiplicity of possible isomers becomes greater. Knowledge of the structure of ions is critical in determining what neutral products result from dissociative recombination. Although some classes of ion–molecule reactions, such as proton transfer reactions, lead to products with relatively well-characterized structures, the problem can be more severe with other classes of reactions.

Consider, for example, the well-studied reaction between C^+ and NH_3, for which one set of products consists of the ion $CH_2N^+ + H$. But what is the structure of the product ion? Based on detailed quantum chemical studies of the very complex potential surface, it is likely that two isomers are produced initially—the linear $HCNH^+$ ion and the T-shaped H_2NC^+ form[89]—although it is also possible that the latter form can subsequently isomerize via a unimolecular path into the more stable linear structure. The product branching fractions for each isomer are very important in determining the HNC/HCN abundance ratio in dense interstellar clouds because the T-shaped ion, when it undergoes dissociative recombination reactions, can produce only HNC, while the linear ion can produce both HCN and HNC. The slightly larger abundance of HNC in dark clouds has been ascribed, probably incorrectly, to a preponderance of H_2NC^+ product in the C^+ and NH_3 reaction.[89]

Association reactions, in particular, seem to present a severe problem for structural determination. In these reactions, an ion and a neutral species form a complex which is stabilized either by collision with a third body or, at especially low pressures, by the emission of radiation. The radiative mechanism, prominent in interstellar chemistry, is discussed below. Although some studies of radiative association have been performed in the laboratory,[30,31,90] most association reactions studied are three-body in nature. It is customarily assumed that the product of three-body association is the same as that of radiative association, although this assumption need not be universally valid.

When an atomic or molecular ion A^+ associates with a neutral B, the stabilized product ion AB^+ can be a very weakly bound "van der Waals" molecule or a more strongly bound isomer (these processes are discussed further by Adams and Fisher in this volume). Unless an experimental determination of the structure of the product ion is made, by, e.g. determining its reactivity or by measuring its collisional induced dissociation (CID) pattern, one must rely on theoretical methods. Quantum chemistry can be used to determine the allowable isomers and their energies and whether or not each structure can be reached from reactants without significant activation energy. Another approach is to use a theoretical determination of the rate of association.[31,91] Statistical theories of association rates, discussed below, are often accurate to an order of magnitude if there is thermodynamic, vibrational, and structural information available concerning the product ion. In general, the stronger the bonding between A^+ and B in the association product, the larger the rate coefficient. A large discrepancy between theory and experiment is one sign that the assumed product ion is not the right isomer. Sometimes, however, the assorted experimental and theoretical approaches are in conflict.

As an example, consider the association reaction,

$$H_3O^+ + C_2H_2 \rightarrow C_2H_5O^+ \tag{39}$$

which has been studied under three-body conditions at assorted temperatures in the laboratory with a SIFT apparatus and found to have a rate coefficient k_{39} at room temperature of 8×10^{-28} cm^6 s^{-1} with He as the third body.[91] The theoretically determined rate coefficient, obtained with the assumption that the product is protonated acetaldehyde, is more than 2 orders of magnitude greater.[91] Although protonated acetaldehyde had been suggested as the product ion based on previous CID studies, the discrepancy between theory and experiment is large enough to suggest that a product ion with less binding energy is being formed, possibly the cyclic O-protonated oxirane. If so, reaction 39 probably does not lead to interstellar acetaldehyde as had previously been thought because it is unlikely that dissociative recombination reactions can lead to a large change in skeletal structure. Recent work on the reactions of the product ions of reaction 39 shows that protonated vinyl alcohol ($CH_2CHOH_2^+$) and either protonated acetaldehyde or a weakly bound cluster ion are formed in a 50–50 mixture[92] so that the situation remains murky.

A clearer case of a mistake being made by interstellar modelers concerns the association reaction,

$$C_2H_3^+ + CO \rightarrow C_3H_3O^+ \tag{40}$$

which had been assumed to lead to protonated propynal, a precursor of propynal (HCCCHO), an observed interstellar molecule. Reactivity and quantum chemistry studies now show, however, that protonated propynal is not produced in the association reaction.[93]

A recent study by Matthews et al.[94] on the reactivity of products of the association reactions,

$$CH_3^+ + CH_3OH \rightarrow C_2H_7O^+ \tag{41}$$

$$H_3O^+ + C_2H_4 \rightarrow C_2H_7O^+ \tag{42}$$

suggests that the products of reactions 41 and 42 under laboratory three-body conditions are *not* fully thermalized by collisions with He, but are high energy forms which can still relax to a variety of structures, including both strongly bound and weakly bound molecules. The data imply that the analogous radiative association reactions under interstellar conditions will produce at least partially $(CH_3)_2OH^+$ (protonated dimethyl ether) in reaction 41 and $C_2H_5OH_2^+$ (protonated ethanol) in reaction 42. Calculations of the branching fractions for these structures and the weakly bound structures have yet to be undertaken, however. Prior to this work, the existence of weakly bound (van der Waals-type) product ions had not been suspected, although the strongly bound products had been correctly identified. Branching fractions produced via relaxation from high energy forms can be treated by master equation methods once the potential surface is known.[95] Perhaps the explanation of incomplete relaxation under three-body conditions is also relevant to reaction 39.

Products in several isomeric forms can occur in systems with fewer atoms than considered above; the association reaction between C_3H^+ and H_2 to produce both cyclic and noncyclic $C_3H_3^+$ is a case in point, although the branching ratio in this instance seems to be noncontroversial.[30] The problem of whether product hydrocarbon ions are cyclic or noncyclic extends to other classes of ion–molecule reactions such as condensation and carbon insertion reactions, where studies of product reactivity have only been undertaken in a few instances. In general, cyclic ion products are less reactive than their noncyclic counterparts. For systems with a large number of atoms, isomers with aromatic, monocyclic, tricyclic, and even fullerene-type structures are possibilities.

C. Radiative Association: Competition

Ion–molecule radiative association reactions have been studied in the laboratory using an assortment of trapping and beam techniques.[30,31,90] Many more radiative association rate coefficients have been deduced from studies of three-body association reactions plus estimates of the collisional and radiative stabilization rates.[91] Radiative association rates have been studied theoretically via an assortment of statistical methods.[31,90,96] Some theoretical approaches use the RRKM method to determine complex lifetimes; others are based on microscopic reversibility between formation and destruction of the complex. The latter methods can be subdivided according to how rigorously they conserve angular momentum; without such conservation the method reduces to a "thermal" approximation—with rigorous conservation, the term "phase space" is utilized.

The best understood class of reactions, which we will term "simple," involves those with no competitive exothermic products; in this case at low densities the temporary reaction complex can either redissociate into reactants or be stabilized by the emission of radiation. In general, both experiment and theory show for simple radiative and unsaturated three-body systems that the rate of association depends strongly on the size of the product, the binding energy (well depth) of the product with respect to reactants, and the temperature.[96] As the size and well depth of the product increase, the rate of association increases dramatically because the reaction complex lives longer and has a greater chance to emit a stabilizing photon. As the temperature decreases, the rate of association increases, typically as $T^{-r/2}$ where r is the total number of rotational degrees of freedom for the separated reactants (e.g. two nonlinear tops possess 6 degrees of freedom so that the rate coefficient goes as T^{-3}). The temperature effect is related to the energy above the dissociation limit in the complex; the less energy, the longer the lifetime. Of course, the upper limit to the rate coefficient is the collision rate. In general, statistical determinations of simple radiative association reactions are in order of magnitude agreement with experimental rates despite the fact that these rates range over many orders of magnitude. In addition, the measured temperature dependence appears to follow the statistical result. With selected reactions, there are theoretical uncertainties depending on the mode of radiative stabilization. Most reactions proceed via the emission of infrared radiation at rates between $10 - 10^3$ s^{-1}, but some proceed more rapidly via the emission of visible and ultraviolet radiation; such emission requires the existence of an excited electronic state accessible to reactants.[97] With other systems, there can be an effective barrier to the formation of a deeply bound complex; this barrier can determine the rate coefficient since the complex, once formed, can no longer redissociate before relaxation.[40,98]

Based on the study of simple radiative association processes that are statistical in nature, one can conclude that even with small binding energies, as the size of the reactants becomes sufficiently large, radiative association becomes 100% efficient. Using the phase-space approach,[96] Herbst and Dunbar[99] have studied the rate of radiative association reactions between hydrogen-rich hydrocarbons of the type,

$$C_nH_m^+ + C_nH_m \rightarrow C_{2n}H_{2m}^+ + h\nu \tag{43}$$

as a function of N, where $N = 2n + 2m$, using generic molecular properties. At a temperature of 10 K, these authors found that for a nonpolar neutral reactant and a product bond energy of 2 eV, 100% stabilization occurs by $N = 14$ atoms, while with a product bond energy of 0.1 eV, 100% stabilization occurs by $N = 106$ atoms. At a temperature of 100 K, the corresponding values of N for 100% stabilization are larger, with $N = 20$ for a bond energy of 2 eV. The results at 10 K are not strongly changed if the neutral species has a permanent dipole.

Before one can utilize these results in models of large molecule formation in interstellar clouds, however, one must take note of a crucial problem: the existence of normal exothermic product channels. It was originally thought that the existence of such channels effectively eliminates the possibility of competitive association channels except for very large systems (Section V)[100]; however, the measured existence of such competition in a variety of reactive systems led to a reevaluation.[53,101] It was then realized that if the exothermic product channel has an activation energy barrier which is not large enough to choke off reaction completely, association can be competitive, especially for collisions with large amounts of relative angular momentum since angular momentum adds a centrifugal barrier contribution to the existing barrier, which eventually closes the exothermic channel. This so-called "series" mechanism for association–reaction competition is most efficient with polar neutral reactants, since ion–polar neutral collisions can occur with very large amounts of angular momentum. Still, the calculated radiative association rates are depressed compared with what they might be in the absence of competition. For accurate calculations, the size of the exit channel barriers must be well determined. Some laboratory measurements on association reactions seem to be understandable in terms of this mechanism,[30,101] which requires that ion–molecule reactions, which tend not to possess activation energy barriers in the entrance channel, possess them in the exit channel.

Herbst and Dunbar[99] have investigated the effects of exit channel barriers on association reactions of type 43 and have shown that, depending on the size of the barrier, the efficiency of radiative association reactions as a function of N can be strongly curtailed. For example, at 10 K and a nonpolar neutral reactant, they found for a system with a well depth of 2 eV and an exothermic channel barrier of 1.0 eV, $N = 130$ atoms for 100% sticking efficiency, approximately 10 times the corresponding value of N in the absence of a competitive exothermic channel.

There is now known to be a second mechanism allowing competition between association and reaction, which can be termed the "parallel" mechanism. In this mechanism, the pathway on the potential energy surface leading to reaction is completely different from the pathway leading to association with the bifurcation between the two pathways occurring at long range in the reactant channel. Perhaps the best studied competition with this mechanism is the $C_2H_2^+ + H_2$ system,

$$C_2H_2^+ + H_2 \rightarrow C_2H_4^+ + h\nu \qquad (44a)$$

$$\rightarrow C_2H_3^+ + H \qquad (44b)$$

where experimental results concerning both channels are available but somewhat contradictory concerning reaction 44b at low temperature.[30,102] Detailed ab initio calculations[44] show that the minimum energy pathway for reaction 44b occurs via a weakly bound entrance channel complex followed by a transition state in which the H_2 attacks perpendicular to the acetylene. Following a weakly bound exit channel complex, the pathway ends with the nonclassical ($HCHCH^+$) form of the $C_2H_3^+$ ion + H, a channel calculated to be endothermic by 2.0 kcal mol^{-1}, in agreement with those measurements that show a low reaction rate at low temperature. Interestingly, the transition state, which lies at 2.5 kcal mol^{-1}, can be tunneled under with high efficiency by low temperature reactants with the standard one-dimensional treatment.[103] The association channel proceeds via the same weakly bound entrance channel complex. However, in this case the complex leads via a complicated pathway to the deep potential well of the ethylene ion. In the exit channel, the ethylene ion would dissociate into the classical (H_2CCH^+) form of $C_2H_3^+ + H$, but these products are calculated to lie 5.4 kcal mol^{-1} higher in energy than the reactants. In other words, here there is no competitive exit channel with association in the sense of the series mechanism; the branching between reaction and association occurs after the weakly bound entrance channel. The competition between association and reaction involving the reactants $C_3H^+ + H_2$ is even more complex than the $C_2H_2^+ + H_2$ system; for $C_3H^+ + H_2$, it is unclear whether a series or parallel mechanism dominates. In addition, there are two isomeric forms of the $C_3H_3^+$ association product ion, and the $C_3H_2^+ + H$ reaction channel may also be slightly endothermic.[30,44]

Since two mechanisms are possible for the competition between association and reaction, detailed ab initio calculations of the potential surface are even more necessary in theoretical determinations of the rates of association channels. More experimental work is also needed; it is possible that as a larger number of competitive systems is studied, our understanding of the competition will increase. Critical systems for interstellar modeling include the association/reactive channels for C^+ and bare carbon clusters, as well as for hydrocarbon ions and H_2.

D. Neutral–Neutral Reactions

As discussed in Section II, neutral–neutral reactions involving atoms and/or small radicals play an uncertain role in the gas-phase chemistry of interstellar clouds. Neutral–neutral reactions may be important in the synthesis of large species if, as measured in a variety of recent studies, reactions involving atoms/radicals and so-called stable species can occur without activation energy and with rate coefficients at or even exceeding the capture model limit.[18] Several model studies have recently attempted to elucidate this role by adding to the standard models a significant number of neutral–neutral reactions with large rate coefficients ($k \approx 10^{-10}$ cm^3s^{-1}) and mainly estimated products, with the systems added based on the limited amount of work in the laboratory.[62,104] The first study[104] showed that the addition of rapid neutral–neutral reactions worsens the agreement between theory and observation by reducing the calculated abundances of many large molecules, some dramatically. The second study looked at the matter more carefully and considered the results when several different classes of neutral–neutral reactions were added to the models.[62] The result was that the addition of neutral-neutral reactions could either help or hurt the synthesis of large molecules depending on which sets of reactions are included. Particularly important in curbing the synthetic power of the models in the standard oxygen-rich case are possible reactions of the sort,

$$O + C_n \rightarrow \text{Products (e.g. } CO + C_{n-1}) \tag{45}$$

since atomic oxygen is a major repository of oxygen in oxygen-rich models (fractional abundance $\approx 10^{-4}$) and unsaturated carbon clusters are calculated to be quite abundant if reactions of type 45 do not occur. At room temperature, atomic oxygen, unlike atomic carbon, is not particularly reactive with stable species, where activation energy barriers appear to be the rule, but does react rapidly with radicals.[17] But should one regard C_n as a radical? The linear carbon clusters lie in singlet and triplet electronic states for odd and even numbers of carbon atoms respectively, not doublet states as radicals normally do. The question is really unanswerable without detailed experimental or theoretical studies. Very recently, Woon and Herbst[105] calculated that the $O + C_3$ reaction possesses an entrance channel transition state barrier of ≈ 2 kcal mol^{-1} (1000 K), and that the reaction rate coefficient at low temperatures is far too low to be competitive, even with tunneling. This result can probably be generalized to reactions between O and linear carbon clusters with an odd number of carbon atoms. The situation for the carbon clusters with an even number of carbon atoms (in ground triplet states) remains unclear. Unlike the $O + C_3$ case, ab initio quantum chemical studies confirm the reactivity of atomic carbon; new ab initio calculations on the $C + C_2H_2$ reaction indicate a rather complex mechanism.[106]

Since the extent of neutral–neutral chemistry in dense interstellar clouds is currently unclear, we have constructed three different interstellar models according to the extent of neutral–neutral reactions incorporated in them.[62] Our normal model, referred to as the "new standard" model, does not have a significant number of atom/radical–stable neutral reactions. Ironically, this model still shows the best agreement with interstellar observations, especially for large molecules. In addition to the "new standard" model, we have constructed a "new neutral–neutral" model, in which large numbers of rapid neutral–neutral reactions have been included, extrapolating from the limited number of new measurements. This model fails to produce sizeable abundances of large molecules at "early time" (see Section III) if the standard elemental abundances, in which the elemental abundance of oxygen exceeds that of carbon, are chosen; presumably the destructive power of O atoms exceeds the synthetic power of C atoms. A third model, known as "Model 4" (the name pertains to the fourth in a wide series of attempted models), contains most of the neutral–neutral reactions in the "new neutral–neutral" model, but eliminates reactions of type 45 as well as $N + C_n$ reactions. (Interestingly, these reactions are included in the "new standard" model with lower rates.) "Model 4" is in almost as good agreement with observation for sources such as TMC-1 as is the "new standard" model (see Table 2), but tends to predict too low abundances for some small species and too large abundances for some large species. The latest UMIST network of reactions is closest to our "new standard" model, but does include the latest measurements.[63]

Even if atomic oxygen–carbon cluster reactions are rapid, as is assumed in the "new neutral–neutral" model, the synthetic power of this model can be recovered if exothermic hydrogen atom abstraction reactions of the sort,

$$C_n + H_2 \rightarrow C_nH + H \tag{46a}$$

$$C_nH + H_2 \rightarrow HC_nH + H \tag{46b}$$

are sufficiently rapid ($k > 10^{-14}$ cm^3 s^{-1}) that the C_n clusters and the C_nH radicals react with abundant H_2 rather than with O, and form species that are not reactive with O, thus eliminating the destructive power of this atom. Reactions between radicals and H_2 are known to possess activation energy barriers, so that the processes can occur at low temperatures via tunneling only, in analogy with the ion–molecule reaction between NH_3^+ and H_2,[36] and possibly $C_2H_2^+ + H_2$.[103] Calculations of the reaction between CCH and H_2 using an ab initio short-range potential and an estimate for the strength of the long-range van der Waals binding suggest that the rate coefficient does indeed fall as the temperature is reduced below 300 K and then starts to rise again at very low temperatures due to tunneling.[107] Still, the estimated rate coefficient at 10 K is only $10^{-14} - 10^{-15}$ cm^3 s^{-1}. The prospect for measuring such small rate coefficients at such low temperatures is not an encouraging one.

At the moment, far more experimental and theoretical studies are needed to get a more coherent picture of the interstellar role of neutral–neutral reactions. A list of important systems for interstellar modeling is given by Bettens et al.[62]

E. Understudied Classes of Ion–Molecule Reactions

Although most of the ion–molecule reactions used in large chemical models of interstellar clouds have not been studied in the laboratory, a few classes of reactions are particularly poorly studied. One such class involves neutral atoms. Although the neutral atoms O, C, N, and H can be abundant in interstellar models, difficulty in studying their reaction rates with assorted ions has led to little information for use by modelers, who typically generalize based on laboratory results. The situation is particularly bleak for C atoms, and somewhat better for H, O, and N atoms, for which several groups have reported limited results.[35,41,108] Particularly exciting has been the recent study of the reaction between N and H_3^+ (reaction 11) as well as a number of reactions between H atoms and hydrocarbon ions.[109] The major concern with C atoms is whether or not they can insert into hydrocarbon ions, while the major concern with O atoms is whether they react with hydrocarbon ions to form oxygenated ions, leading to new organo–oxygen molecules, or whether they form mainly HCO^+ and smaller neutral hydrocarbons.

A second major class of ion–molecule reactions that is relatively poorly studied consists of systems involving very unsaturated hydrocarbon neutrals, especially radicals. The unsaturated nature of the organic chemistry in interstellar clouds leads to sizeable abundances of very unsaturated hydrocarbons such as the polyacetylenes HC_nH, the carbenes H_2C_n, the radicals C_nH, and the clusters C_n. Although some work has been done on the chemistry of such species, much of the relevant ion–molecule chemistry involving ions such as C^+, CH_3^+, and even $C_2H_2^+$ must be guessed at from generalizations based on a small number of studied systems.

A third major class of reactions involves the formation of fairly saturated organo–oxygen species from methanol and ions. Such reactions may occur in hot cores (Section VI), in which the onset of star formation drives large amounts of methanol from grain surfaces into the gas phase where it is possibly converted into more complex species such as ethanol, ether, and methyl formate.[110] Although suggestions have been made concerning this conversion of methanol by reactions such as 41 under prevailing conditions ($n \approx 10^{6-7}$ cm^{-3}, $T \approx 100–300$ K), the lack of corroborating laboratory evidence is striking. A synthesis of the highly abundant methyl formate is particularly needed; the only suggested reaction leading to its protonated precursor is that between $CH_3OH_2^+$ and H_2CO.

Reactions between molecular ions and H_2 at low temperature constitute another class of understudied ion–molecule reactions. Laboratory and/or theoretical evidence for a variety of ions—NH_3^+, $C_2H_2^+$, HOC^+—suggests that tunneling can lead to significant rate coefficients at low temperatures even for systems that react rather slowly at room temperature. By "significant" in an interstellar context, we mean only that the reaction possesses a rate coefficient of $\geq 10^{-13}$ cm^3 s^{-1} since there is 10^4 times more molecular hydrogen than any other neutral species. Rate coefficients for reactions between unsaturated hydrocarbon ions and H_2 are particularly important in molecular synthesis (as are their neutral counterparts); null results at room temperature for such systems may not lead to null results at low temperatures, if the reason for the null result is an activation energy barrier rather than endothermicity.

Finally, the amount of laboratory information available to modelers seems to become more sparse as the reactants become larger. An extension of current models to include species with more than 10 atoms (as discussed below) is rendered highly speculative by the lack of experimental information. Particularly crucial are reactions leading to the formation and destruction of species in the same classes as observed molecules but somewhat larger in size.

F. Photodestruction

Although external visible and ultraviolet photons do not penetrate dense clouds appreciably, there are two mechanisms by which such photons may be present in cloud interiors and cause photodestruction. First, a small residual photon flux is maintained via cosmic ray bombardment; secondary electrons from the primary ionization excite H_2 which subsequently fluoresces.[42] Second, the formation of a star inside a cloud can lead to a sufficiently large radiation field that, even with the protection of dust particles, the photon flux can be significant at large distances from the internal star, leading to so-called "photon-dominated regions" (PDRs; see Section VI).[111] In addition, photodestruction is prominent in outer layers of clouds and in diffuse clouds, which have low gas densities even in their interiors.[58,112] It is therefore necessary to include photodestruction in interstellar models, and summaries of suggested rates have appeared as functions of cloud depth based on both measured and calculated cross sections for assorted channels.[113] In general, although several groups are measuring relevant photodestruction rates,[114] the rates and the branching fractions for polyatomic species are poorly determined. The basic problem is that the radiation field extends down to 13.6 eV (912 Å), requiring vacuum UV experiments.

An extremely interesting question in the photochemistry of interstellar molecules concerns the size at which classes of molecules become resistant to the interstellar radiation field. Up to now, only statistical theories have been brought to bear on the question (see below).[83,115]

V. THE PRODUCTION OF VERY LARGE MOLECULES

A. Chains, Rings, and Fullerenes

The larger molecules seen via radioastronomical techniques (≈ 10 atoms in size) are customarily referred to as "large" by the astronomical community; molecules larger than this in size, the existence of which is inferred only from low resolution spectra, are referred to as "complex" or "very large." Are complex molecules produced in significant abundance in the interstellar medium? Several attempts have been made to estimate the efficiency of ion–molecule chemistry in producing very large hydrocarbons and carbon clusters in interstellar clouds. Herbst[116] estimated the abundances of hydrocarbons through 30 carbon atoms in size that could be produced in dense clouds via radiative association, carbon insertion, and condensation reactions. To simplify the equations, he considered neither the state of hydrogenation of the hydrocarbons nor their structure. Major routes for destruction included reactions with generic ions as well as oxygen atoms. In general, it was found that the abundances of very large molecules depend quite sensitively on the unknown rate coefficients utilized in the model, but that 20–30 carbon-atom hydrocarbons could conceivably be produced with a total fractional abundance of $\approx 10^{-7}$ in perhaps 10^6 years in the most optimistic case.

Thaddeus[117] suggested subsequently that the growth of very large clusters can occur more efficiently in diffuse clouds (where the so-called diffuse interstellar bands, or DIBs, originate[9,118]) than in dense clouds, because the degree of ionization is much higher. In particular, the radiation field is strong enough that most carbon is in the form of C^+; thus, carbon insertion reactions (see e.g. reactions 25–26) as well as radiative association between C^+ and C_n (reaction 28) can produce complex unsaturated species as long as these species do not photodissociate rapidly. Since small species do photodissociate relatively rapidly in the unshielded interstellar radiation field, one cannot build up complex molecules from atoms in diffuse clouds. Rather, it is necessary to consider "seed" molecules of a size large enough that their photodissociation rates are not too rapid to compete with reaction with C^+. These seed molecules can form in the gas phase of dense interstellar clouds which eventually disperse as stars form and blow much of the material surrounding them away. Alternately, the seed molecules can condense onto dust particles in the dense cloud environment and then photodesorption can bring them back into the gas once the radiation field increases.

In the simple steady-state model of Thaddeus,[117] bare carbon cluster seed molecules with 12 carbon atoms are used with reaction 28 to produce large linear carbon clusters with sizeable abundances since it is assumed that the C^+_{n+1} ions produced in reaction 28 do not dissociate when they recombine with electrons if $n \geq 12$. Rather, neutral C_{n+1} clusters are formed which either photodissociate (slowly) or recombine further with C^+. In this limited system, cluster growth would be catastrophic were it not for photodissociation. The large abundances of carbon clusters with $20 \leq n \leq 40$ suggests that such molecules may well be the carriers of the well-known DIBs.[118]

A much more detailed and time-dependent study of complex hydrocarbon and carbon cluster formation has been prepared by Bettens and Herbst,[83,84] who considered the detailed growth of unsaturated hydrocarbons and clusters via ion–molecule and neutral–neutral processes under the conditions of both dense and diffuse interstellar clouds. In order to include molecules up to 64 carbon atoms in size, these authors increased the size of their gas-phase model to include approximately 10,000 reactions. The products of many of the unstudied reactions have been estimated via simplified statistical (RRKM) calculations coupled with ab initio and semiempirical energy calculations. The simplified RRKM approach posits a transition state between complex and products even when no obvious potential barrier exists. In this sense, it is equivalent to modern variational transition state theories.[96] Analogous simplified RRKM calculations have been undertaken for determining photodissociation, radiative association, and recombination rate coefficients and product branching fractions. For these three processes, the statistical treatment allows us to make crude estimates as to when complex dissociative ("series") channels are no longer competitive with radiative stabilization channels. Since bond energies are often large, the number of atoms required for the latter processes to dominate ($\approx$ 30) is significantly smaller than for the species studied by Herbst and Dunbar[99] (Section IV.C) using the phase-space technique. In addition, the existence of a potential barrier in the exit channel is not needed. Experimental confirmation would be welcome.

The mechanism for the formation of complex hydrocarbons through fullerenes is loosely taken from Helden et al.[119] and Hunter et al.,[120] and is depicted in Figure 2. As in the work of Thaddeus,[117] linear carbon clusters grow via carbon insertion and radiative association reactions, although in this case a large number of additional reactions involving neutral atoms such as C, O, and H and neutral molecules such as H_2 are also included. Reactions with H and H_2 serve to produce

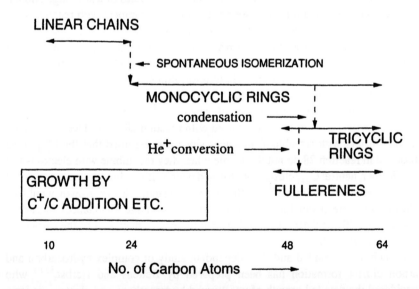

Figure 2. The new species added to our chemical models of interstellar clouds. The species range in complexity from 10–64 carbon atoms and comprise the following groups of molecules: linear carbon chains, monocyclic rings, tricyclic rings, and fullerenes. The synthetic pathways are also indicated. See ref. 83. Reproduced from the *International Journal of Mass Spectrometry and Ion Processes*, vol. 149/150, R.P.A. Bettens, Eric Herbst "The interstellar gas phase production of highly complex hydrocarbons: construction of a model", pp 321–343 (1995) with kind permission from Elsevier Science-NL, Sara Burgerhartstraat 25, 1055 KV, Amsterdam, The Netherlands.

very unsaturated hydrocarbons so that not only carbon clusters are produced. As opposed to continual growth of linear chains, spontaneous isomerization of linear chains into monocyclic rings is assumed to occur when monocyclic rings become so much more stable than chains that ring strain is unable to prevent the isomerization. This is assumed to occur for species with 24 carbon atoms. Monocyclic rings grow in a similar manner to the linear chains. Tricyclic rings (*t*-, often referred to incorrectly as "bicyclic" rings) are produced by association and condensation reactions between ionic and neutral monocyclic rings. Finally, fullerenes (*f*-) are formed from tricyclic rings via reactions with ions that can overcome the activation energy barriers known to be present. An example is the process:

$$C^+ + t\text{-}C_{48} \rightarrow f\text{-}C_{46}^+ + C_3 \tag{47}$$

Another example, as shown in Figure 2, is charge exchange with helium ions followed by internal rearrangement of the energized tricyclic ion. Once fullerenes are produced, they too can grow via insertion and radiative association reactions involving C and C^+, except for the cases of C_{60} and $C_{60}H^+$ which are assumed to be stable. The growth of monocyclic and tricyclic rings, and of fullerenes, is terminated at 64 carbon atoms; growth of larger species undoubtedly occurs and leads to small grains. The role of negative ions in molecular synthesis is considered for linear hydrocarbons with large electron affinities; negative ions are formed via simple addition reactions between the hydrocarbons and free electrons; these reactions are assumed to proceed at the collision rate coefficient[86] and not to possess barriers.[87]

Given that uncertainties in the rates of neutral–neutral reactions strongly affect predicted large molecule abundances, it would not be surprising for such uncertainties to have even stronger consequences for the abundances of very large molecules. At the current time, Bettens and Herbst[83,84] have extended two of their models to include complex hydrocarbons. One model, the "new standard" model, has been used to obtain the excellent agreement between observation and theory for TMC-1 shown in Table 2. This model does not contain a wide sample of rapid neutral–neutral reactions. The other model, "Model 4," contains a wide sample of rapid neutral–neutral reactions but excludes reactions between carbon clusters and oxygen and nitrogen atoms.[62]

In their cloud models, Bettens and Herbst have investigated both static and dispersing clouds.[84] The term "dispersing" here refers to a cloud with gas density decreasing with increasing time as the material spreads out. For a static diffuse cloud, complex molecules cannot be synthesized in a purely gas-phase model because no "seed" molecules can ever be grown. For a static dense cloud, the extension of "Model 4" shows a much higher degree of complex molecule production than does the extension of the "new standard model," presumably due to the additional synthetic power of rapid neutral–neutral reactions and the removal of destructive O atom reactions. In the case of 64-carbon-atom species, the largest considered, and a class representative of still larger molecules and small grains, a total fractional abundance of 10^{-7} for all molecules this size with respect to the overall gas density can be produced in a time of 10^6 years. The comparable fractional abundance using the extension of the "new standard" model is only 10^{-15}. These numbers do not decrease as steady-state conditions are achieved because the large molecules cannot be broken down into CO.

Dispersing clouds would seem to represent the most likely habitat for the growth of very large molecules, because seed molecules can be formed under dense cloud conditions and then grow under diffuse cloud conditions. In general, it is found that the rate of dispersal has a much stronger effect on chemical models based on the "new standard" model than on "Model 4." In particular, for clouds with dispersion velocities exceeding 1 km s^{-1} (this reduces a density of 10^4 cm^{-3} to one near 0.1 cm^{-3} in 10^6 years), a fractional abundance for all 64-carbon-atom species of 10^{-9} can be achieved in a time which is dependent on the exact dispersion velocity with the "new standard" model extension. The extension of "Model 4" shows little difference between the already efficient results for a static dark cloud and the results for dispersing clouds.

Large abundances of individual fullerenes (especially f-C_{60}) can be produced using the extension of "Model 4" for both dark and dispersing clouds, but the degree of hydrogenation of these molecules is uncertain so that our results are certainly not in conflict with the hypothesis of Webster[9] that the DIBs are caused by "fulleranes," or partially hydrogenated fullerenes. Nor do our results for f-C_{60}^+ strongly contradict the tentative identification of this molecule as the carrier of two DIBs by Foing and Ehrenfreund,[9] but do show that if f-C_{60}^+ is abundant, so is the heretofore undetected f-C_{60}.

Interestingly, our purely gas-phase models predict small abundances in dispersing clouds for linear clusters and hydrocarbons, due mainly to their relatively rapid photodissociation rates (the molecules are too small to be stable against photodissociation). This result is different from that of Thaddeus, who used smaller photodestruction rates, and a larger (assumed) abundance of seed molecules. If we assume that seed molecules do exist on the mantles of interstellar dust particles and that these species can be photodesorbed once the radiation field can penetrate the dispersing clouds, then we too calculate large abundances for linear (and also for monocyclic) hydrocarbons. In dense clouds, we predict sizeable abundances for such species in purely gas-phase models only if the "Model 4" network of reactions is utilized. For both dispersive and static dense cloud models, our results are typically time-dependent and are best shown with three-dimensional diagrams, in which abundances are plotted against both number of carbon atoms and time.

B. PAHs

The synthesis of PAHs is generally thought not to occur within the clouds themselves, but in the envelopes of old carbon-rich stars. These stars possess large envelopes in which material expands outward and cools as it expands. In the inner layers thermodynamic equilibrium pertains and some "parent" molecules such as acetylene form with high abundances. In the outer layers, external photons and cosmic rays initiate a low temperature acetylenic chemistry with a prominent role for radicals and ions.[13,121] Intermediate between these parts of the envelope is an area where the temperature is around 1100 K and PAHs may form via condensation reactions between aromatic (Ar) radicals.[14,122] The radicals are formed by abstraction reactions with atomic hydrogen:

$$Ar\text{-}H + H \rightarrow Ar + H_2 \qquad (48)$$

These reactions do not occur at lower temperatures because of activation energy barriers and because H_2 becomes the dominant form of hydrogen. Aromatic species are produced initially from acetylene via Diels–Alder type processes, in which a two-carbon and a four-carbon hydrocarbon condense into an aromatic species. Once PAHs are synthesized, they may continue to grow to form carbonaceous small grains.

What is the ultimate fate of the molecular material formed in the envelopes of carbon-rich stars as it heads out into space? The dust grains will be processed only slowly by the interstellar radiation field and survive almost intact until they become part of an interstellar cloud. The survival of individual PAHs depends on their size; the larger ones withstand radiation much better than do the smaller ones.[115] By survival we are referring to the aromatic skeleton; the interstellar radiation field will efficiently break H bonds and cause ionization so that unsaturated, ionized PAHs are likely to dominate those found in the diffuse interstellar medium. Such species have been suggested as a source of the DIBs.[118,123] Small molecules photodissociate in the interstellar radiation field before the material becomes part of an interstellar cloud.

VI. CHEMISTRY AND STAR FORMATION

The formation of stars in the interiors of dense interstellar clouds affects the chemistry of the immediate environment in a variety of ways depending on many factors such as the stage in the evolution of star formation, the mass of the star or protostar, and the density and temperature of the surrounding material. In general, the dynamics of the material in the vicinity of a newly forming star are complex and show many manifestations. Table 3 contains a list of some of the better studied such manifestations, which tend to have distinctive chemistries. These are discussed individually below.

A. Dense Condensates

Most stars that are forming or already formed in interstellar clouds are of the low mass variety; these objects are relatively cool (by stellar standards)[3] and do not heat up the area surrounding them greatly. Since the influence of gravity results in higher

Table 3. Some Types of "Active" Sources

Region	Physical Conditions	Characteristics
1) Dense condensates	enhanced density	
a) near low mass star formation	"low" temperature ($T < 100$ K)	adsorption onto dust
b) near high mass star formation ("hot core")	"high" temperature ($T = 300$ K)	desorption from dust
2) PDR (photon-dominated)	high radiation field	highly structured and layered
3) Shock		
a) J-type	1 fluid, $T \approx 3,000$–$10,000$ K	reactions with barriers; endothermic reactions
b) C-type	2 fluid, ion streaming	endothermic ion–molecule reactions

gas ($n \approx 10^7$ cm^{-3}) and dust densities surrounding these objects, however, the time scale for adsorption onto grains shortens. In the absence of an efficient mechanism for nonthermal desorption, one expects that the gas phase will become depleted in molecules containing heavy elements in a relatively short period of time. There is ample observational evidence for such depletion, especially involving CO.[124] Some of the cloud material around a newly forming low mass star becomes a disk of gas and dust.[125] Such disks are of extreme interest since they may lead to the formation of planets.

If the star that forms is a massive one, which puts out much more heat, the scenario is very different because the dust particles can warm up to such an extent that the mantles are desorbed. The newly released material in the gas phase can subsequently undergo a gas-phase chemistry under high density ($n \approx 10^7$ cm^{-3}), "high" temperature (300 K) conditions. Objects with these conditions, known as hot cores or "molecular heimats," have been studied in some detail both observationally and theoretically. Generally, they are much richer in more saturated species such as NH_3, CH_2CHCN, CH_3CH_2CN, CH_3OH, $HCOOCH_3$, CH_3OCH_3, and C_2H_5OH, but chemical differences among the cores can be large. The observation of large abundances of ethanol, in particular, was noted with amusement by the press, with headlines such as "pub at the end of the universe" and "scotch on the galactic rocks." The high degree of saturation is universally ascribed to dust chemistry occurring in a previous low temperature era, since hydrogenation (via H atoms) is assumed to be much more facile on dust surfaces than in the gas. Once the molecules are released into the gas, the gas-phase chemistry they undergo is very similar to the standard low temperature ion–molecule chemistry except that more neutral–neutral reactions become important. Detailed chemical models of hot cores have been run.[110,126] The differences between chemical abundances among these objects can be ascribed to two causes: differences in dust chemistry, which, unlike ion–molecule chemistry, is exponentially dependent on temperature, and differences in the amount of time that has passed since the dust mantles were released, since the gas-phase chemistry will eventually return the chemical abundances to steady-state values.

Two well-studied objects in the Orion Nebula, known as the Hot Core and the Compact Ridge sources, show widely varying chemistries though lying in close proximity to one another.[110,126] In the Hot Core, nitrogen-containing saturated species show high abundances, whereas in the Compact Ridge, oxygen-containing species are more prominent. The model of Caselli et al.[110] ascribes these differences mainly to temperature in the low temperature era; the hydrogenation of CO into methanol on grain surfaces can only occur in the Compact Ridge because the grain temperature in the Hot Core is too high for CO to remain on grain surfaces. Once warming begins, methanol desorbs and leads to the other oxygen-containing organic species via ion–molecule reactions, e.g.:

$$CH_3OH_2^+ + CH_3OH \rightarrow (CH_3)_2OH^+ + H_2O \tag{49}$$

$$(CH_3)_2OH^+ + e^- \rightarrow CH_3OCH_3 + H \tag{50}$$

As discussed in previous sections, however, the routes from methanol to more complex species are not well mapped out, and those reactions that have been utilized by modelers may not have the assumed products.

B. Photon-Dominated Regions (PDRs)

Once a massive star or group of stars reaches a certain stage of evolution, it ionizes the area around it, forming what is known as an HII region in which the dominant form of hydrogen is H^+. These very bright objects are seen in many dense interstellar clouds; perhaps the most famous is the Orion Nebula, which is a region of hot, ionized gas surrounding a group of newly born stars known as the Trapezium stars. In this and other HII regions, there is little in the way of molecular development. As the distance from the star increases, however, the temperature and the intensity of the radiation decrease. Still, the chemistry is quite different from that of the ambient medium. Originally known as photodissociation regions and now known as photon-dominated regions (both labels abbreviated as PDRs), these regions in between intense sources of radiation and more ambient interstellar cloud conditions are of chemical interest.[111] Their distinctiveness resides in the strength of the radiation field, which weakens as it traverses through the medium.

The standard approach to modeling PDRs is to use a one-dimensional approach in which the radiation strikes perpendicularly. The region is divided into slabs, so that the equations of radiative transfer and chemical kinetics can be solved conveniently. The slabs can be homogeneous, or can have different gas densities. The radiation is scattered and absorbed by dust particles, but, in addition, both H_2 and

CO absorb discrete frequencies of radiation when they photodissociate.[61,111,112] Since the photodissociation is not continuous, the molecules are self-shielded. In some models, the temperature is actually calculated as a function of depth into the PDR by equating heating and cooling rates. For example, in the model of Sternberg and Dalgarno,[111] which represents rather dense gas subjected to a radiation field χ $= 2 \times 10^5$ times the standard radiation field, the temperature rapidly decreases from 3000 K at the border of the HII region to 30 K by the time the optical depth reaches ≈ 2. Most models assume (a) that only gas-phase chemistry occurs, and (b) that steady-state conditions are reached sufficiently rapidly that time-dependent effects need not be considered. The second assumption may not be justified in all cases. The results are presented as a function of depth into the cloud, normally in terms of optical depth or an astronomical parameter closely related to optical depth called the "visual extinction," A_v. If the exciting stars lie such that the radiation traverses the PDR perpendicular to the line of sight to the observer, radioastronomers can actually observe the change in fractional abundance as a function of distance to the star. The best-known source of this type is the so-called Orion Bar which borders the Orion Nebula on one side. On the other hand, if the PDR lies between the exciting source and the observer, then radioastronomers can only measure the so-called total column densities (concentrations times distance) integrated over the total line of sight.

The models show a variety of interesting chemical transitions between ions and neutrals, such as the $C^+/C/CO$ structure. Close to the exciting source, C^+ is the dominant form of carbon, whereas far removed from the radiation, CO becomes dominant. At steady state, there is no region where neutral atomic C dominates, since the transition between C^+ and CO is rather sharp. Different molecules show different behavior in plots of fractional abundance versus depth; many radicals, for example, tend to peak at intermediate depths, whereas stable molecules are at their greatest abundances at large depth. One possible problem with the models is that they don't take clumpiness into account. If the gas is clumpy, the penetration of radiation is far more extensive than if the gas is homogeneous. Another shortcoming of the models is that, up to now, they have not included large molecules.

One can view a quiescent molecular cloud as a one-dimensional PDR with $\chi = 1$. Here, instead of spherical shells representing outer and inner layers, one has one-dimensional slabs. The advantage of such "shell" models[65] over homogeneous models of the inner portions of clouds is that the roles of the outer layers can be accounted for; such roles are especially important for atoms (e.g. C) and radicals (e.g. OH, CH). Small dense clouds, known as "translucent" clouds, have particularly salient outer portions which should be included in models.

C. Shocks

As stars form, they occasionally send out shock waves that, as they traverse through a medium, result in transient very high temperatures which are related to the velocity of propagation of the shock.[127] Temperatures of a few thousand K or more can be attained. Obviously, the chemistry is strongly affected by the passage of a shock wave, with both exothermic reactions that contain activation energy barriers, and even endothermic reactions possible. Shocks are deduced observationally based on the large Doppler linewidths and shifts in radioastronomical and visible spectra.

Shocks were first invoked to explain the unusually high abundance of the radical CH^+ in diffuse clouds ($n \approx 100$ cm^{-3}; $T \approx 100$ K), where standard chemical models predict far too little of this substance which is reactive both with electrons and with atomic and molecular hydrogen.[128] In a high temperature medium, however, the reaction,

$$C^+ + H_2 \rightarrow CH^+ + H \tag{51}$$

which is endothermic by ≈ 0.4 eV, can produce CH^+ and narrow, if not remove, the disagreement with observation.[129] One problem with diffuse cloud shock models is that in seeking to explain the abundance of CH^+, they must not destroy the agreement between regular models and observation for other species such as OH. It may well be that a more promising mechanism in diffuse clouds is turbulence, which allows endothermic ion–molecule reactions to occur due to ion streaming, but not does allow endothermic neutral–neutral reactions and so does not affect the calculated OH abundance.[130]

Shocks in dense clouds have also been investigated[131]; at sufficiently high shock velocities, the molecules can be dissociated and must reform after the shock passes.[132] The best known shock in a dense region probably occurs in the Orion Nebula, along the same line of sight as the Hot Core and the Compact Ridge; the source is known as the Orion Plateau. In this source, the chemical abundances are different from those found in hot cores and in quiescent clouds. In particular, high abundances of sulfur-containing molecules are observed.[110]

The type of shock wave in which the temperature rises dramatically to thousands of degrees and then cools is now known as a J-shock, where "J" stands for jump. Depending on the unknown strength of the relatively weak magnetic fields in interstellar clouds, the shock behavior can be quite different and, in general, more mild. More moderate peak temperatures are reached in these so-called C-shocks, where "C" is for continuous, but an interesting two-phase, or two "fluid" situation results, in which the ions are actually streaming with respect to the neutrals. Therefore, as with turbulence,[130] endothermic ion–molecule reactions (e.g. 51) can be powered by this streaming, whereas neutral–neutral reactions cannot.

Even in J-type shock models, it is not appropriate to use thermal rate coefficients because the internal degrees of freedom will cool rapidly (via radiation) in the low density medium, whereas the translational degree of freedom will cool much more slowly. Appropriate rate coefficients are then those in which only translation is strongly excited; such rate coefficients can be considerably lower than thermal rates for systems in which vibrational energy is the most efficient at inducing reaction.

In general, rate coefficients determined in drift tube studies, where the internal degrees of freedom are cooled, are most appropriate.[133]

VII. SUMMARY

The field of molecular astronomy is, in reality, only 25 years of age, having started with the observations of polyatomic molecules in interstellar space in the late 1960s and early 1970s.[1-2] During these 25 years, a large amount of information, both chemical and physical, has been learned about interstellar clouds. The study of astrochemistry has inspired some important laboratory work, especially in the areas of microwave spectroscopy of transient species and of ion–molecule reactions. Both normal ion–molecule exothermic reactions and association reactions at low temperature have been studied by a variety of groups, and the results used to improve interstellar chemical models. In addition to ion–molecule studies, astrochemistry has also spawned the study of low temperature neutral–neutral reactions, as well as laboratory studies on grains and their heterogeneous chemistry.

There remain many outstanding problems to solve, without which models of interstellar chemistry will never move to a secure basis. In the area of ion–molecule reactions, some crucial reactions involving neutral atoms such as C and O remain to be studied. Reactions involving radicals and carbon clusters as the neutral reactants are also of the greatest interest. Detailed studies of the ion–molecule reactivity of bare clusters are particularly needed to determine how rapidly these species can grow in the interstellar medium. Even slow processes involving ions and molecular hydrogen, possibly proceeding via tunneling under an activation energy barrier, are critical since the abundance of molecular hydrogen in dense interstellar clouds is so dominant. Detailed studies of ion–molecule reactions involving likely interstellar species (e.g. carbon chains) with more than 10 atoms in size are also desirable. In the area of radiative association reactions, the mechanisms for competition between associative and reactive channels are not fully understood. It is currently far from clear whether radiative association rate coefficients will approach collisional rates for many large systems if there are exothermic competitive reactive channels. The products of dissociative recombination processes must still be estimated based on a small amount of laboratory information, and even this limited amount of information is conflicting. Moreover, the molecular size at which radiative recombination begins to compete with dissociative recombination is still unknown. The structures of ionic products of many types of reactions need to be better characterized. In addition to ionic reactions, many more studies are needed in the areas of low temperature neutral–neutral reactions, reactions occurring on the surfaces of cold dust grains, and in nonthermal desorption from dust grains.

The existence of molecules in interstellar space has stimulated the field of ion–molecule reactions for almost a quarter of a century; it is reasonable to expect that a high degree of stimulation will remain for many years to come.

ACKNOWLEDGMENTS

Support by the Division of Astronomical Sciences of the National Science Foundation is gratefully acknowledged. The model calculations were performed on the Cray YMP8 computer of the Ohio Supercomputer Center; we are grateful for the award of time on this machine.

REFERENCES AND NOTES

1. Winnewisser, G.; Herbst, E.; Ungerechts, H. In *Spectroscopy of the Earth's Atmosphere and Interstellar Medium*; Rao, K. N.; Weber, A., Eds.; Academic Press: New York, 1992, p 423.
2. Wilson, T. L.; Johnston, K. J. (Eds.). *The Structure and Content of Molecular Clouds*; Lecture Notes in Physics 439; Springer-Verlag: New York, 1994.
3. Abell, G. O.; Morrison, D.; Wolff, S. C. *Exploration of the Universe*; 6th ed.; Saunders: Philadelphia, 1993.
4. Irvine, W. M.; Goldsmith, P. F.; Hjalmarson, Å. In *Interstellar Processes*; Hollenbach, D. J.; Thronson, Jr., H. A., Eds.; Reidel: Dordrecht, 1987, p 561; Ohishi, M.; Irvine, W. M.; Kaifu, N. In *Astrochemistry of Cosmic Phenomena*; Singh, P. D., Ed.; Kluwer: Dordrecht, 1992, p 171; Jansen, D. J.; van Dishoeck, E. F.; Black, J. H.; Spaans, M.; Sosin, C. *Astron. Astrophys.* **1995**, *302*, 223.
5. Lacy, J. H.; Carr, J. S.; Evans II, N. J.; Baas, F.; Achtermann, J. M.; Arens, J. F. *Astrophys. J.* **1991**, *376*, 556.
6. Mathis, J. S. *Ann. Rev. Astron. Astrophys.* **1990**, *28*, 37.
7. Whittet, D. C. B. In *Dust and Chemistry in Astronomy*; Millar, T. J.; Williams, D. A., Eds.; Institute of Physics: Philadelphia, 1993, p 9.
8. Allamandola, L. J.; Tielens, A. G. G. M.; Barker, J. R. *Astrophys. J. Supp.* **1989**, *71*, 733.
9. Foing, B. H.; Ehrenfreund, P. *Nature* **1994**, *369*, 296; Herbig, G. H. *Astrophys. J.* **1993**, *407*, 142; Webster, A. *Mon. Not. R. Astron. Soc.* **1993**, *263*, 385; Fulara, J.; Lessen, D.; Freivogel, P.; Maier, J. P. *Nature* **1993**, *366*, 439.
10. Solomon, P. M.; Werner, M. W. *Astrophys. J.* **1971**, *165*, 41.
11. Winnewisser, G.; Herbst, E. *Rep. Prog. Phys.* **1993**, *56*, 1209.
12. Omont, A. *J. Chem. Soc., Faraday Trans.* **1993**, *89*, 2137.
13. Millar, T. J.; Herbst, E. *Astron. Astrophys.* **1994**, *288*, 561.
14. Frenklach, M.; Feigelson, E. D. *Astrophys. J.* **1989**, *341*, 372.
15. Rowe, B. R. In *Rate Coefficients in Astrochemistry*; Millar, T. J.; Williams, D. A., Eds., Kluwer Publishing: Dordrecht, 1988, p 135.
16. Clary, D. C. *Ann. Rev. Phys. Chem.* **1990**, *41*, 61.
17. Baulch, D. L.; Cobos, C. J.; Cox, R. A.; Esser, C.; Frank, P.; Just, Th.; Kerr, J. A.; Pilling, M. J.; Troe, J.; Walker, R. W.; Warnatz, J. *J. Phys. Chem. Ref. Data* **1992**, *21*, 411.
18. Sims, I. R.; Queffelec, J.-L.; Defrance, A.; Rebrion-Rowe, C.; Travers, D.; Bocherei, P.; Rowe, B. R.; Smith, I. W. M. *J. Chem. Phys.* **1994**, *100*, 4229; Sims, I. R.; Smith, I. W. M.; Bocherei, P.; Defrance, A.; Travers, D.; Rowe, B. R. *J. Chem. Soc., Faraday Trans.* **1994**, *90*, 1473; Sims, I. R.; Smith, I. W. M. *Ann. Rev. Phys. Chem.* **1995**, *46*, 109; Clary, D. C.; Haider, N.; Husain, D.; Kabir, M. *Astrophys. J.* **1994**, *422*, 416.
19. Watson, W. D. *Rev. Mod. Phys.* **1976**, *48*, 513; Herbst, E. In *Dust and Chemistry in Astronomy*; Millar, T. J.; Williams, D. A., Eds.; Institute of Physics: Philadelphia, 1993, p 183.
20. Somorjai, G. A. *Introduction to Surface Chemistry and Catalysis*; Wiley: New York, 1994.
21. Tielens, A. G. G. M.; Hagen, W. *Astron. Astrophys.* **1982**, *114*, 245.
22. Hasegawa, T. I.; Herbst, E.; Leung, C. M. *Astrophys. J. Supp. Ser.* **1992**, *82*, 167.

23. Williams, D. A. In *Dust and Chemistry in Astronomy*; Millar, T. J.; Williams, D. A., Eds.; Institute of Physics: Philadelphia, 1993, p 143.

24. Leung, C. M.; Herbst, E.; Huebner, W. F. *Astrophys. J. Supp. Ser.* **1984**, *56*, 231; Herbst, E.; Leung, C. M. *Astrophys. J. Supp. Ser.* **1989**, *69*, 271; Millar, T. J.; Herbst, E. *Astron. Astrophys.* **1990**, *231*, 466; Herbst, E. *Ann. Rev. Phys. Chem.* **1995**, *46*, 27.

25. Geballe, T. R.; Oka, T. *Nature* **1996**, *384*, 334.

26. Adams, N. G. In *Advances in Gas Phase Ion Chemistry*; Adams, N. G.; Babcock, L. M., Eds.; JAI Press: Greenwich, CT, 1992, Vol. 1, p 271; Herd, C. R.; Adams, N. G.; Smith, D. *Astrophys. J.* **1990**, *349*, 388; Adams, N. G.; Herd, C. R.; Geoghegan, M.; Smith, D.; Canosa, A.; Gomet, J. C.; Rowe, B. R.; Queffelec, J. L.; Morlais M. *J. Chem. Phys.* **1991**, *94*, 4852.

27. Datz, S.; Larsson, M.; Stromholm, C.; Sundström, G.; Zengin, V.; Danared, H.; Källberg, A.; af Ugglas, M. *Phys. Rev. A* **1995**, *52*, 901; Larsson, M.; Lepp, S.; Dalgarno, A.; Stromholm, C.; Sundström, G.; Zengin, V.; Danared, H.; Källberg, A.; af Ugglas, M.; Datz, S. *Astron. Astrophys.* **1996**, *309*, L1.

28. Williams, T. L.; Adams, N. G.; Babcock, L. M.; Herd, C. R.; Geoghegan, M. *Mon. Not. Roy. Astron. Soc.* **1996**, *282*, 413.

29. Vejby-Christensen, L.; Andersen, L. H.; Heber, O.; Kella, D.; Pederssen, H. B.; Schmidt, H. T.; Zaifman, D. *Astrophys. J.* **1997**, *483*, 531.

30. Gerlich, D.; Horning S. *Chem. Rev.* **1992**, *92*, 1509.

31. Bates, D. R.; Herbst, E. In *Rate Coefficients in Astrochemistry*; Millar, T. J.; Williams, D. A., Eds., Kluwer Publishing: Dordrecht, 1988, p 17.

32. Galloway, E. T.; Herbst, E. *Astrophys. J.* **1991**, *376*, 531; Bates, D. R. *Astrophys. J.* **1989**, *344*, 531.

33. Herbst, E.; Lee, H.-H. *Astrophys. J.* **1997**, *485*, 689.

34. Herbst, E.; DeFrees, D. J.; McLean, A. D. *Astrophys. J.* **1987**, *321*, 898.

35. Scott, G. B. I.; Fairley, D. A.; Freeman, C. G.; McEwan, M. J. *Chem. Phys. Lett.* **1997**, *269*, 88.

36. Herbst, E.; DeFrees, D. J.; Talbi, D.; Pauzat, F.; Koch, W.; McLean, A. D. *J. Chem. Phys.* **1991**, *94*, 7842.

37. Ziurys, L. M.; Apponi, A. J. *Astrophys. J.* **1995**, *455*, L73.

38. Herbst, E.; Woon, D. E. *Astrophys. J.* **1996**, *463*, L113; Herbst, E.; Woon, D. E. *Astrophys. J.* **1996**, *471*, L73.

39. Le Bourlot, J. *Astron. Astrophys.* **1991**, *242*, 235; Galloway, E. T.; Herbst, E. *Astron. Astrophys.* **1989**, *211*, 413; Adams, N. G.; Smith, D.; Millar, T. J. *Mon. Not. Roy. Astron. Soc.* **1984**, *211*, 857.

40. Herbst, E.; DeFrees, D. J.; Koch, W. *Mon. Not. R. Astron. Soc.* **1989**, *237*, 1057.

41. Anicich, V. G. *J. Phys. Chem. Ref. Data* **1993**, *22*, 1469.

42. Prasad, S. S.; Tarafdar, S. P. *Astrophys. J.* **1983**, *267*, 603; Gredel, R.; Lepp, S.; Dalgarno, A.; Herbst, E. *Astrophys. J.* **1989**, *347*, 289.

43. Giles, K.; Adams, N. G.; Smith, D. *Int. J. Mass Spectrom. Ion Proc.* **1989**, *89*, 303.

44. Maluendes, S. A.; McLean, A. D.; Herbst, E. *Chem. Phys. Letters* **1994**, *217*, 571; Maluendes, S. A.; McLean, A. D.; Yamashita, K.; Herbst, E. *J. Chem. Phys.* **1993**, *99*, 2812.

45. Freed, K. F.; Oka, T.; Suzuki, H. *Astrophys. J.* **1982**, *263*, 718.

46. Clary, D. C.; Haider, N.; Husain, D.; Kabir, M. *Astrophys. J.* **1994**, *422*, 416.

47. Kaiser, R. I.; Ochsenfeld, C.; Head-Gordon, M.; Lee, Y. T.; Suits, A. G. *J. Chem. Phys.* **1997**, *106*, 1729; Kaiser, R. I.; Ochsenfeld, C.; Head-Gordon, M.; Lee, Y. T.; Suits, A. G. *Science* **1996**, *274*, 1508.

48. Kaiser, R. I.; Stranges, D.; Bevsek, H. M.; Lee, Y. T.; Suits, A. G. *J. Chem. Phys.* **1997**, *106*, 4945.

49. Liao, Q.; Herbst, E. *Astrophys. J.* **1995**, *444*, 694.

50. Klippenstein, S. J.; Kim, Y.-W. *J. Chem. Phys.* **1993**, *99*, 5790.

51. Sims, I. R.; Queffelec, J.-L.; Travers, D.; Rowe, B. R.; Herbert, L. B.; Karthäuser, J.; Smith, I. W. M. *Chem. Phys. Letters* **1993**, *211*, 461.

52. Woon, D. E.; Herbst, E. *Astrophys. J.* **1997**, *477*, 204; Fukuzawa, K.; Osamura, Y. *Astrophys. J.* **1997**, in press.
53. Herbst, E. *Astrophys. J.* **1987**, *313*, 867.
54. Walmsley, C. M. In *Chemistry and Spectroscopy of Interstellar Molecules*; Bohme, D. K., Herbst, E., Kaifu, N., Saito, S., Eds.; Tokyo University Press: Tokyo, 1992, p 267; Millar, T. J. In *Dust and Chemistry in Astronomy*; Millar, T. J.; Williams, D. A., Eds.; Institute of Physics: Philadelphia, 1993, p 249.
55. Herbst, E.; Leung, C. M. *Astrophys. J.* **1986**, *310*, 378; Moran, T. F.; Hamill, W. H. *J. Chem. Phys.* **1963**, *39*, 1413; Su, T.; Chesnavich, W. J. *J. Chem. Phys.* **1982**, *76*, 5183.
56. Lehfaoui, L.; Rebrion-Rowe, C.; Laubé, S.; Mitchell, J. B. A.; Rowe, B. R. *J. Chem. Phys.* **1997**, *106*, 5406.
57. Graedel, T. E.; Langer, W. D.; Frerking, M. A. *Astrophys. J. Supp. Ser.* **1982**, *48*, 321.
58. Van Dishoeck, E. F.; Black, J. H. In *Rate Coefficients in Astrochemistry*; Millar, T. J.; Williams, D. A., Eds.; Kluwer: Dordrecht, 1988, p 209.
59. Herbst, E.; Klemperer, W. *Astrophys. J.* **1973**, *185*, 505.
60. Prasad, S. S.; Huntress, Jr., W. T. *Astrophys. J. Supp. Ser.* **1980**, *43*, 1; Prasad, S. S.; Huntress, Jr., W. T. *Astrophys. J.* **1980**, *239*, 151.
61. Stecher, T. P.; Williams, D. A. *Astrophys. J.* **1967**, *149*, L29.
62. Bettens, R. P. A.; Lee, H.-H.; Herbst, E. *Astrophys. J.* **1995**, *443*, 664; Lee, H.-H.; Bettens, R. P. A.; Herbst, E. *Astron. Astrophys. Suppl.* **1996**, *119*, 114.
63. Millar, T. J.; Farquhar, R. P. A.; Willacy, K. *Astron. Astrophys.* **1997**, 121, 139.
64. Prasad, S. S.; Heere, K. R.; Tarafdar, S. P. *Astrophys. J.* **1991**, *373*, 123.
65. Bergin, E. A.; Langer, W. D.; Goldsmith, P. F. *Astrophys. J.* **1995**, *441*, 222; Xie, T.; Allen, M.; Langer, W. D. *Astrophys. J.* **1995**, *440*, 674; Lee, H-H.; Herbst, E.; Pineau des Forêts, G.; Roueff, E.; Le Bourlot, J. *Astron. Astrophys.* **1996**, *311*, 690.
66. Green, S.; Herbst, E. *Astrophys. J.* **1979**, 229, 121.
67. Herbst, E. *Astron. Astrophys.* **1982**, *111*, 76; Herbst, E. In *Isotope Effects in Gas-Phase Chemistry*; Kaye, J. A., Ed.; American Chemical Soc.: Washington, DC, 1992, p 358.
68. Adams, N. G.; Smith, D. *Astrophys. J.* **1981**, *248*, 373.
69. Henchman, M. J.; Paulson, J. F.; Smith, D.; Adams, N. G.; Lindinger, W. In *Rate Coefficients in Astrochemistry*; Millar, T. J.; Williams, D. A., Eds., Kluwer Publishing: Dordrecht, 1988, p 201; Maluendes, S. A.; McLean, A. D.; Herbst, E. *Astrophys. J.* **1992**, *397*, 477.
70. Millar, T. J.; Bennett, A.; Herbst, E. *Astrophys. J.* **1989**, *340*, 906.
71. Pineau Des Forêts, G.; Roueff, E.; Flower, D. *Mon. Not. Roy. Astron. Soc.* **1992**, *258*, 45; Flower, D. R.; Le Bourlot, J.; Pineau Des Forêts, G.; Roueff, E. *Astron. Astrophys.* **1994**, *282*, 225; Le Bourlot, J.; Pineau Des Forêts, G.; Roueff, E. *Astron. Astrophys.* **1995**, *297*, 251.
72. Westley, M. S.; Baragiola, R. A.; Rohnson, R. E.; Baratta, G. A. *Nature* **1995**, *373*, 405.
73. Dzegilenko, F.; Herbst, E. *Astrophys. J.* **1995**, *443*, L81. For the role of translation, see Dzegilenko, F.; Herbst, E. *J. Chem. Phys.* **1996**, *104*, 6330.
74. Charnley, S. B.; Tielens, A. G. G. M.; Rodgers, S. D. *Astrophys. J.* **1997**, *482*, L203.
75. Caselli, P.; Hasegawa, T. I.; Herbst, E. *Astrophys. J.* **1998**, in press.
76. Hasegawa, T. I.; Herbst, E. *Mon. Not. Roy. Astron. Soc.* **1993**, *263*, 589; Hasegawa, T. I.; Herbst, E. *Mon. Not. Roy. Astron. Soc.* **1993**, *261*, 83.
77. Mitchell, J. B. A.; Guberman, S. (Eds.) *Dissociative Recombination: Theory, Experiment, and Applications*; World: Singapore, 1989.
78. Bates, D. R. *Astrophys. J.* **1986**, *306*, L45; Bates, D. R. *J. Phys. B: At. Mol. Opt. Phys.* **1991**, *24*, 3267.
79. Bates, D. R. *Mon. Not. R. Astron. Soc.* **1993**, *263*, 369.
80. Adams, N. G.; Babcock, L. M. *J. Phys. Chem.* **1994**, *98*, 4564; Butler, J. M.; Babcock, L. M.; Adams, N. G. *Mol. Phys.* **1997**, *91*, 81.
81. Talbi, D.; Ellinger, Y., in preparation.

82. Hiraoka, K.; Ohashi, N.; Kihara, Y.; Yamamoto, K.; Sato, T.; Yamashita, A. *Chem. Phys. Lett.* **1994**, *229*, 408.

83. Bettens, R. P. A.; Herbst, E. *Int. J. Mass Spectrom. Ion Proc.* **1995**, *149/150*, 321.

84. Bettens, R. P. A.; Herbst, E. *Astrophys. J.* **1996**, *468*, 686; Bettens, R. P. A.; Herbst, E. *Astrophys. J.* **1997**, *478*, 585.

85. Lepp, S.; Dalgarno, A. *Astrophys. J.* **1988**, *324*, 553.

86. Herbst, E. *Nature* **1981**, *289*, 656.

87. Smith, D.; Spanel, P.; Märk, T. D. *Chem. Phys. Letters* **1993**, *213*, 202; Spanel, P.; Matejcik, S.; Smith, D. *J. Phys. B* **1995**, *28*, 2941.

88. Bates, D. R.; Herbst, E. In *Rate Coefficients in Astrochemistry*; Millar, T. J.; Williams, D. A., Eds.; Kluwer Publishing: Dordrecht, 1988, p 41.

89. Allen, T. L.; Goddard, J. D.; Schaefer III, H. F. *J. Chem. Phys.* **1980**, *73*, 3255; Talbi, D.; Herbst, E., *Astron. Astrophys.*, submitted.

90. Dunbar, R. C. In *Unimolecular and Bimolecular Ion–Molecule Reaction Dynamics*; Ng, C. Y.; Baer, T.; Powis, I., Eds.; Wiley: New York, 1994, p 270; McMahon, T. B. In *Advances in Gas Phase Ion Chemistry*; Adams, N. G.; Babcock, L. M., Eds.; JAI Press, Greenwich, CT, 1996, Vol. 2, p 41; Dunbar, R. C. In *Advances in Gas Phase Ion Chemistry*; Adams, N. G.; Babcock, L. M., Eds.; JAI Press, Greenwich, CT, 1996, Vol. 2, p 87.

91. Herbst, E.; Smith, D.; Adams, N. G.; McIntosh, B. J. *J. Chem. Soc., Faraday Trans.* **1989**, *85*, 1655.

92. Fairley, D. A.; Scott, G. B. I.; Freeman, C. G.; Maclagan, R. G. A. R.; McEwan, M. J. *J. Chem. Soc., Faraday Trans.* **1996**, *92*, 1305.

93. Maclagan, R. G. A. R.; McEwan, M. J.; Scott, G. B. I. *Chem. Phys. Letters* **1995**, *240*, 185; Scott, G. B. I.; Fairley, D. A.; Freeman, C. G.; Maclagan, R. G. A. R.; McEwan, M. J. *Int. J. Mass Spectrom. Ion Proc.* **1995**, *149/150*, 251.

94. Matthews, K.; Adams, N. G.; Fisher, N. D. *J. Phys. Chem. A* **1997**, *101*, 2841. See also Fairley, D. A.; Scott, G. B. I.; Freeman, C. G.; Maclagan, A. R.; McEwan, M. J. *J. Phys. Chem. A* **1997**, *101*, 2848, in which the strongly-bound protonated ethanol product is found for reaction (42).

95. DeFrees, D. J.; McLean, A. D.; Herbst, E. *Astrophys. J.* **1985**, *293*, 236.

96. Herbst, E. In *Atomic, Molecular, & Optical Physics Handbook*; Drake, G., Ed.; AIP Press: New York, 1996, p 429; Adams, N. G. In *Atomic, Molecular, & Optical Physics Handbook*; Drake, G., Ed.; AIP Press: New York, 1996, p 441. For three-body systems, a slightly more complex temperature dependence is observed. For saturated systems, more complex treatments are needed —see Gilbert, R. G.; Smith, S. C. *Theory of Unimolecular and Recombination Reactions*; Blackwell: Oxford, 1990.

97. Herbst, E.; Bates, D. R. *Astrophys. J.* **1988**, *329*, 410.

98. Herbst, E.; Yamashita, K. *J. Chem. Soc., Faraday Trans.* **1993**, *89*, 2175.

99. Herbst, E.; Dunbar, R. C. *Mon. Not. R. Astron. Soc.* **1991**, *253*, 341.

100. Bates, D. R. *Astrophys. J.* **1983**, *267*, L121.

101. Herbst, E. *J. Chem. Phys.* **1985**, *82*, 4017; Herbst, E. *Astrophys. J.* **1985**, *292*, 484; Herbst, E.; McEwan, M. J. *Astron. Astrophys.* **1990**, *229*, 201.

102. Hawley, M.; Smith, M. A. *J. Chem. Phys.* **1992**, *96*, 1121; Smith, M. A. *J. Chem. Soc., Faraday Trans.* **1993**, *89*, 2209; Gerlich, D. *J. Chem. Soc., Faraday Trans.* **1993**, *89*, 2210.

103. Yamashita, K.; Herbst, E. *J. Chem. Phys.* **1992**, *96*, 5801.

104. Herbst, E.; Lee, H.-H.; Howe, D. A.; Millar, T. J. *Mon. Not. R. Astron. Soc.* **1994**, *268*, 335.

105. Woon, D. E.; Herbst, E. *Astrophys. J.* **1996**, *465*, 795.

106. Takahashi, J.; Yamashita, K. *J. Chem. Phys.* **1996**, *104*, 6613; Ochsenfeld, C.; Kaiser, R. I.; Lee, Y. T.; Suits, A. G.; Head-Gordon, M. *J. Chem. Phys.* **1997**, *106*, 4141.

107. Herbst, E. *Chem. Phys. Letters* **1994**, *222*, 297.

108. Federer, W.; Villinger, H.; Lindinger, W.; Ferguson, E. E. *Chem. Phys. Letters* **1986**, *123*, 12; Viggiano, A. A.; Howarka, F.; Albritton, D. L.; Fehsenfeld, F. C.; Adams, N. G.; Smith, D. *Astrophys. J.* **1980**, *236*, 492.

109. Scott, G. B. I.; Fairley, D. A.; Freeman, C. G.; McEwan, M. J.; Adams, N. G.; Babcock, L. M. *J. Phys. Chem.*, **1997**, *A101*, 4973.
110. Blake, G. A.; Sutton, E. C.; Masson, C. R.; Phillips, T. G. *Astrophys. J.* **1987**, *315*, 621; Millar, T. J.; Herbst, E., Charnley, S. B. *Astrophys. J.* **1991**, *369*, 147; Caselli, P.; Hasegawa, T. I.; Herbst, E. *Astrophys. J.* **1993**, *408*, 548.
111. Sternberg, A.; Dalgarno, A. *Astrophys. J. Supp. Ser.* **1995**, *99*, 565; Tielens, A. G. G. M.; Hollenbach, D. J. *Astrophys. J.* **1985**, *291*, 722; Le Bourlot, J.; Pineau des Forêts, G.; Roueff, E.; Flower, D. R. *Astron. Astrophys.* **1993**, *267*, 233; Jansen, D. Ph.D. Thesis, Leiden University, Leiden, Holland; 1995; Jansen, D. J.; van Dishoeck, E. F.; Black, J. H.; Spaans, M.; Sosin, C. *Astron. Astrophys.* **1995**, *302*, 223.
112. van Dishoeck, E. F.; Black, J. H. *Astrophys. J.* **1988**, *334*, 771.
113. van Dishoeck, E. F. In *Rate Coefficients in Astrochemistry*; Millar, T. J.; Williams, D. A., Eds.; Kluwer Publishing: Dordrecht, 1988, p 49; Roberge, W. G.; Jones, D.; Lepp, S.; Dalgarno, A. *Astrophys. J. Supp. Ser.* **1991**, *77*, 287.
114. Lee, L. C. *Astrophys. J.* **1984**, *282*, 172.
115. Léger, A.; Boissel, P.; Désert, F. X.; d'Hendecourt, L. *Astron. Astrophys.* **1989**, *213*, 351; Jochims, H. M.; Rühl, E.; Baumbärtel, H.; Tobita, S.; Leach, S. *Astrophys. J.* **1994**, *420*, 307.
116. Herbst, E. *Astrophys. J.* **1991**, *366*, 133.
117. Thaddeus, P. In *Molecules and Grains in Space*; Nenner, I.; Trojanowski, L., Eds.; American Institute of Physics Conf. Proc. *312*, Am.. Inst. Phys.: New York, 1994, p 711.
118. Tielens, A. G. G. M.; Snow, T. P. (Eds.). *The Diffuse Interstellar Bands*, Kluwer: Dordrecht, 1995.
119. von Helden, G.; Gotts, N. G.; Bowers, M. T. *Nature* **1993**, *363*, 60.
120. Hunter, J. M.; Fye, J. L.; Roskamp, E. J.; Jarrold, M. F. *J. Phys. Chem.* **1994**, *98*, 1810.
121. Cherchneff, I.; Glassgold, A. E. *Astrophys. J.* **1993**, *419*, L41; Cherchneff, I.; Glassgold, A. E.; Mamon, G. A. *Astrophys. J.* **1993**, *410*, 188.
122. Cherchneff, I.; Barker, J. R.; Tielens, A. G. G. M. *Astrophys. J.* **1992**, *401*, 269.
123. Crawford, M. K.; Tielens, A. G. G. M.; Allamandola, L. J. *Astrophys. J.* **1985**, *293*, L45.
124. Blake, G. A.; Sandell, G.; van Dishoeck, E. F.; Groesbeck, T. D.; Mundy, L. G.; Aspin, C. *Astrophys. J.* **1995**, *441*, 689; Walmsley, C. M. In *CO: Twenty-Five Years of Millimeter-Wave Spectroscopy (IAU Symposium 170)*; Latter, W. B.; Radford, S. J. E.; Jewell, P. R.; Mangum, J. G.; Bally, J., Eds.; Kluwer Publishing: Dordrecht, 1996, p 79.
125. Aikawa, Y.; Miyama, S. M.; Nakano, T.; Umebayashi, T. *Astrophys. J.* **1996**, *467*, 684.
126. Brown, P. D.; Charnley, S. B.; Millar, T. J. *Mon. Not. R. Astron. Soc.* **1988**, *243*, 65; Charnley, S. B.; Tielens, A. G. G. M.; Millar, T. J. *Astrophys. J.* **1992**, *399*, L71.
127. Draine, B. T.; McKee, C. F. *Ann. Rev. Astron. Astrophys.* **1993**, *31*, 373.
128. Elitzur, M.; Watson, W. D. *Astrophys. J.* **1978**, *222*, L141; Elitzur, M.; Watson, W. D. *Astrophys. J.* **1980**, *236*, 172.
129. Flower, D. R.; Monteiro, T. S.; Pineau des Forêts, G.; Roueff, E. In *Rate Coefficients in Astrochemistry*; Millar, T. J.; Williams, D. A., Eds., Kluwer Publishing: Dordrecht, 1988, p 271; Hawkins, I.; Craig, N. *Astrophys. J.* **1991**, *375*, 642.
130. Federman, S. R.; Rawlings, J. M. C.; Taylor, S. D.; Williams, D. A. *Mon. Not. Roy. Astron. Soc.* **1996**, *279*, L41.
131. Mitchell, G. F. In *Astrochemistry*; Vardya, M. S.; Tarafdar, S. P., Eds.; Reidel Publishing: Dordrecht, 1987, p 275; Pineau des Forêts, G.; Roueff, E.; Schilke, P.; Flower, D. R. *Mon. Not. R. Astron. Soc.* **1993**, *262*, 915.
132. Neufeld, D. A.; Dalgarno, A. *Astrophys. J.* **1989**, *340*, 869; Neufeld, D. A.; Dalgarno, A. *Astrophys. J.* **1989**, *344*, 251.
133. Smith, D.; Adams, N. G. In *Rate Coefficients in Astrochemistry*; Millar, T. J.; Williams, D. A., Eds.; Kluwer Publishing: Dordrecht, 1988, p 153.

COMPLEX FORMATION IN
ELECTRON–ION RECOMBINATION OF
MOLECULAR IONS

R. Johnsen and J. B. A. Mitchell

Advances in Gas-Phase Ion Chemistry
Volume 3, pages 49–80
Copyright © 1998 by JAI Press Inc.
All rights of reproduction in any form reserved.
ISBN: 0-7623-0204-6

ABSTRACT

Recent colliding-beam experiments have shown that the low-energy recombination cross section of H_3^+ ions is surprisingly sensitive to electric fields that are applied in the post-collision region. Several plasma afterglow experiments indicate that H_3^+ recombination coefficients vary with electron density. It is shown that these findings can be reconciled by assuming that the recombination proceeds through an intermediate long-lived complex, rather than by the fast direct mechanism that is usually invoked in discussions of molecular–ion recombination. This article examines the experimental evidence that has been adduced in support of the binary dissociative recombination mechanism for molecular ions. It is concluded that the evidence is perhaps not as firm as is often assumed.

I. INTRODUCTION

Experimental and theoretical studies of dissociative recombination (DR) of molecular ions,

$$AB^+ + e \rightarrow A + B \tag{1}$$

have been pursued for many years, but, with the exception of the recombination of H_2^+ ions, a quantitative "convergence" of theory and experiment has not really been achieved. One problem, of course, is that ab initio calculations of recombination cross sections are exceedingly difficult. On an even more basic level, recent theoretical work[1] casts some doubt on our qualitative understanding of the recombination mechanisms. It is no longer quite clear if crossings between ionic potential and repulsive states are really as essential as originally thought. If dissociative recombination can occur without curve crossings, then many arguments that are commonly adduced to rationalize experimental observations are no longer very persuasive. The recombination of H_3^+ ions has become an important test case: It is an important ion in astrophysical environments[2,3] and it is one of the few polyatomic ions that can be treated theoretically. Calculations of the molecular states clearly indicate that the curve-crossing mechanism cannot be operative but there is substantial experimental evidence for sizable recombination coefficients. Do these experiments really lend support to proposed recombination mechanisms without curve crossings? The answer depends, in part, on the interpretation of the experimental data, which is the subject of this review.

Since the experimental work on H_3^+ recombination provided much of the impetus for this work, we will compare two recent experiments, a merged-beam and a plasma afterglow experiment. We conclude that the results can be reconciled by postulating the formation of long-lived Rydberg states and that these are quite sensitive to the environment in which the recombination occurs. In beam experiments (Section IV. B) the observed cross sections are shown to be sensitive to the presence of electric fields. In the afterglow experiments, the concentration of third bodies (electrons, ions, and neutrals) seems to affect the rate at which electrons recombine.

The mechanisms that we propose to explain the H_3^+ data has led us to revisit and to reexamine earlier data on the recombination of other ions (see Section V). The goal was to see to what extent the experimental observations in afterglow plasmas support the assumption that only binary as opposed to third-body assisted recombination is important. While this is most likely true in many cases, the evidence is not as firm as one might think. Often, the original authors stated their conclusions rather cautiously, but over the years tentative conclusions have often assumed a status that is not justified by the original data. We will point out some gaps in our knowledge which should be closed by more detailed studies.

We will be rather brief in describing the principles of the experimental methods that are most commonly used in studies of dissociative recombination, afterglows and colliding beams (merged beams and ion storage rings). It should suffice to recount some of their generic properties.

All afterglow experiments, whether the plasma is "stationary" or "flowing," have in common that the recombination occurs in the presence of a substantial density of third particles, such as neutral atoms, electrons, and ions. One measures the volume loss rate of free electrons and derives a "deionization coefficient,"

$$\beta = -\frac{1}{n_e^2} \frac{\partial n_e}{\partial t} \tag{2}$$

where n_e is the electron density (equal to the total positive ion density) and t denotes afterglow time. We will use the term "deionization coefficient" for this phenomenological quantity to distinguish it from the binary recombination coefficient of a single ion type defined as,

$$\alpha = \int_0^\infty f(v)\sigma(v)v\,dv \tag{3}$$

where $\sigma(v)$ is the velocity-dependent recombination cross section and $f(v)$ is the velocity distribution of the electrons. The quantities α and β should be equal when only binary recombination occurs, but this should not be assumed without supporting evidence.

In addition to measuring total recombination coefficients, experimentalists seek to determine absolute or relative yields of specific recombination products by emission spectroscopy, laser induced fluorescence, and optical absorption. In most such measurements, the products suffer many collisions between their creation and detection and nothing can be deduced about their initial translational energies. Limited, but important, information on the kinetic energies of the nascent products can be obtained by examination of the widths of emitted spectral lines and by time-of-flight analysis of products that effuse from the plasma into a high-vacuum region (see Section V.A).

Beam experiments provide much information that cannot be obtained from afterglows. Most important of all, they permit measurements of the detailed structure of recombination cross sections with good energy resolution. After some ingenious technical improvements (electron "coolers" and adiabatic expansion of the electron beam), ion storage rings now achieve energy resolutions on the order of 10 meV. A further major achievement is the recent introduction of position-sensitive and gridded detectors which permit determinations of the kinetic energies of products and their relative yields.

Since beam experiments are carried out under high-vacuum conditions, third-particle interactions are essentially absent. However, recombination in merged beams and in ion storage rings often takes place in the presence of electric or magnetic fields. In addition to undesired "stray fields," strong electric or magnetic fields are employed to separate charged particles from neutrals before they reach the detector. The effect of fields on measured cross sections may be negligible in most cases, but perhaps not when the electron–ion collisions form long-lived autoionizing complexes. For instance, if the complexes reach the detector before autoionizing they will be counted as neutral recombination products. If they become field-ionized in the deflection field, they will not be detected. Experimental observations by Mitchell et al. (see Section IV.B) show that the magnitude of electric fields in the post-collision region can have a pronounced effect on measured cross sections.

II. THREE-BODY RECOMBINATION MECHANISMS

A. Collisional–Radiative Recombination

Third-body stabilized recombination,

$$AB^+ + e + M \rightarrow \text{products} \tag{4}$$

where M may be an electron, an ion, or a neutral, will play a role in the discussion of the afterglow experiment. We introduce it here to distinguish the types of recombination that may occur. The best known mechanisms are those in which an electron is captured by transferring part of its recombination energy to a second electron or to a neutral atom.[4] In both cases, the electron is captured into a (non-autoionizing) Rydberg state which can either be collisionally reionized or can be stabilized by further collisions or radiation. Since only highly excited electronic states are involved, the internal structure of the recombining ion is not important. These mechanisms are often referred to as "collisional radiative recombination". For the electron-stabilized collisional–radiative process, a "working formula" that agrees quite well with experiment has been derived by Stevefelt et al.[5]

$$\beta_{ecr}[\text{cm}^3/\text{s}] = 3.8 \times 10^{-9} T_e^{-4.5} n_e + 1.55 \times 10^{-10} T_e^{-0.63} + 6 \times 10^{-9} T_e^{-2.18} n_e^{0.37} \tag{5}$$

where T_e is the electron temperature and n_e is the electron density. At $T_e = 300$ K and at the highest electron density that is commonly used in flowing afterglow studies of recombination ($\sim 10^{11}$ cm^{-3}) β_{ecr} takes the value 3×10^{-9} cm^3/s, which is quite small compared to typical recombination coefficients of simple diatomic ions (a few $\times 10^{-7}$ cm^3/s).

Neutral stabilized collisional-radiative recombination is also quite slow under typical afterglow conditions. In his review, Flannery[4] gives the simple formula,

$$\beta_{ncr} \, [\text{cm}^3/\text{s}] = [10^{-26}/M \, (\text{amu})] \, (300/T)^{2.5} \, N_{\text{neutral}} \tag{6}$$

for estimating the recombination coefficient, where N_{neutral} is the neutral gas density. At $T = 300$ and $M = 4$ (helium) β_{ncr} will be less than 10^{-9} cm^3/s at pressures below 10 torr.

Thus, collisional–radiative recombination will not make an important contribution to the deionization in typical low-density afterglow experiments, except in cases where dissociative recombination is extremely slow, e.g. for He$_2^+$. However, collisional–radiative recombination may not be slow compared to the partial dissociative recombination coefficient for a particular product branch. For instance, a product channel that accounts for only 1% of the total may very well arise from collisional–radiative recombination.

B. Three-Body Recombination Involving Long-Lived Complexes

Neither of the two three-body mechanisms discussed above involves long-lived complexes. The situation is quite different with a recombining molecular ion which possesses internal energy states (electronic, vibrational, and rotational degrees of freedom) that can temporarily absorb part of the recombination energy. For instance, the electron can become temporarily bound in a Rydberg state by exciting the ion core to a vibrational or rotational state. The lifetime with respect to decay of the complex into a free electron and an ion (autoionization) is thereby greatly lengthened and more time is available for third-body stabilization.

Complex formation is well known to occur in chemical reactions, ion–molecule association, and electron attachment to large molecules. Why then is it not commonly invoked in discussion of electron–ion recombination? After all, the existence of long-lived intermediate complexes is quite familiar from studies of thermal energy electron attachment to polyatomic molecules.[6] A well-studied case is the attachment to SF$_6$. Here, the auto-detaching complex (SF$_6^-$)** lives long enough (10^{-5} s or longer) to be detected in a mass spectrometer or to be collisionally stabilized at low gas densities. There is however an important distinction between attachment and recombination: The energy gained by attaching a thermal electron to a neutral molecule is often insufficient to allow dissociation of the molecule. By contrast, electron capture by a molecular ion invariably (with the possible exception of the oxides of heavy metals and large polyaromatic molecular ions) results in a complex that has enough energy to dissociate, i.e. dissociative recombination occurs. Unless dynamical constraints prevent dissociation, complexes formed by electron–ion recombination will usually not live long enough to be collisionally stabilized. Therefore, third-body stabilization of complexes will play a role only when the ion does *not* recombine rapidly by dissociative recombination. In the case of H_3^+ recombination, experimental and theoretical evidence indicates that dissociation of autoionizing Rydberg states is quite slow. For this reason, it is a good test case for recombination by complex formation. For most other polyatomic ions, the lifetimes of highly excited complexes are not known and it is difficult to assess the role of complex formation.

It is worth noting that electron–ion complexes can be stabilized by several mechanisms that are not available to the complexes formed by electron–molecule collisions. Since electronically excited states are involved, radiative stabilization will be faster. Also, collisional stabilization can be enhanced by the effect of "*l*-mixing." The basic idea is that the angular momentum of the Rydberg electron rapidly spreads out over all available *l* and *m* states and that the lifetime of the complex is thereby further increased. It is known from studies of dielectronic recombination[7] and zero-electron-kinetic-energy (ZEKE) spectroscopy [8] that redistribution of angular momenta of Rydberg electrons (*l*-mixing) due to electric microfields and collisions can increase predissociation and autoionization lifetimes by several orders of magnitude (for an overview of the ZEKE technique see the chapter by Richard et al.). The reason is that the wave functions of high-l states do not interact with the ion core. In a simple classical picture given by Chupka,[9] autoionization and dissociation occur by short-range interactions ("collisions") of low-*l* electrons with the ion core. High-*l* electrons in nearly circular orbits collide with the core less often and thus remain bound sufficiently long to lose energy by radiation or further collisions. Lundeen[10] apparently was the first to suggest that *l*-mixing should be considered in electron–ion recombination.

III. FIELD EFFECTS IN ELECTRON-ION RECOMBINATION

Field effects in *dielectronic* recombination of atomic ions have been observed, for instance by Müller et al.,[11] but they are thought to be less important in *dissociative* recombination. The sensitivity to fields arises from the effects of fields on the rate of autoionization. Both, dielectronic and dissociative recombination have in common that an electron is temporarily captured into an autoionizing core-excited Rydberg state, but the stabilizing mechanisms differ greatly in efficiency. In dielectronic recombination, only a small fraction of the Rydberg states are radiatively stabilized while most autoionize. Hence, the rate of autoionization and its field dependence directly affect the recombination cross section.[11] In the case of most molecular ions, dissociation occurs before autoionization and the sensitivity to fields is thus greatly reduced. This can be seen if one expresses the recombination coefficient in the form,

$$\alpha = k_c \frac{A_r + A_{diss}}{A_r + A_{diss} + A_a} \tag{7}$$

where k_c is the capture rate and A_r, A_{diss}, and A_a denote the rates of radiative transitions, dissociation, and autoionization, respectively. The derivation of this expression makes use of the steady-state assumption. If either or both A_{diss} and A_r are far larger than the competing decay rates, then a reduction of the autoionization rate A_a due to ambient fields is of no consequence. This is probably the case for most molecular ions. However, if neither A_{diss} or A_r are large compared to A_a, then any dependence of A_a on external fields propagates to α. The latter condition may exist in recombination of H_3^+ and perhaps other polyatomic ions. Additional field effects, which have no counterpart in dielectronic recombination, may arise if the electric field enhances or suppresses the rate of dissociation of the core-excited state.

IV. RECOMBINATION OF H_3^+ IONS

A. History

As has been mentioned before, there are good theoretical reasons [12,13] to believe that the recombination of H_3^+ ($v = 0$) ions,

$$H_3^+ + e^- \rightarrow H + H + H$$

$$\rightarrow H_2 + H \tag{8}$$

by the "direct" and also by the "indirect" recombination processes[14] should be slow due to a lack of suitable crossings between the ionic and neutral potential surfaces. The term "direct," as opposed to "indirect" recombination, means that the electron is thought to be captured directly into a dissociating state of the neutral. In indirect recombination, the electron is thought to be captured into a vibrationally excited Rydberg state which subsequently predissociates along a repulsive potential surface of the neutral. Both mechanisms require that a suitable repulsive potential curve intersects the ionic ground state near its minimum. The problem is that the early microwave–afterglow experiments [15,16] and some later experiments (see Bates et al.[17] for a review) showed that H_3^+ recombines with a fairly "normal" recombination coefficients ($\sim 2 \times 10^{-7}$ cm^3/s) even though suitable potential curve-crossings seem to be absent.

One resolution of the problem that gained acceptance for a while was proposed by Adams et al.[18] In their flowing afterglow measurements, the apparent recombination coefficient of H_3^+ fell off to a small value in the "late afterglow". This finding could be interpreted by assuming the presence of two species of H_3^+ ions, one of which recombined fairly fast while the other recombined slowly ($\alpha < 2 \times 10^{-8}$

cm^3/s). Theory and experiment could be reconciled by identifying the fast and slowly recombining ions as, respectively, $v > 3$ or $v = 0$ ions. This solution of the problem became less plausible when afterglow studies in hollow-cathode discharge by Amano[19,20] showed that spectroscopically identified H_3^+ ions in $v = 0$ recombined readily with electrons ($\alpha \sim 1.8 \times 10^{-7}$ cm^3/s). Also, a repeat of the flowing-afterglow measurements by Canosa et al.[21,22] failed to show evidence for the existence of more than one species of H_3^+ ions.

Merged-beam measurements [23-26] have consistently shown that the measured recombination cross section depends on conditions in the ion source. The authors have ascribed the effect to differing vibrational distributions. In one of the later measurements,[16] the H_3^+ vibrational state was inferred from the threshold energy for electron–ion dissociative excitation,

$$e + H_3^+ (v) \rightarrow H + H_2 \tag{9}$$

and the results imply that $v = 0$ ions recombine only slowly with electrons. This finding, however, is in conflict with results of ion–storage-ring experiments. Here, vibrationally excited ions have time to relax to the ground state by radiative transitions. Larsson et al.[27] and Sundström et al.,[28] using the CRYRING storage ring at Stockholm, obtained recombination cross sections for H_3^+ ($v = 0$) from 0.0025 to 30 eV. The analysis of the experimental cross sections yielded a recombination coefficient of $(1.15 \pm 0.13) \times 10^{-7}$ cm^3/s at an electron temperature of 300 K.

After publication of the ion–storage-ring data, new recombination mechanisms were proposed[1] that do not require crossings between ionic and neutral potential curves. However, no detailed calculations have been made for H_3^+ and it is not yet clear if such mechanisms are capable of explaining the experimental findings.

A careful repeat of the flowing-afterglow measurement of Adams et al. [18] was subsequently carried out by Smith and Spanel.[29] The results confirmed the Adams' et al. observation that the recombination coefficient appears to fall off to a small value in the late afterglow. The authors concluded that the small recombination coefficient observed in the late afterglow is the proper value for $v = 0$ ions and that the initial rapid plasma decay in the early afterglow should be ascribed to vibrationally excited ions.

The afterglow experiment described in Section IV. C, in part, was undertaken to check the experimental findings of Smith and Spanel's work.

B. Colliding-Beam Experiments on H_3^+ Recombination

Figure 1 shows three configurations for colliding-beam experiments that have been used for studies of dissociative recombination. These are the inclined-beam apparatus at the University of Newcastle-upon-Tyne,[30] the merged-electron–ion-beam experiment (MEIBE) at the University of Western Ontario,[31] and an electron-cooler apparatus at the Manne Siegbahn laboratory in Stockholm.[32] In the inclined-beam method (Figure 1a), the ion beam is accelerated to an energy of 30

KeV and is mass analyzed so that the ions under study can be not only preselected but also identified. The beam enters the interaction region where it intersects an electron beam, produced from a hot cathode gun. Following this intersection, the ion beam is separated from the neutrals using an electrostatic deflector, and neutrals that pass undeflected through this analyzer are detected using a surface barrier detector that is sensitive to the energy of the incoming particles. This feature is very important as it allows dissociation products with different masses (and therefore different kinetic energies) to be distinguished and to be counted individually. This means that dissociative excitation reactions that produce both neutral and charged fragments e.g.,

$$e + H_2^+ \rightarrow H + H^+ + e \tag{10}$$

can be distinguished from dissociative recombination reactions, producing only neutrals, e.g.,

$$e + H_2^+ \rightarrow H + H \tag{11}$$

since in the former case, the charged H^+ would be deflected away and would not reach the detector which would then only register a single particle (H-atom) carrying half of the total H_2^+ beam energy. In the latter case, the two H-atoms arrive essentially simultaneously at the detector and appear as a single, full-beam-energy particle. Since the intensities of the initial ion and electron beams are measured and their energies are well defined, and since the production rate of the products is determined, the inclined-beam experiment is capable of measuring absolute collision cross sections for recombination reactions. These are determined by the formula,[23]

$$\sigma = \frac{e^2 C_n}{I_e I_i} \frac{v_i v_e F \sin(\vartheta)}{v_i^2 + v_e^2 - 2v_i v_e \cos(\vartheta)} \tag{12}$$

where I_i, I_e, v_i, and v_e are the ion and electron beam currents and velocities, respectively, e is the electronic charge, and ϑ is the angle of intersection of the two beams. F is the effective height of the region of beam overlap which is determined by measuring the geometrical current profile of the two beams. C_n is the count rate of neutral products.

The merged-beam and electron-cooler techniques operate in essentially the same fashion, namely that an electron beam of well-defined energy is made to intersect a high-energy ion beam and the resulting neutral species are detected. The fundamental difference between these techniques and the inclined-beam method is that in these cases, the angle of intersection of the beams is zero so that collision energies close to zero (limited only by energy resolution considerations) can be achieved. In the inclined-beam case, there is a definite lower limit to the collision energy. This

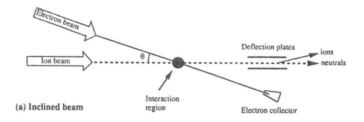

(a) Inclined beam

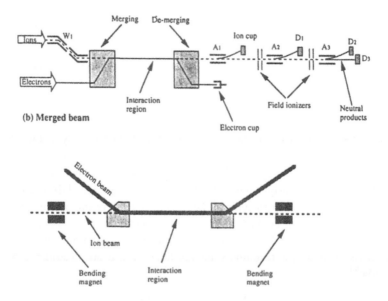

(b) Merged beam

(c) Electron-cooler section of ion storage ring.

Figure 1. Comparison of the three types of colliding beam experiments. The top figure (**a**) shows the geometry of the inclined beam apparatus (as used in ref. 30). Here the beams intersect at an angle θ in the interaction region and the residual ions are separated from the neutral recombination products by deflection plates. In the merged-beam apparatus (**b**) (ref. 31) the angle between the electron and ion beams is reduced to zero by using a merging magnet. At the end of the interaction region, the beams are separated and residual ions are removed by deflection plates (A_1). The neutral products of recombination are detected by detector D_3. The field ionizers (A_1 and A_2) in conjunction with ion detectors D_1 and D_2 are used to detect neutral products in high Rydberg states. As in the merged-beam apparatus, the colliding beams in an ion-storage ring (**c**) (ref. 32). also intersect at zero angle. Only the interaction region and the "electron cooler" are shown. The ion beam circulates in the storage ring and traverses the interaction region many times.

can be seen from the equation describing the collision energy in the center-of-mass frame of reference,

$$E_{cm} = E_i + E_e - 2\sqrt{E_i}\cos\vartheta, \tag{13}$$

where E_i and E_e are the ion and electron energies in the laboratory frame. The equation used to determine the cross section is somewhat different from that used in the inclined-beam experiment, namely,

$$\sigma = \frac{e^2 C_n}{I_i I_e L} \frac{v_i v_e}{|v_i - v_e|} F, \tag{14}$$

where the variables are the same as those in Eq. 12. In this case, F is a two-dimensional quantity, representing the effective collision area and L is the length of the interaction region.

The primary experimental differences between the three techniques are in the beam energies and currents and in the physical scale of the apparatuses, as will be discussed below.

Let us start by examining some reaction processes that seem to be rather well understood and where there is good agreement between the various techniques. The recombination of electrons with H_2^+ ions has been studied by a number of groups and is such a case. Figure 2 shows results for the recombination of ions that have a vibrational distribution close to that predicted by a Franck–Condon analysis of the ionization process, i.e. by the assumption of vertical ionization from the ground vibrational level of the H_2 molecule. The vibrational levels of H_2^+ are strongly metastable and this population distribution seems to be very robust, varying little with different ion source conditions and configurations. It is possible to alter the distribution by adding gases to the ion source that selectively remove excited levels[33] or by using a laser to de-excite vibrational states (as has been demonstrated recently at the Aarhus storage ring[34]). Such measures allow one to determine cross sections that can be compared directly with sophisticated quantum mechanical calculations and this has produced results that give one great confidence that this particular reaction is well understood.

In the case of H_2^+ recombination, the electron is captured into a doubly excited neutral state that crosses the ion state close to the $v = 1$ level. This direct recombination mechanism dominates the reaction. The presence of neutral Rydberg states lying just underneath the ion curve interferes with the direct capture producing window resonances that show up as narrow dips in the cross section[35] (see the solid curve in Figure 2). The dips arise because vibrationally excited levels of these states lie above the $v = 0$ level of the H_2^+ ion and so appear as resonances in the electron-scattering process. Capture into these states is unlikely because of their weak vibronic coupling to the electron–ion scattering state. Being bound states, however, the wave function for the system is localized in these configurations at the resonant energies and this has the effect of turning off the direct process, leaving only the weak indirect capture. The end result is a sharp dip in the recombination cross section.

Figure 3 shows another example where colliding-beam experiments have produced results in excellent agreement with each other. This particular reaction is the recombination of H_3^+ ions with electrons producing ion-pair products, i.e.:

$$e + H_3^+ \rightarrow H_2^+ + H^- \tag{15}$$

This reaction proceeds via a direct transition from the ground electronic state of the molecular ion to the doubly excited $^1A_1'$ state of the H_3 molecule which dissociates to the ion-pair limit. The process is endothermic by 5 eV and so displays an onset threshold as demonstrated in the figure. The fact that the inclined-beam results of Peart et al.[36] have a threshold somewhat lower than the merged-beam results of Yousif et al.[37] simply shows that the former experiment used ions which had a small amount of v = 1 states populated. In the merged-beam experiment the ions were completely relaxed. Unfortunately, there has not been a storage ring measurement of this reaction as of the time of writing but it would be very interesting to make this measurement as a check of all three experimental techniques.

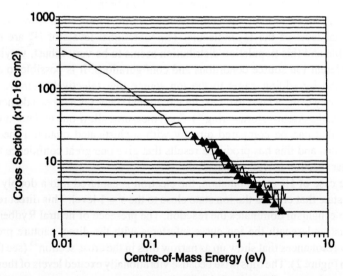

Figure 2. Dissociative recombination cross section for H_2^+ (all v) measured using the inclined-beam method (solid triangles, from ref. 30) and by the merged-beam method (curve, from ref. 31).

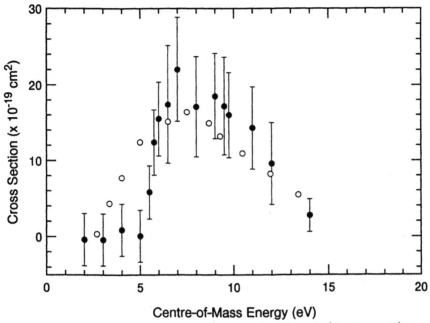

Figure 3. Cross sections for the dissociative recombination of H_3^+ leading to $H_2^+ + H^-$ formation. Inclined-beam results: open circles (from ref. 36). Merged-beam results: closed circles (from ref. 37).

If we examine the dissociative recombination of H_3^+ to produce neutral products, we see that the situation regarding agreement between various experimental techniques is far from satisfactory (see Figure 4). The figure shows results for cold ions obtained using the merged-beam and the inclined-beam[38] methods, respectively. It can be seen that the latter results are 2 orders of magnitude larger than the merged-beam results for energies in excess of 0.5 eV! The MEIBE results are in reasonable agreement with the storage ring results in this energy region although they are lower by about a factor of 5 than the storage ring data at low energies.[28] Clearly there is an experimental factor that is producing this surprising discrepancy. All of the experiments were conducted under ultrahigh vacuum conditions so third-body effects can be discounted. What is different between the three experiments is the magnitude of electric fields experienced by the ions and neutral products in each case. In the inclined-beam experiment, the primary ion beam had an energy of about 30 keV and so the deflection field used to separate out the primary ion beam from the recombination products had a magnitude of about 200 V/cm. The merged-beam experiment uses a 400 keV ion beam and a deflection field of 3000 kV/cm separates the ion beam from the products. In the storage-ring case, the ion beam is maintained in its periodic orbit by means of strong magnetic fields that produce a Lorentz field of about 500 kV/cm, the beam having an energy of 39 MeV.

Since the formation of high Rydberg states is not a major mechanism in H_2^+ recombination, the presence of electric fields has little or no effect on the total cross section. This is also the case with the H_3^+ ion-pair reaction since this proceeds via a direct electronic transition from the initial ion ground state to a doubly excited

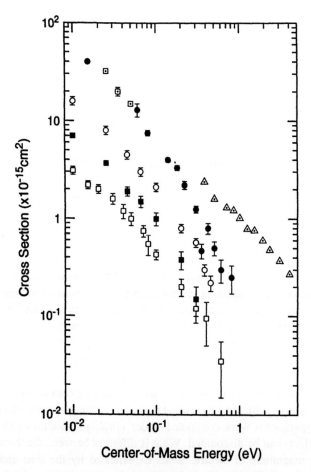

Figure 4. Comparison of measured dissociative recombination cross sections for H_3^+. Inclined-beam results for cold ions (triangles with dots, from ref. 38). Afterglow results for cold ions (open squares with dots, from ref. 15). Merged-beam results (from ref. 23) for excited ions (solid circles), for cold ions (open squares) (from ref. 23), and for ions with intermediate vibrational excitation (solid squares and open circles).

neutral state that has the ion pair as its asymptote. H_3^+ recombination to form neutral products is a very different situation since this process does not proceed via direct electron capture. This is because the only doubly excited neutral state that can participate in the reaction intersects the ion ground state far from the $v = 0$ level [12] (see Figure 5). Two theoretical arguments have been advanced to explain the large recombination cross section observed in experiments. Guberman[39] has proposed that recombination can occur with substantial rates even when there is no curve-crossing but when a dissociating neutral state lies close the ion ground state. He has estimated that this can explain the large cross sections although detailed calculations have not been performed. An alternative model, proposed by Bates[40] and further developed by Flannery[41] pictures the initial electron capture as occurring into an autoionizing, vibrationally excited level of a neutral Rydberg state that goes on to make a number of horizontal transitions to lower n, but more vibrationally excited Rydberg states. As these transitions occur, the molecule expands until

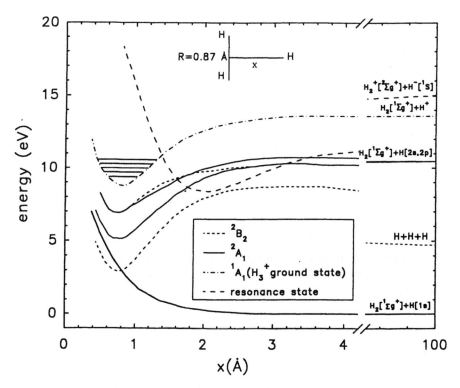

Figure 5. Potential energy curves for H_3 and H_3^+ in C_{2v} symmetry (redrawn from ref. 12).

overlap with the ground state (or some other dissociating state) is sufficiently large to produce a stabilizing transition. In essence, these two models are not so incompatible; the only difference being in the need for the participation of intermediate bound Rydberg states in the latter picture.

If we suppose that these Rydberg states have non-negligible lifetimes against autoionization, as could occur via dispersion of the potential energy of the molecule into various vibrational modes, then these states would be very sensitive to electric fields in the measuring apparatus. It is suggested here that this is the reason for the large discrepancies between cross sections measured in the various beam experiments.

In a recent experiment performed using the MEIBE apparatus, Mitchell et al.[42] varied the deflection field used to separate out the primary ion beam (see Figure 6) following the interaction region and found that when this field strength was reduced from 3000 V/cm to about 200 V/cm, the measured recombination signal increased by about a factor of 5, bringing the measured cross section in line with that obtained in storage ring [27,28] and afterglow measurements.[15,16] Careful experimental checks were performed to ensure that true signals were being measured and not artifacts due to space charge interaction between the electron and ion beams. This measurement provides strong evidence that the recombination of H_3^+ proceeds via the formation of molecular Rydberg states that subsequently predissociate to form the final recombination products. Evidence for long lived H_3^* molecules has also been found in previous MEIBE experiments [43] in which the branching ratios for the final reaction channels, i.e.,

$$e + H_3^+ \rightarrow H_2 + H \tag{16a}$$

$$\rightarrow H + H + H \tag{16b}$$

$$\rightarrow H_3^* \tag{16c}$$

were determined. In this experiment, which employed ground-state ions, it was found that channels (16a) and (16b) contributed about 40 and 52%, respectively, to the total cross section while channel (16c) accounted for about 8%. Recent storage ring measurements[44] of these branching ratios produced very different results. These indicated that channel (16b) dominated channel a by a factor of 2 to 1 and that channel (16c) was negligible. If the recombination proceeds via the formation of an intermediate Rydberg state, this could explain this discrepancy given the very different electric fields experienced when the ions and neutrals are separated in the two apparatuses. What is difficult to explain, however, is why the storage ring experiments, which have much higher separation fields, do not show evidence for signal loss due to the field ionization of Rydberg states.

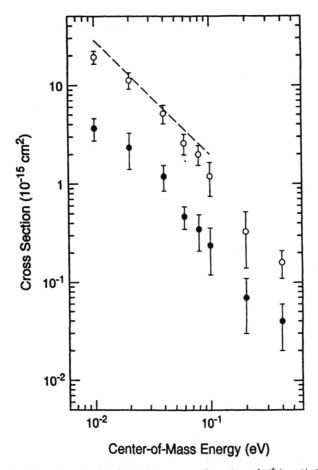

Figure 6. Cross sections for the dissociative recombination of H_3^+ ($v = 0$). Storage-ring results (ref. 28): dashed line. Merged beam results (ref. 42) for a deflector field of 3kV/cm (closed circles) and a deflector field of 200 V/cm (open circles).

A possible explanation for the above is that both high ($n > 30$) and low Rydberg states ($n = 7$ and below) are created in the recombination process. The former could arise via a mechanism such as that proposed by Guberman,[39] and the latter via Bates' mechanism.[40] The former would be field ionized in the MEIBE apparatus and in a storage ring experiment. Low Rydberg states would pass through the fields in the MEIBE apparatus without effect but such states could be subject to field-induced dissociation[45] in the high Lorentz fields, encountered as the neutral particles exit the storage rings through the bending magnet. This could give rise to an enhanced measured recombination signal and explain why a larger recombination rate is found in these experiments. This is certainly speculative but experiments are being planned for the MEIBE in which such an effect will be searched for.

C. Flowing-Afterglow Experiments in Recombining H_3^+ Plasmas

Since the flowing-afterglow method is quite well established, a brief review of the basic features of this technique should be sufficient. The experiments of Gougousi et al.[46] on H_3^+ recombination and those of Adams et al.[18] and of Smith and Spanel[24] used nearly the same experimental method, but there are significant differences in the data analysis and in the interpretation of the results.

In such flow-tube experiments, helium carrier gas is ionized and excited to the metastable state by passing it through a microwave discharge (see Figure 7). Argon (at a pressure of 10 to 15% of that of helium) is added downstream to convert metastable helium atoms to argon ions by Penning ionization. The electron density after addition of argon is typically 4×10^{10} cm^{-3}.

To create H_3^+ ions, high concentrations of H_2 are added (from 10^{14} to 10^{15} cm^{-3}). The ion–molecule reactions that occur are,

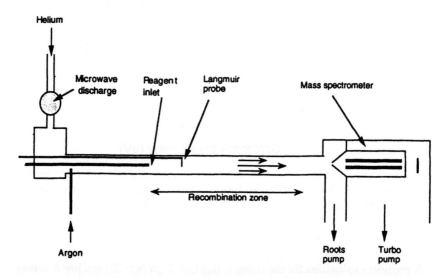

Figure 7. Schematic diagram of a flowing-afterglow electron-ion experiment. The diameter of flow tubes is typically 5 to 10 cm and the length is 1 to 2 meters. The carrier gas (helium) enters through the discharge and flows with a velocity of 50 to 100 m/s towards the downstream end of the tube where it exits into a fast pump. Recombination occurs mainly in the region 10 to 20 cm downstream from the movable reagent inlet, at which the ions under study are produced by ion-molecule reactions. The Langmuir probe measures the variation of the electron density in that region. A differentially pumped mass spectrometer is used to determine which ion species are present in the plasma.

$$Ar^+ + H_2 \rightarrow Ar\,H^+ + H\ (+\,1.53\ eV) \tag{17}$$

and:

$$ArH^+ + H_2 \rightarrow Ar + H_3^+\ (+\,0.57\ eV) \tag{18}$$

The decay of the electron density downstream from the hydrogen inlet is then measured using a movable Langmuir probe. To obtain the volume loss rate of electrons, Gougousi et al. constructed numerical models that include diffusion in both the axial and the radial directions. This is an improvement over the simpler form of analysis used by Smith and Spanel.[29] However, the latter authors' methods should have been sufficiently accurate for the large flow tube that they used. Both groups of authors tested their procedures in recombining O_2^+ plasmas and were able to reproduce the well-known recombination coefficient of O_2^+. Hence, there are no reasons to suspect serious problems in separating the volume loss of electrons from the wall loss.

The results obtained in H_3^+ plasmas by Gougousi et al. confirmed the most important experimental observations of Adams et al. and of Smith and Spanel; namely that the apparent recombination (or "deionization") coefficient decreases with afterglow time (or position) to a low value in the late afterglow (large distances from the H_2 inlet). An example taken from Gougousi et al. is shown in Figure 8. The graphs shown by Smith and Spanel[29] are remarkably similar. There appears to be little doubt that the plasma decay does not follow a simple recombination law.

There is, however, one point where the different experiments do not agree: Gougousi et al. found that the plasma decayed *faster* when the H_2 concentration was increased. They concluded this from a large set of data over a significant range of H_2 densities and they found the same to be true when D_3^+ ions were studied in the presence of D_2. Smith and Spanel carried out tests over a smaller range of H_2 concentrations. Their data show the opposite dependence on H_2 concentration, but unfortunately the authors discontinued this set of measurements since they became concerned about an increase in the concentration of impurity ions.

The afterglow data of Gougousi et al. have two features that need to be explained: The first is the observation (a), also made by Adams et al. and Smith and Spanel, that the "deionization coefficient β declines with increasing time. The other observation (b) is that hydrogen addition enhances the rate of deionization.

The interpretation of observation (a) given by Smith and Spanel [29] relies on the assumption that the dominant fraction of the H_3^+ ions are vibrationally excited to $v > 3$ and that those recombine first. The remaining, slowly recombining ions then should be $v = 0$ ions (with perhaps an admixture of $v = 1$). There is, however, one serious problem with this interpretation that has been noted earlier.[21,22] Reactions 18 and 20 that are used to create H_3^+ ions release sufficient energy to produce H_3^+ in vibrational states up to $v = 5$. However, even if highly excited ions were produced

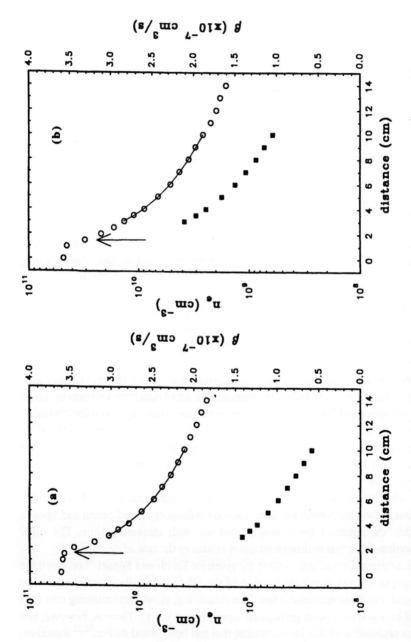

Figure 8. Measured values of the deionization coefficient β(z) (filled squares, right-hand scale) in H_3^+ plasmas derived from measured $n_e(z)$ data (open circles, left-hand scale). The H_2 concentration was 1.3×10^{14} cm^{-3} for (**a**) and 1.7×10^{15} cm^{-3} for (**b**).

(which is not at all obvious), those with $v > 1$ would be destroyed rapidly by proton transfer to Ar:

$$H_3^+ + Ar \rightarrow Ar\,H^+ + H_2 \tag{19}$$

This leaves only H_3^+ ions in the vibrational ground state, in the $v_2 = 1$ bending-mode vibration at 0.3126 eV, and in the $v_1 = 1$ breathing mode vibration at 0.394 eV (see Lie and Frye [47] or Oka and Jagod[6]). Since the argon density in these experiments is quite high ($\sim 5 \times 10^{15} \text{cm}^{-3}$), $v > 1$ ions would be destroyed in less than 1 µs. This is an important point, since Smith and Spanel's proposed reconciliation of theory and experiment rests on the assumption that vibrationally excited ions dominate the plasma.

One can go a little further in narrowing down the expected vibrational distribution. While radiative lifetimes of the $v_1 = 1$ and $v_2 = 1$ levels (1.2 s and 4 ms, respectively) [19] are too long for radiative relaxation to be important, quenching of $v_2 = 1$ ions in collisions with H_2 seems to be quite fast. Amano's [48] absorption studies of the v_2 hot band indicate that the $v_2 = 1$ level is quenched by H_2 with a rate coefficient of approximately 3×10^{-10} cm^3/s. This eliminates $v_2 = 1$ ions since they will be destroyed in 30 µs at $[H_2] = 1 \times 10^{14} \text{cm}^{-3}$. Experimental evidence [49–51] suggests a smaller quenching coefficient ($k \sim 10^{-12}$ cm^3/s) for the $v_1 = 1$ state. Thus, the only vibrationally excited state that may survive long enough to undergo recombination should be the $v_1 = 1$ state.

Gougousi et al. first attempted to explain their findings by assuming the presence of H_3^+ ions in two vibrational states ($v = 0$ and $v_1 = 1$) with two different recombination coefficients, and quenching of the $v_1 = 1$ ions by H_2. This two-state model did not produce a consistent, quantitative fit to the data and this interpretation was abandoned.

The anomalous plasma decay suggests that the deionization coefficient is larger at higher electron densities and higher H_2 densities. The second interpretation proposes that electrons and H_2 molecules can act as stabilizing third bodies and that this process involves the same long-lived intermediate complexes that are observed in merged beam experiments (see Section IV.B).

A multistep mechanism may occur in the following form: First, an electron of low angular momentum (s or p) is captured into a rovibrational level of a Rydberg state of H_3;

$$e^- + H_3^+ \rightarrow H_3^{**} \ (v = 1, n, s) \tag{20}$$

This is the same step that is invoked in the indirect recombination process. The $n = 7$ Rydberg state would be a good candidate since excitation to the $v_2 = 1$ state (0.3126 eV) in $n = 7$, $l = 0$ is resonant at an electron energy of +0.0309 eV (1.2 kT at 300 K).

In the second step the angular momentum of the Rydberg molecule may be randomized in collisions with ambient electrons (l-mixing and possibly m-mixing), thus further enhancing the lifetime against auto-ionization. Since the l-mixing rate coefficient of thermal electrons is exceedingly large ($>10^{-3}$ cm³/s),[52] l-mixing should occur in less than 10^{-7} s at electron densities near 4×10^{10} cm⁻³. This time is comparable to the lifetimes of H_3^{**} autoionizing states ($\sim 10^{-7}$ s).[53]

Once formed, the long-lived H_3^{**} ($n = 7$, $v = 1$, $l > 1$) molecules can be returned to the autoionizing state (s or p) by further electron collisions. Also, they may be quenched to lower n-states by neutral atoms or electrons, they may undergo slow predissociation, radiate to a lower state, or react with an H_2 molecule. The reaction scheme is shown in Figure 9. Several of the decay rates can be estimated with some confidence. Collisional energy loss (n-changing) with helium atoms is not impor-tant[54] compared to radiative decay. The rate coefficient for n-changing collisions of electrons with $n = 7$ Rydberg atoms (electron quenching) has been measured by Devos et al.[43] to be $\sim 10^{-5}$ cm³/s. Radiative decay rates for $n = 7$ hydrogenic states are close to 1×10^6 s⁻¹ for s-states, 2×10^7 s⁻¹ for p-states, and about 2×10^6 s⁻¹ for a statistical mixture of l-values up to $l = 6$.[55] The reaction with H_2,

$$H_3^{**} + H_2 \rightarrow H_2 + H_3^* \tag{21}$$

will compete with radiative decay only if it occurs on a time scale of < 1 μs. At $[H_2] = 1 \times 10^{15}$ cm⁻³, a rate coefficient of $\sim 1 \times 10^{-9}$ cm³/s would be sufficient.

It is a straightforward task to construct a rate model for the reaction scheme shown in Figure 9. Electron capture and its inverse (autoionization),

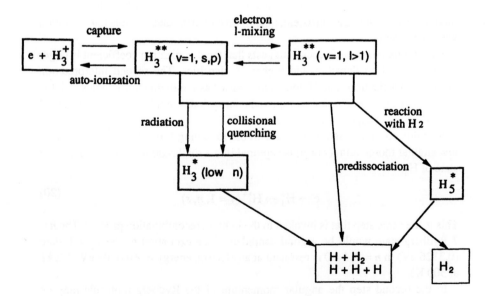

Figure 9. Schematic representation of the reactions in H_3^+ recombination.

$$H_3^+ + e^- \leftrightarrow H_3^{**} \ (n = 7, s, p) \tag{22}$$

are described by a capture coefficient k_c and an autoionization rate ν_a. Stabilization by l-mixing due to electrons,

$$H_3^{**} \ (n = 7, s, p) + e^- \leftrightarrow H_3^{**} \ (n = 7, l > 1) + e^- \tag{23}$$

and its reverse are described by rate coefficients k_s and k_i, respectively. The destruction mechanisms for H_3^{**} are:

$$H_3^{**} \ (n = 7) \qquad \rightarrow \qquad H_3^{**} \ (n < 7) + h\nu \ (\text{radiation, rate } \nu_{rad}) \tag{24}$$

$$H_3^{**} \qquad \rightarrow \qquad H_2 + H \ (\text{or 3H}) \ (\text{predissociation, rate } \nu_p) \tag{25}$$

$$H_3^{**} \ (n = 7) + e^- \quad \rightarrow \qquad H_3^* \ (n < 7) + e^- \ (\text{rate coefficient } k_{qe}) \tag{26}$$

$$H_3^{**} + H_2 \qquad \rightarrow \qquad H_2 + H_3 \ (\text{low } n) \ (\text{rate coefficient } k_H). \tag{27}$$

The rate coefficients for H_3^{**} in the autoionizing states (s, p) may differ from those in higher l-states and are treated differently (subscripts "0" and "1"). A steady-state analysis of the rate equations gives a deionization coefficient of:

$$\beta = -\frac{1}{n_e^2}\frac{dn_e}{dt} = \frac{k_c}{1 + \dfrac{1}{\dfrac{n_e k_s \nu_l}{(n_e k_i + \nu_l)\nu_a} + \dfrac{\nu_0}{\nu_a}}} \tag{28}$$

where $\nu_0 = \nu_{rad,0} + \nu_{p,0} + k_{qe}\, n_e + k_H\,[H_2]$
and $\nu_l = \nu_{rad,1} + \nu_{p,1} + k_{qe}\, n_e + k_H\,[H_2]$

Gougousi et al. showed that it is possible to find a set of plausible rate coefficients so that Eq. 28 gives a reasonable fit to their data for different H_2 concentrations. Figure 10 shows a comparison to experimental values of $\beta \ (n_e)$ for the following set of rate coefficients:

$$k_c = 4.4 \times 10^{-7} \ \text{cm}^3/\text{s}, \ \nu_a = 1.3 \times 10^8 \ \text{s}^{-1}, \ k_s = 1.25 \times 10^{-2} \ \text{cm}^3/\text{s}$$

$$k_i = 3.1 \times 10^{-4} \ \text{cm}^3/\text{s}, \ \nu_{rad,0} = \nu_{rad,1} = 1.1 \times 10^6 \ \text{s}^{-1}$$

$$\nu_{p,0} = \nu_{p,1} = 0, \ k_H = 4.5 \times 10^{-9} \ \text{cm}^3/\text{s}, \ k_{qe} = 1 \times 10^{-5} \ \text{cm}^3/\text{s}$$

It is clear that this set of model parameters is not uniquely determined by experiment, but the fitting parameters appear to be reasonable. The model proposed

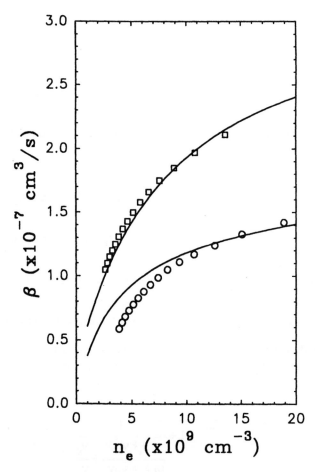

Figure 10. Comparison of measured values of $\beta(n_e)$ for $[H_2] = 1.3 \times 10^{14}$ cm^{-3} (circles) and $[H_2] = 1.7 \times 10^{15}$ cm^{-3} (squares) to the model described in the text.

by Gougousi et al. has the virtue that it reconciles the results of merged-beam experiments (see Section IV.B) with the afterglow observations. Also, the postulated three-body mechanism (Figure 9) eliminates the need to explain recombination without curve crossing since no such crossings are required. The binary recombination coefficient of H_3^+ ions may be far smaller than the deionization coefficients measured in afterglow experiments.

The experiment, however, does not provide conclusive proof for three-body recombination. Perhaps, spectroscopic studies of emissions in recombining H_3^+ plasmas may shed further light on the mechanism. Two interesting observations were made by Amano [56,57] and by Miderski and Gellene.[58] They found that H$_3$ (or

D_3) absorption and emission lines (from $n = 3$ states) in H_2 (D_2) plasmas were strongly Doppler-broadened which seems to indicate high, nonthermal energies (about 0.3 eV) of the absorbing or emitting H_3 molecules. The energy is close to that expected if the excited ($n = 3$) H_3 molecules were formed by recombination of H_5^+, but in Amano's work no H_5^+ ions should have been present. Perhaps, the fast H_3 molecules are produced from H_3^{**} + H_2 collisions , and the spectroscopic observations provide indirect evidence for the existence of H_3^{**} molecules. The conjecture needs to be examined by more detailed work.

One may ask why some experiments, for instance those done by microwave–afterglow technique [15,16] and the experiments by Canosa et al., [21,22] gave no indications of an anomalous decay. In part, the answer may be that small variations of the deionization coefficients are not easily detected in the presence of ambipolar diffusion. They were detected in the work of Adams et al. and of Smith and Spanel only because the diffusion losses were unusually slow in their large flow tube.

V. THREE-BODY EFFECTS IN EARLY RECOMBINATION MEASUREMENTS

In the following section, we will reexamine some earlier recombination measurements to see if third-body effects may have played a role. The rare gas dimer ions provide a good starting point.

A. Rare-Gas Dimer Ions

Not surprisingly, a rather extensive set of experimental data has been accumulated on the recombination of rare-gas dimer ions, e.g. He_2^+, Ne_2^+, etc., with electrons. Here, the binding energy of the molecule is sufficiently small (1 to 1.5 eV) that the atomic fragments of dissociative recombination can be formed in a variety of excited states, many of which decay by radiation. With the exception of He_2^+, the thermal energy recombination coefficients for Ne_2^+, Ar_2^+, Kr_2^+, and Xe_2^+ are very large (from 1.75×10^{-7} cm^3/s for Ne_2^+ to 2.3×10^{-6} cm^3/s for Xe_2^+). Bates[59] placed them into the category of "super dissociative recombination." He argued that the capture of an electron into a repulsive dissociating state is more efficient when the slope of the repulsive potential curve is small. This situation exists in the rare-gas dimer ions, but Bates' proposal suffers somewhat from the problem that dissociation along a weakly repulsive potential takes a longer time and hence more time will be available for ejecting the captured electron by autoionization. Bates apparently was aware of the problem but did not pursue it.

Many recombination and spectroscopic studies have been carried out in decaying rare-gas plasmas. Invariably, the buffer gas was the parent gas of the ion under study. In addition to recording the emitted spectra, Frommhold and Biondi [60] examined the line shapes of many afterglow lines in neon and in argon. The intent was to find spectroscopic evidence for Doppler broadening of the lines which would prove that kinetic energy is released in the dissociative mechanism. In some cases, the measured line shapes exhibited broadening and this is often quoted as evidence for the binary dissociative recombination mechanism. The evidence, however, is not really as convincing as one might wish it to be. Let us examine the results of those measurements.

The afterglow data for Ne_2^+,[60] Ar_2^+,[61] Kr_2^+,[62] and Xe_2^+ [63] showed that the total recombination coefficients, as determined from the decay of the electron density, were sensibly independent of the gas pressure. Typically, the pressure was varied by at least a factor of 3 and no systematic variation of the total recombination coefficient was found. The observed plasma decay was compatible with the assumption that the total recombination coefficients are independent of the electron density (in the range from approximately 10^9 to 10^{10} cm^{-3}).

Numerous visible atomic spectral lines were observed during the afterglow. The decay of the intensities of these lines followed approximately the n_e^2 dependence that is expected for lines produced by recombination. However, the absolute intensities of the lines were not measured and no partial recombination coefficients were obtained. Since only a limited spectral range was examined, many lines that might be expected to come from recombination were therefore not observed.

Spectral line shape studies by Frommhold and Biondi[60] (using Fabry–Perot interferometry) indicated measurable broadening of several lines coming from Ne_2^+ and Ar_2^+ recombination. The authors showed that dissociative recombination,

$$X_2^+ + e^- \rightarrow X + X^* + \text{kinetic energy} \tag{29}$$

from a single ionic state would lead to a Doppler-broadened line of trapezoidal shape. However, only one line shape (Ne 585.2 nm, $2p_1 - 1s_2$) was reasonably close to the expected trapezoidal profile. Other lines showed a more rapid fall-off of the intensity away from line center and the lines exhibit barely discernible structures (called "shoulders" by Frommhold and Biondi) which were tentatively ascribed to recombination of vibrationally excited ions. A repetition of the experiment by Edwin and Turner [64] confirmed the broadening of the Ne 585.2 nm but failed to show evidence for broadening of other afterglow lines in neon and krypton.

In later work, Malinovsky et al.[65] found that the relative intensities of spectral lines observed in neon afterglows vary with gas pressure. Rather surprising observations were made by Ramos et al.[66] and by Barrios et al.[67] who used a time-of-flight method to determine the kinetic energy spectrum of atoms from discharges in neon, argon, and krypton. The authors arrived at the conclusion that most recombination events produce states that are not spectroscopically observable. They also report an unexpectedly large production of kinetically excited ground-state atoms.

The various observations leave little doubt that recombination of rare-gas dimer ions can produce excited atoms, but it is quite possible that formation of radiating excited states occurs only in a small fraction of recombination events. If this is true, then the observed line shapes do not reflect the dominant recombination mechanism. One explanation of the observed line shapes may be that the excited atomic fragments suffer excitation transfer or state-mixing collisions before radiating. This may be one way to explain the observed pressure dependence. Another explanation may be that the atomic excitation results from a two-step process in which three-body electron–ion recombination produces a Rydberg molecule that subsequently collides with an atom:

$$e^- + X_2^+ + M \rightarrow X_2^{**} + M \tag{30}$$

$$X_2^{**} + X \rightarrow X^* + X + X \tag{31}$$

The second step will impart kinetic energy to the excited atom and thus lead to Doppler broadening of the atomic lines. As will be discussed later, the analogous mechanism is thought to account for the optical emissions in helium afterglows. Perhaps it also occurs in other rare gases. The spectral lines should have similar widths as those produced by direct dissociative recombination since nearly the same energy is released. The line shapes, however, will be different since energy and momentum are shared by three rather than by two particles.

There is another puzzle in the recombination of rare-gas dimer ions that casts some doubt on the interpretation by the standard curve-crossing mechanism. Shock-tube measurements[68] have shown that the recombination coefficients decline much faster with gas temperature ($T_g^{-3/2}$) than with electron temperature ($T_e^{-1/2}$). The authors of the shock-tube work explained this by a model proposed by O'Malley[69] which assumes that vibrationally excited dimer ions recombine only very slowly. Such a model can be made to fit the shock-tube data fairly well, but as Bardsley[70] pointed out, it would be surprising if many dissociating potential curves intersected the vibrational ground state while very few crossed vibrationally excited states. Why should this be so for all rare-gas dimer ions except He_2^+? There is no good resolution of this puzzle: It is hard to see which effect other than vibrational (or perhaps rotational) excitation could have suppressed the recombination at elevated gas temperatures in the shock-tube experiments. Bardsley's criticism assumes that the curve-crossing mechanism is dominant. Perhaps the recombination of vibrationally excited is indeed slow, but a recombination mechanism other than the curve-crossing mechanism occurs in these ions.

The recombination of He_2^+ is a special case. We include it here because of the similarities with H_3^+ and because it is the only known example where three-body recombination of a diatomic molecular ion dominates over the binary process. The literature on the helium afterglow is quite large and we will not be able to do justice to all aspects of this problem. Mulliken[71] had predicted that fast dissociative recombination of He_2^+ should not occur due to a lack of a suitable curve crossing between the ionic potential curve and repulsive curves of He_2^*. Afterglow experiments in pure helium, at sufficient pressure to enable formation of He_2^+ ions, have confirmed this expectation. It does not appear that the true binary recombination coefficient of He_2^+ has ever been established reliably, but there is a wealth of data that supports the three-body mechanism of recombination. The observed pressure dependences show that both electrons and atomic ions can act as stabilizing third bodies. The dependence on electron density was found to be well described by the collisional–radiative mechanism, but the surprisingly strong dependence on the helium density initially was hard to explain. Following a suggestion of Collins and Hurt,[72] Bates[73] proposed that (electron-stabilized) recombination of He_2^+ ions produces a population of highly excited He_2^* Rydberg molecules (with quantum numbers $n > 3$) and that these are stabilized by reactions of the form:

$$He_2^* + H_e \leftrightarrow He_3^* \rightarrow He^* + 2\,He \tag{32}$$

The second step in the reaction, dissociation of the He_3^* Rydberg molecule, is similar to dissociative recombination of He_3^+ with a free electron. For this reason, Bates[73] called this recombination mechanism "Rydberg dissociative recombination." It enhances the overall loss rate of free electrons because the stabilization of He_2^* prevents the return of "weakly bound" electrons to the population of free electrons. The reaction plays the same role as the reaction of H_3^{**} with H_2 that we discussed in Section IV.C. As has been discussed by Bates, the mechanism also provides an explanation for spectroscopic observations of atomic and molecular emissions in helium afterglows. There is direct evidence for the existence of a substantial population of weakly bound electrons in helium afterglows.[74] Most likely, the weakly bound electrons are Rydberg electrons in He_2^* molecules.

B. Diatomic Atmospheric Ions

There is little reason to believe that complex formation plays a rate-controlling role in the recombination of atmospheric diatomic ions (e.g. O_2^+, N_2^+, NO^+). For instance, in the case of O_2^+ ions, the agreement between afterglow,[75] ion-trap[76] and merged-beam experiments,[77] and ionospheric satellite observations[78] strongly suggests that the total recombination coefficient is due to simple binary recombination. This does not mean, however, that the yield for specific product channels is likewise independent of the environment (electron and neutral densities). It is interesting to note that the best theoretical calculations of Guberman and Giusti-Suzor[79] indicate that $O(^1S)$ should arise in only 0.26% of O_2^+ $(v = 0) + e^-$ recombination events. While this is a "minor" channel, it is nevertheless geophysically important since it produces the airglow emission at 557.7 nm. Laboratory measurements in afterglow plasmas (electron densities from 10^{10} to 10^{12} cm^{-3}) have consistently given yields of $O(^1S)$ up to 10% (see Bates[80] for a very detailed review of experimental data and ionospheric implications). The reasons for differing $O(^1S)$ yields are by no means clear, but it is conceivable that the $O(^1S)$ yield measured in the laboratory may in part result from three-body recombination.

C. Three-Body Effects in Recombination of Polyatomic Ions

Complex formation may be most important in the recombination of large polyatomic ions but there is presently no direct experimental support for this conjecture. We also do not have a clear qualitative understanding how polyatomic ions recombine. Should one view the process as a direct process in which a particular bond is rapidly broken? Bates[81] attempted to rationalize experimental observations using this picture. An alternative view is that the capture of the electron is followed by rapid intramolecular energy transfer. If the complex is sufficiently long-lived, many dissociative decay channels are available and the probability of a given one should depend largely on its statistical weight. This approach has been taken by Herbst.[82] Long-lived complexes, however, can also transfer energy to third particles and can become stabilized. Abouelaziz et al.[83] recently suggested such a three-body (or radiation-stabilized) mechanism for aromatic cyclic ions that they studied using the afterglow method. If this indeed occurs, then one would expect that total recombination coefficients and yields for specific recombination products depend on gas density or electron density.

Large cluster ions (e.g. $H_3O^+ (H_2O)_n$) provide good test cases for recombination mechanisms of polyatomic species. Some early experiments[84] seemingly showed that recombination rates of water-cluster ions were nearly independent of temperature and it was first thought that this indicated a different recombination mechanism for cluster ions. Later work[85] showed that water-cluster ions actually exhibit the usual $T^{-1/2}$ temperature variation that is expected for the curve-crossing mechanism. This, of course, does not prove that this is the dominant mechanism. Smirnov[86] has pointed out that even a simple classical recombination model yields a $T^{-1/2}$ variation.

VI. CONCLUDING REMARKS

Our discussion of complex formation in electron–ion recombination, field effects, and three-body recombination has perhaps posed more questions than it has answered. In the case of H_3^+ recombination, the experimental observations suggest but do not prove that complex formation is an important mechanism. Three-body recombination involving complex formation is not likely to have much effect on the total recombination coefficients of diatomic ions, but it may alter the yield of "minor" product channels. Complex formation may be most prevalent in the case of large polyatomic ions, but there is a serious lack of experimental data and theoretical calculations that can be adduced for or against complex formation.

Our lack of knowledge concerning the mechanisms of molecular recombination has serious consequences for plasma modeling. If the measured data are subject to three-body and field effects, then their application to low-density plasmas (e.g. space plasmas) is questionable. For this reason, studies that focus on recombination mechanisms probably deserve more attention that they have received in recent years.

ACKNOWLEDGMENTS

This work was, in part, supported by NASA under Grant NAGW 1764 and by the Canadian Natural Sciences and Engineering Research Council.

REFERENCES

1. Guberman, S. L. *Phys. Rev. A* **1994**, *49*, R4277.
2. Dalgarno, A. *Advances in Atomic, Molecular and Optical Physics* **1993**, Vol. 32, p. 57.
3. Tennyson, J.; Miller, S.; Schild, H. *J. Chem., Faraday Trans.* **1993**, *89*, 2155.
4. Flannery, M. R. *Advances in Atomic, Molecular and Optical Physics* **1994**, Vol. 32, 117.
5. Stevefelt, J.; Boulmer, J.; Delpech, J. F. *Phys. Rev. A* **1975**, *12*, 1246.
6. Compton, R. N.; Christophorou, L. G.; Hurst, G. S.; Reinhardt, P. W. *J. Chem. Phys.* **1966**, *45*, 4634.
7. Müller, A.; Belic', D. S.; DePaola, B. D.; Djuric', N.; Dunn, G. H.; Mueller, D.W.; Timmer, C. *Phys. Rev. A* **1987**, *36*, 599.
8. Merkt, F.; Zare, R. N. *J. Chem. Phys.* **1994**, *101*, 3495.
9. Chupka, W. A. *J. Chem. Phys.* **1993**, *98*, 4520.
10. Lundeen, S. R. In *Dissociative Recombination: Theory, Experiment, and Applications;* Mitchell, J. B. A.; Guberman, S. L., Eds.; World Scientific: Singapore, 1989.
11. Müller, A.; Belic, D. S.; Paola, B. D.; Djuric, N.; Dunn, G. H.; Mueller, D. W.; Timmer, C. *Phys. Rev. A* **1987**, *36*, 599.
12. Michels, H. H.; Hobbs, R. H. *Ap. J. (Letters)* **1984**, *286*, L27.
13. Kulander, K. C.; Guest, M. F. *J. Phys. B* **1979**, *12*, L501.
14. Bardsley, J. N. *J. Phys. B* **1968**, *1*, 365.
15. Leu, M. T.; Biondi, M. A.; Johnsen, R. *Phys. Rev. A* **1973**, *8*, 413.
16. Macdonald, J. A.; Biondi, M. A.; Johnsen, R. *Planet. Space Sci.* **1984**, *32*, 651.
17. Bates, D. R.; Guest, M. F.; Kendall, R. A. *Planet. Space Sci.* **1993**, *41*, 9.
18. Adams, N. G.; Smith, D.; Alge, E. *J. Chem. Phys.* **1984**, *81*, 1778.
19. Amano, T. *Astrophysi. J.* **1988**, *329*, L121.
20. Amano, T. *J. Chem. Phys.* **1990**, *92*, 6492.
21. Canosa, A.; Gomet, J. C.; Rowe, B. R.; Mitchell, J. B. A.; Queffelec, J. L. *J. Chem. Phys.* **1992**, *97*, 1028.
22. Canosa, A.; Rowe, B. R.; Mitchell, J. B. A.; Gomet, J. C.; Brion, C. *Astron. Astrophys.* **1991**, *248*, L19.
23. Hus, H.; Yousif, F. B.; Sen, A.; Mitchell, J. B. A. *Phys. Rev. A* **1988**, *38*, 658.
24. Mitchell, J. B. A.; Forland, J. L.; Ng, C. T.; Levac, D. P.; Mitchell, R. E.; Mul, P. M.; Claeys, W.; Sen, A.; McGowan, J. Wm. *Phys. Rev. Lett.* **1983**, *51*, 885.
25. Mitchell, J. B. A.; Ng, C. T.; Forland, J. L.; Janssen, R.; McGowan, J. Wm. *J. Phys. B* **1984**, *17*, L909.
26. Yousif, F. B.; Van der Donk, P. J. T.; Orakzai, M.; Mitchell, J. B. A. *Phys. Rev. A* **1991**, *44*, 5653.
27. Larsson, M.; Danared, H.; Mowat, J. R.; Sigray, P.; Sundström, G.; Broström, L.; Filevich, A.; Källberg, A.; Mannervik, S.; Rensfelt, K. G.; Datz, S. *Phys. Rev. Lett.* **1993**, *70*, 430.
28. Sundström, G.; Mowat, J. R.; Danared, H.; Datz, S.; Broström, L.; Filevich, A.; Källberg, A.; Mannervik, S.; Rensfelt, K. G.; Sigray, P.; af Ugglas, M.; Larsson, M. *Science* **1994**, *263*, 785.
29. Smith, D.; Spanel, P. *Int. J. Mass Spec. Ion Proc.* **1987**, *81*, 67.

30. Peart, B.; Dolder, K. T. *J. Phys. B* **1974**, *7*, 236.
31. Auerbach, D.; Cacak, R.; Caudano, R.; Gaily, T. D.; Keyser, C. J.; McGowan, J. Wm.; Mitchell, J. B. A.; Wilk, S. F. *J. Phys. B.* **1977**, *10*, 3797.
32. Larsson, M.; Stromholm, C. In *Dissociative Recombination: Theory, Experiment and Applications, III*; Zajfman, D.; Mitchell, J. B. A.; Schwalm, D.; Rowe, B. R., Eds.; World Scientific: Singapore, 1996.
33. Hus, H.; Yousif, F. B.; Noren, C.; Sen, A.; Mitchell, J. B. A. *Phys. Rev. Lett.* **1988**, *67*, 1006.
34. Andersen, L. H. In *Dissociative Recombination: Theory, Experiment and Applications, III*; Zajfman, D.; Mitchell, J. B. A.; Schwalm, D.; Rowe, B. R., Eds.; World Scientific: Singapore, 1996.
35. Schneider, I. F.; Dulieu, O.; Giusti-Suzor A. *J. Phys. B* **1991**, *24*, L289.
36. Peart, B.; Forrest, R. A.; Dolder, K. T. *J. Phys. B.* **1979**, *12*, 3441.
37. Yousif, F. B.; Van der Donk, P.; Mitchell, J. B. A. *J. Phys. B.* **1993**, *26*, 4249.
38. Peart, B.; Dolder, K. T. *J. Phys. B* **1974**, *7*, 1948.
39. Guberman, S. L. In *Electronic and Atomic Collisions, Invited Papers of the XIXth ICPEAC*; Dube, L. J.; Mitchell, J. B. A.; McConkey, J. W.; Brion, C. E., Eds; AIP Press: 1996.
40. Bates, D. R. *Proc. Roy. Soc. Lond.* **1993**, *443*, 257.
41. Flannery, R. Private communication, 1995.
42. Mitchell, J. B. A.; Rogelstad, M.; Yousif, F. B. In preparation.
43. Yousif, F. B.; Rogelstad, M.; Mitchell, J. B. A. In *Atomic and Molecular Physics: Proc. 4th US/Mexico Symp.* Alvarez, I.; Cisneros, C.; Morgan, T. J. World Scientific: Singapore, 1995, p. 343.
44. Datz, S.; Sundstrom, G.; Biederman, Ch.; Brostrom, L.; Danared, H.; Mannervik, S.; Mowat, J. R.; Larsson, M. *Phys. Rev. Lett.* **1995**, *74*, 896.
45. Bordas, C.; Helm, H. *Phys. Rev. A* **1991** *43*, 3645.
46. Gougousi, T.; Johnsen, R.; Golde, M. F. *Int. J. Mass Spectr. Ion Proc.* **1995**, *145/150*, 131.
47. Lie, G. C.; Frye, D. *J. Chem. Phys.* **1992**, *96*, 6784.
48. Amano, T. In *Dissociative Recombination: Theory, Experiment and Applications*; Rowe, B. R.; Mitchell, J. B.; Canosa, A., Eds.; Plenum Press: New York, 1993.
49. Lee, H. S.; Drucker, M.; Adams, N. G. *Int. J. Mass Spectrom. Ion Proc.* **1992**, *117*, 101.
50. Blakley, C. R.; Vestal, M. L.; Futrell, J. H. *J. Chem. Phys.* **1977**, *66*, 2392.
51. Bawendi, M. G.; Rehfuss, B. D.; Oka, T. *J. Chem. Phys.* **1990**, *93*, 6200.
52. Devos, F.; Boulmer, J.; Delpech, J.-F. *J. de Physique* **1979**, *40*, 215; *Phys. Rev. Lett.* **1977**, *39*, 1400.
53. Berardi, V.; Spinelli, N.; Velotta, R.; Armenante, M.; Zecca, A. *Phys. Rev. A* **1993**, *47*, 986.
54. Bates, D. R.; Khare, S. P. *Proc. Phys. Soc.* **1965**, *85*, 231.
55. Hiskes, J. R.; Tarter, C. B. Report UCRL-7088; Lawrence Radiation Laboratory: Livermore CA, 1964.
56. Amano, T. In *Dissociative Recombination: Theory, Experiment and Applications*; Rowe, B. R.; Mitchell, J. B.; Canosa, A., Eds.; Plenum Press: New York, 1993.
57. Amano, T. SPIE, Vol. 1858, 1993.
58. Miderski, C. A.; Gellene, G. I. *J. Chem. Phys.* **1988** *88*, 5331.
59. Bates, D. R. *J. Phys. B* **1991**, *24*, 703.
60. Frommhold, L.; Biondi, M. A. *Phys. Rev.* **1969**, *185*, 244.
61. Shiu, Y.-J.; Biondi, M. A. *Phys. Rev. A* **1978**, *17*, 868.
62. Shiu, Y.-J.; Biondi, M. A. *Phys. Rev. A* **1977**, *16*, 1817.
63. Shiu, Y.-J.; Biondi, M. A.; Sipler, D. P. *Phys. Rev. A* **1977**, *15*, 494.
64. Edwin, R. P.; Turner, R. *J. Chem. Phys.* **1969**, *50*, 4388.
65. Malinovsky, L.; Lucac, P.; Hong, C. J. *Czech J. Phys. B* **1986**, *36*.
66. Ramos, G. B.; Schlamkowitz, M.; Sheldon, J.; Hardy, K. A.; Peterson, J. R. *Phys. Rev. A* **1995**, *51*, 2945.

67. Barrios, A.; Sheldon, J. W.; Hardy, K. A.; Peterson, J. R. *Phys. Rev. Lett.* **1992**, *69*, 1348.
68. O'Malley, T. F.; Cunningham, A. J.; Hobson, R. M. *J. Phys. B* **1972**, *5*, 2126.
69. O'Malley, T. F. *Phys. Rev.* **1969**, *185*, 101.
70. Bardsley, J. N. *Phys. Rev. A* **1970**, *2*, 1359.
71. Mulliken, R. S. *Phys. Rev. A* **1964**, *136*, A 926.
72. Collins, C. B.; Hurt, W. *Phys. Rev.* **1969**, *179*, 203.
73. Bates, D. R. *J. Phys. B* **1984**, *17*, 2363.
74. Kaplafka, J. P.; Merkelo, H.; Goldstein, L. *Phys. Rev. Lett.* **1968**, *14*, 970.
75. Mehr, F. J.; Biondi, M. A. *Phys. Rev.* **1969**, *181*, 264.
76. Walls, F. L.; Dunn, G. H. *J. Geophys. Res.* **1974**, *79*, 1911.
77. Mul, P. M.; McGowan, J. Wm. *J. Phys. B* **1979**, *12*, 1591.
78. Abreu, V. J.; Salomon, S. C.; Sharp, W. E.; Hays, P. B. *J. Geophys. Res.* **1983**, *88*, 4140.
79. Guberman, S. L.; Giusti-Suzor, A. *J. Chem. Phys.* **1991**, *95*, 2602.
80. Bates, D. R. *Planet. Space Science* **1990**, *38*, 889.
81. Bates, D. R. *J. Phys. B* **1991**, *24*, 3276.
82. Herbst, E. *Astrophys. J.* **1978**, *222*, 508.
83. Abouelaziz, H.; Gomet, J. C.; Pasquerault, D.; Rowe, B. R.; Mitchell, J. B. A. *J. Chem. Phys.* **1993**, *99*, 237.
84. Huang, C.-M; Whitaker, M.; Biondi, M. A.; Johnsen, R. *Phys. Rev. A* **1978** *18*, 64.
85. Johnsen, R. *J. Chem. Phys.* **1993**, *98*, 5390.
86. Smirnov, B. M. *Sov. Phys.* **1977**, JETP *45*, 4.

REACTIVE PROBING OF THE POTENTIAL ENERGY SURFACES OF ISOMERIC IONS

Nigel G. Adams and Neyoka D. Fisher

Advances in Gas-Phase Ion Chemistry
Volume 3, pages 81–123
Copyright © 1998 by JAI Press Inc.
All rights of reproduction in any form reserved.
ISBN: 0-7623-0204-6

ABSTRACT

Experimental studies are reviewed which provide information to distinguish between the different forms of the isomeric ions of HCN^+, $C_3H_3^+$, $CH_3O_2^+$, CH_5O^+, $C_2H_5O^+$, $C_2H_7O^+$, $C_3H_3O^+$, $C_3H_6N^+$, and $C_2H_5O_2^+$. Ternary and binary reactions which generate stable isomers are considered, and these isomers have been probed and identified by their reactivities with a series of molecules. The data are integrated with earlier studies: of binary reactions where the intermediate complex is an excited form of the accessible isomers, of collision induced dissociation of specific stable isomers, and of unimolecular fragmentation of excited metastable isomeric forms of stable ions. The insights from these data are combined with theoretical deductions about the form of the potential energy surfaces where such data on energetics are available.

I. INTRODUCTION

As reacting chemical systems AB^+ + CD become more complex, there is an increasing possibility of intermediate complexes, $ABCD_i^+$ (with excitation much above the minimum of the well), which have different isomeric forms. Indeed, isomeric forms of the reactants and products will also become possible. For such reactions, the multidimensional potential surfaces (or more restrictively, reaction coordinate diagrams) may be represented by a series of potential minima, usually of unequal depth, separated by transition states, i.e. by barriers of unequal heights. Such a surface can terminate in several directions in dissociation products (in principle, in all directions, but generally only channels leading to thermally accessible products are of interest and this makes directions unimportant where there is a great deal of fragmentation). Such a potential surface is represented schematically in Figure 1.

Reactants AB^+ + CD are considered to associate to form a weakly bonded intermediate complex, $AB^+ \cdot CD$, the ground vibrational state of which has a barrier to the formation of the more strongly bound form, $ABCD^+$. The reactants, of course, have access to both of these isomeric forms, although the presence of the barrier will affect the rate of unimolecular isomerization between them. Note that the minimum energy barrier may not be accessed in a particular interaction of AB^+ with CD since the dynamics, i.e. initial trajectories and the detailed nature of the potential surface, control the reaction coordinate followed. Even in the absence (left hand dashed line in Figure 1) of a formal barrier (i.e. of a local potential maximum), the intermediate will resonate between the conformations having $AB^+ \cdot CD$ or $ABCD^+$ character. These complexes only have the possibilities of unimolecular decomposition back to AB^+ + CD or collisional stabilization. In the stabilization process,

$$(ABCD^+)^{**} + M \rightarrow (ABCD^+)^* + M^* \tag{1}$$

only a small amount of energy needs to be taken away by the collider M for the lifetime against unimolecular decomposition to be increased to such an extent that

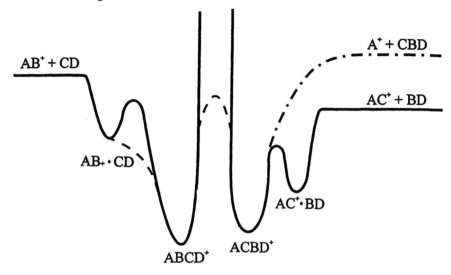

Figure 1. Schematic diagram of a hypothetical potential energy hyper-surface for (ABCD)⁺, constructed to illustrate the main features which relate and control the production of isomeric forms and the isomerization between those forms.

further stabilizing collisions are almost inevitable. The statement is made in this form because such collisions can also impart energy, thus making unimolecular decomposition again possible. With complex reactants, there is a great deal of energy contained within the vibrational modes of the reactants, and thus in the vibrational modes of the intermediate complex (in addition to the weak cluster ion bond), and some of this may be channeled into the AB⁺·CD cluster bond increasing the dissociation rate. For example, even for the four atom intermediate illustrated here, there are six vibrational modes ($3N - 6$, assuming that it is nonlinear) on the six-dimensional potential hypersurface, giving an additional reservoir of ~6 kT of energy. Of course, in general much of this energy will be inaccessible because of the incompatibility of the vibrational frequencies; the weak cluster bond will have low-frequency vibrations, whereas those of the stronger chemical bonds within the molecules will be of higher frequency (this will obviously be less restricting if the more strongly bonded ABCD⁺ is formed). This energy will also be inaccessible because of lack of resonances between the levels of different vibrational modes. True, there will be a large density of such levels close to the dissociation limits of the various modes, but these levels will be unoccupied in all but the cluster bond and thus the transfer of energy away from this bond will be more likely.

Remember that it is not the direct energy resonance between the vibrational levels in the two modes that is important. For cluster bond excitation, it is a resonance between the energy of the unoccupied vibrational levels in the weak cluster bond relative to the occupied excited level and quanta of thermal vibrational energy in the other modes. For cluster bond de-excitation, it is a resonance of the energy of an unoccupied level in these other modes relative to the occupied level with the energy of the occupied level of the cluster bond relative to a lower vibrational level.

As the intermediate becomes more de-excited due to successive collisions with M, it will eventually be stabilized into either the $AB^+ \cdot CD$ or $ABCD^+$ wells (assuming a barrier exists; solid line on the left-hand side of Figure 1). In these collisions, the position on the $(ABCD)^+$ surface is altered on a time scale of a vibrational period and thus the transition need not be vertical (as it is in the case of radiative stabilization). Indeed, it is actually a motion on an $(ABCDM)^+$ potential surface. Such an interaction could cause a different part of the $(ABCD)^+$ surface to be accessed by altering the subsequent dynamics of the motion over the surface. This may result in different degrees of excitation (rovibrational and translational) in the products or the appearance of a different reaction channel (if one is available). It is important to realize that the stable forms, $AB^+ \cdot CD$ and $ABCD^+$, may not be cleanly isolated even below the barrier to isomerization because, for small thin barriers, quantum tunneling of one of the atoms may be significant if the tunneling species has low mass. This will be especially significant if there is a resonance between the energy levels.[1] For $AB^+ \cdot CD$, there is an interaction with the barrier every vibrational period ($\sim 10^{-13}$ to 10^{-14} s) and thus barrier tunneling from this direction (especially for the ground vibrational level) may be quite likely even for a small tunneling probability. When there is no formal isomerization barrier, de-excitation to $ABCD^+$ is of course inevitable. However, during the initial part of this process, the association complex may still have $AB^+ \cdot CD$ character for a considerable proportion of the time. This occurs since in the vibration (or motion) on this surface, the species will spend more of the time where the potential is shallow rather than where it is deep.

For the situation denoted by the solid lines in Figure 1, the only possibilities for $AB^+ + CD$ are denoted by the left-hand side of the potential surface since this is completely decoupled from the right-hand side. On the right-hand side, as illustrated, the association of $AC^+ + BD$ to produce $AC^+ \cdot BD$ and $ACBD^+$ is analogous to the left-hand side. However, an additional part of the potential surface is illustrated where the reactants $A^+ + CBD$ (dot-dashed line) are also shown to access the $ACBD^+$ well. Thus, these reactants are directly coupled to $AC^+ + BD$ and the binary reaction,

$$A^+ + CBD \rightarrow (ACBD^+)^* \rightarrow (AC^+ \cdot BD)^* \rightarrow AC^+ + BD \qquad (2)$$

is likely (unlike that to $AB^+ + CD$ for which there is a barrier). However, association to form $ACBD^+$ may be able to compete if the binary channel requires a high degree of bond breaking and bond making. This is equivalent to requiring a complex path across the potential surface, which even if overall efficient, may take a long time giving collisional stabilization a chance to occur.

If the barrier between the left- and right-hand sides is lower (as depicted by the dashed line), then all of the possible reactants have greater access to the entire potential surface, but to differing degrees depending on the energy contained within the reactants. Then the relative importance of the various reaction channels and their occurrence will give information on that accessibility and thus on the structure of the surface.

For some real systems, structures and relative energies of the stable isomers have been calculated and, in some cases, the transition states have also been determined.[2-5] Many of these calculations have recently been reviewed by Radom[2,3] and some of the features of the calculations of Radom and coworkers are presented in Figures 2 and 6. In general, however, the detailed form of the surface is not known, with the exception of the separated reactant entrance channels where the potential is governed by long-range ion/induced dipole, ion/permanent dipole, etc. forces. In any chemical reaction, it is the detailed form of the surface that governs the progress of the reaction and determines the relative importance of the various processes. Motions across such surfaces were first discussed by Brauman[6] in the restricted (relative to the present situation) form of the double-minimum potential surface, essentially the part of the surface in Figure 1 depicted by reaction 2. Using this restricted surface, RRKM and Marcus formalisms have been used to interpret gas-phase reaction rate coefficients as expounded by Dodd and Brauman.[7] Note that it is often not necessary to know the form of the whole potential surface to perform such calculations, or even all of the details of the potential along the whole of the reaction path. Only the parts of the surface controlling the rate of the reaction and the product channels are important.

With these basic principles, we are now in a position to consider how the potential surface can be probed experimentally. As discussed above, the surface can be accessed either permanently by association,

$$\overset{M}{AB^+ + CD \rightarrow ABCD^+ + M} \tag{3}$$

or

$$\overset{M}{AC^+ + BD \rightarrow ACBD^+ + M} \tag{4}$$

with the surface being probed erratically due to random collisions with M, or in a transient crossing of the $(ABCD^+)^*$ potential surface in binary reactions such as that depicted by reaction 2. The surface can also be permanently accessed by other, more complicated, binary reactions which give $ABCD^+$ or $ACBD^+$ as one of the binary products. For example (see Section III.B), the binary reactions $O_2^+ + CH_4$ and $HCO_2H^+ + HCO_2H$ access the $CH_3O_2^+$ potential surface via the more complicated surfaces of the systems $CH_4O_2^+$ and $C_2H_4O_4^+$. The ion $CH_3O_2^+$ can also be accessed by the ternary association of CH_3^+ with O_2 or transiently in the binary reaction:

$$OH^+ + H_2CO \rightarrow (CH_3O_2^+)^* \rightarrow CH_3O^+ + O \tag{5}$$

When a stable isomer of $ABCD^+$ is produced, its isomeric form can be probed in a series of ways: (a) by its reactivity with other species, the main topic of this chapter, (b) by thermal dissociation, applied mainly to weakly bonded isomers of the cluster form in equilibrium studies (these are used to determine, cluster bond strengths; see, for example, the review by Kebarle[8]), and (c) by nonthermal collisional dissociation in a drift tube[9] or by injection at high energy into a collision gas.[10,11] In these cases, the isomer under investigation is in a stable vibrationally relaxed form. In the transient production of the isomers, excited intermediates are probed. The observed reaction channels show some of the parts of the potential surface that are coupled at an energy equivalent to the energy of the reactants. By isotopic labeling, information on the isomeric forms accessible is obtained from the degree of scrambling, i.e. how statistical the isotopic distribution of the products is and how many of the available sites are involved in successive reactions.[12] The former gives information on the accessibility of the sites and the time for scrambling relative to the lifetime of the intermediate. The latter gives information on which sites are structurally similar. Excited ions (above the dissociation limit) are also probed when metastable ions are observed in mass spectrometers.[13] Examples of all of the above are given in the specific isomeric systems discussed in the following sections. Those reviewed compose only the tip of the iceberg in a complex web of isomeric species.

In specific applications, it is critically important to know which isomer is produced in a particular situation in order to ascertain its further reactivity. Indeed, further reactivity, in the form of rate coefficients and product ion distributions, both identifies which reactions generate the same isomeric forms and gives information to enable the isomeric forms to be identified (often by determining the energetics and comparing them with theoretical calculations). One such application is to molecular synthesis in interstellar gas clouds. In the synthesis of the >115 molecules (mainly neutral; ~85%) detected in these clouds,[14] a major production route is via the radiatively stabilized analog of the collisional association discussed above,[15] viz.:

$$AB^+ + CD \rightarrow ABCD^+ + h\nu \qquad (6)$$

The process of radiative association has recently been reviewed by McMahon[16] and by Dunbar.[17] The observed neutral molecules are then produced by the dissociative recombination of these product ions with electrons, e.g.:

$$ABCD^+ + e \rightarrow A + BCD \qquad (7)$$

Recombination reactions of this type have been reviewed by Adams[18] and by Johnsen and Mitchell[19] (this volume) from the experimental viewpoint and by Bates from a theoretical viewpoint.[20] The recombination products will be critically dependent on the isomeric form of $ABCD^+$, i.e. the weakly bonded form, $AB^+ \cdot CD$, is more likely to produce $AB + CD$, whereas other products will be more likely for the more strongly bonded isomeric ions. Relatively little information is currently available on recombination products.[18] Indeed, only recently has the first full product distribution been obtained for a polyatomic ion recombination ($H_3O^+ + e$) with multiple product channels.[21] In addition, binary ion-neutral reactions are often competitive with recombination and the rate coefficients and ion product distributions will depend on the isomeric form of the product of reaction 6. The interstellar significance of reactions of the type 6 and 7 are considered by Herbst[22] (this volume).

The reactions and identification of small isomeric species were reviewed by McEwan in 1992.[23] Since that time, additional experimental data have been obtained on more complex systems. In the present review, smaller systems will only be mentioned where there has been an advance since the previous review and emphasis here will be concentrated on the correlation between reactivity, the form of the potential surface, and the isomeric forms. There is also a wealth of kinetic data (rate coefficients and product ion distributions) for ion–molecule reactions in the compilations of Ikezoe et al.[24] and Anicich,[25,26] some of which refer to isomeric species. Thermochemical data relevant to such systems, and some isomeric information, is contained in the compilations of Rosenstock et al.,[27] Lias et al.,[28,29] and Hunter and Lias.[30]

Following this introduction, the chapter is divided as follows. There is a description of the technique used to distinguish reactively between the isomers and identify their structures and reactivities. Various isomeric systems are then discussed in detail. The sections consist of *simple systems* (HCN^+/HNC^+, $C_3H_3^+$) followed by $CH_3O_2^+$, CH_5O^+, $C_2H_5O^+$, $C_2H_7O^+$, $C_3H_3O^+$, $C_3H_6N^+$, and $C_2H_5O_2^+$.

II. EXPERIMENTAL TECHNIQUES FOR IDENTIFYING ISOMERIC IONS BY THEIR REACTIVITY

In obtaining experimental information about the isomeric forms of ions, a variety of techniques have been used. These include ion cyclotron resonance (ICR),[31] flow tube techniques, notably the selected ion flow tube (SIFT),[32] and the selected ion flow drift tube (SIFDT)[32] (and its simpler variant[33]), collision induced dissociation (CID),[10,11] and the decomposition of metastable ions in mass spectrometers.[13] All of these techniques are mentioned in the text of Section III where they have provided data relevant to the present review.

In identifying isomeric ions by their reactivity, the main effort recently has been using selected ion flow tubes, in particular those at the University of Georgia and the University of Canterbury (the latter instrument has been described by McEwan[23]). The SIFT has been thoroughly described in the literature[32,33] and thus it is only necessary to give the details pertinent to the present studies of isomer reactivity. Here the University of Georgia instrument will be briefly described, since this was used to obtain our data reported in Section III.

Ions are generated in a remote ion source: a microwave discharge, low-pressure electron impact, or high-pressure electron impact. The former two were used to produce precursors to the isomeric ions, which were then generated in the flow tube. The microwave source and the last mentioned source were used to produce the isomeric ions directly in the ion source by ion–molecule reactions and then introduce them into the flow tube. After production in the source, the ion type of interest is selected by a quadrupole mass filter and focused through a 1 mm orifice into the flow tube. The ions are then entrained in flowing He gas at a pressure of ~0.5 torr, and carried to a downstream mass spectrometer for detection. For these reaction studies, the injection energies of the ions into the flow tube were kept small to minimize collisional dissociation and rearrangement. This was particularly important for the isomeric ions because they are prone to such processes. If collisional fragmentation occurs, then impurity ions will be present in the flow tube, complicating the determination of the product ion distributions of the reactions of the isomeric ion. Where minor amounts of fragment ions (a few percent) were a problem, their reactions were studied separately and their contribution accounted for in the product distribution of the isomeric ion reaction. Note that these fragmentation ions are not always a problem; they are products of the dissociation of the isomeric ions (a form of CID) and thus provide additional information about the isomeric potential surface. Such multicollision CID has only been used in a qualitative way, but of course could be made quantitative. The existence of vibrational excitation in the injected ions is also a possibility that has to be considered.

In the studies of isomeric ion reactivity detailed in Section III, vibrational excitation was searched for; there is little evidence for the survival of vibrational excitation into the reaction region of the flow tube. For the larger ions, where there are many low-frequency vibrational modes, vibrational to translational energy transfer in the 10^4 to 10^5 collisions with He before reaction would ensure de-excitation even with small energy transfer cross sections. When strongly bonded isomeric ions were directly injected, their reactions were studied by adding the reactant gas and obtaining rate coefficients and product ion distributions using standard well-developed techniques.[34,35] Different isomeric forms of the ions were studied by using different gases or gas mixtures in the high-pressure ion source.

When isomeric ions were produced by ternary association in the flow tube, thus allowing the potential surface to be accessed, the chemistry was more complicated. The gas with which the ions associated was added upstream and sufficient time was allowed for the association reaction to proceed before the reactant gas was added. When the associated ions are initially produced, they will be stable against dissociation if ternary collisions with the He remove sufficient energy to take them below the dissociation limits. However, they will still be internally excited and this excitation needs to be removed before the reactivity is probed. Again, the bulk of the evidence suggests that this de-excitation has occurred before the reactant gas is added. In a few cases there is some indication of residual excitation (see Section

III.E on $C_2H_7O^+$ isomers), but this seems to be a special situation. In parallel with association, there are often binary channels. This provides additional information on the potential surface; however, it does result in some complications. In this situation, the further reactions of the ions, generated in the binary reaction channels, with the reactant gas have to be studied separately and accounted for in determining the product distribution for the reaction of the associated ion. Also, the gas which is used to produce the associated ion can potentially cause some problems due to its presence in the reaction region. Here it could react with the product ions of the reaction which is used to probe the associated ion structure. We know of only one case where this may be a problem (in the reaction of H_3O^+/C_2H_4 with $H_2C_3H_2$); however, it is necessary to be continuously vigilant for such effects. The use of a variety of probing reactant gases considerably reduces the possibility of such effects being overlooked.

In the following sections, studies of isomeric ions are reported in which the ions are reactively probed. Where calculations are available, information on potential energy surfaces is given. This is usually the structure of the stable isomeric forms and transition states and their relative energies; thus only points on the potential surface are known. The detailed form of the potential surface is almost never available nor is the connectivity between the various states usually established theoretically (chemical intuition is often used to connect the states). Pertinent experimental data on CID and metastable ions, isomers produced in binary reactions, and potential surfaces probed by binary reactions (with the excited isomeric ion as the reaction intermediate) are also given.

III. ISOMERIC SYSTEMS

A. Simple Systems

Since the review by McEwan in 1992, there have been several advances for simple isomeric systems.[23]

HCN⁺/HNC⁺

This is one of the simplest systems for which isomers can exist. Ab initio calculations of the energetics for this system have shown that HCN^+ is 17.7 kcal mol^{-1} less stable than HNC^+ and has an energy barrier to isomerization of 30.9 kcal mol^{-1}. Even though this barrier is substantial, isomerization occurs in the reactions with CO and CO_2, in parallel with a small proton transfer channel,[23] viz.,

		CO	CO_2	
$HCN^+ + CO/CO_2$	$\rightarrow HNC^+ + CO/CO_2$	4.6(−10)	5.0(−10)	(8a)
	$\rightarrow HCO^+/HCO_2^+ + CN$	~3(−12)	~2(−12)	(8b)

(all rate coefficients quoted are in $cm^3 s^{-1}$ and were measured at room temperature, i.e. ~300 K, unless otherwise stated). Because it is so surprising that the exothermic proton transfer channel does not dominate (proton transfer usually occurs at the collisional rate when it is exothermic[36]), a double-proton transfer mechanism was proposed,[37] e.g.:

$$NCH^+ + CO_2 \rightarrow (NC...H^+\text{--}CO_2)^* \rightarrow HCO_2^+ + CN \qquad (9a)$$

$$\downarrow$$

$$(CN...H^+\text{--}CO_2)^* \rightarrow CNH^+ + CO_2 \qquad (9b)$$

The efficiency of the double-proton transfer depends to a large degree on the lifetime of the excited intermediate complex. For long lifetimes, double-proton transfer is likely giving isomerization, whereas for short lifetimes this process will be less significant and single-proton transfer to give HCO_2^+ will dominate. This mechanism has recently been confirmed by Hansel et al.[38] who used a selected ion flow drift tube to vary the HCN^+/CO_2 interaction energy and thus the lifetime of the intermediate complex. At low energies, as discussed above, isomerization (double-proton transfer) dominates; however, as the energy is increased, thus decreasing the lifetime, the HCO_2^+ channel (single-proton transfer) increases in importance, becoming the only channel at higher energies. This confirms the operation of a double-proton transfer mechanism. Such a mechanism is expected to occur generally where isomerization is exothermic but where there is an inter-vening barrier. For this, the major requirements are that the lower energy ion species has the larger proton detachment energy, that the catalyst molecule has a proton affinity between the proton detachment energies of the two isomeric ion species, and that the lifetime of the intermediate catalyst complex is sufficiently long for multiple-proton transfers to occur.

$C_3H_3^+$

The two isomeric forms of $C_3H_3^+$—the acyclic propargyl ion, $HC_3H_2^+$, and the cyclopropenylium ion, $c\text{-}C_3H_3^+$—are well known and their relative energies, and those of the dissociation limits, are well established (see Figure 2). The energetic information on the potential surface has been corrected for zero-point energies in this and some of the other cases discussed below. Where this has been done, it is noted in the figure caption for the particular potential surface. These species have been distinguished experimentally in an ICR by Smythe et al.[45] by their reactivities with a series of alkanes, alkenes, and aromatic molecules, and later by Baykat et al. using reactions with aromatic hydrocarbons and simple alcohols.[46] Mixtures of these ions are produced following electron impact on methyl acetylene, allene, and propargyl bromide. The two isomers were also shown to be produced in about equal proportions in a SIFT experiment by Smith and Adams[47] by the reaction,

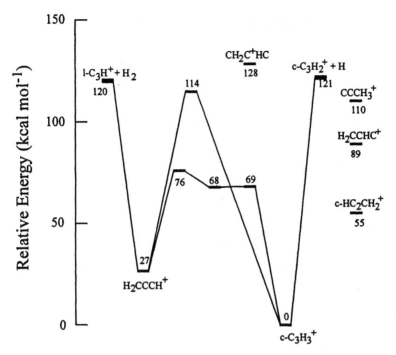

Figure 2. Potential energy surface for $C_3H_3^+$ system. The data are from the theoretical calculations of Maluendes et al.[39,40] with the energies of four additional stable isomers from Radom et al.[41] The values are corrected for zero-point energies.

$$C_3H^+ + H_2 + He \rightarrow C_3H_3^+ + He \qquad (10)$$

and the radiatively stabilized analog was cited as the mechanism for the production (following dissociative electron–ion recombination) of c-C_3H_2[48] which is observed ubiquitously in interstellar gas clouds.[49,50] Indeed, the prediction from this reaction that l-C_3H_2 would exist in the interstellar clouds preceded its detection.[51] In the study of reaction 10, the isomers $C_3H_3^+$ were identified by their reactivities with CO and C_2H_2.[47] At about the same time, Oztuk et al. studied the reactions of the $C_3H_3^+$ isomers with C_2H_2 and C_2H_4 in an FT-ICR.[52] The rate coefficients determined in the low-pressure ICR and the higher pressure SIFT for the reaction of $HC_3H_2^+$ with C_2H_2 were $\leq 5(-13)$ and $1.1(-9)$ cm^3 s^{-1}, respectively, at room temperature (the major product in the SIFT experiment was collisional association with a minor $C_5H_3^+$ product). Ostuk et al.[52] also noticed that some isomerization to c-$C_3H_3^+$ occurred, viz.:

$$HC_3H_2^+ + C_2H_2 \rightarrow c\text{-}C_3H_3^+ + C_2H_2 \qquad (11)$$

To explain these behaviors, Wiseman et al.[53] suggested the scheme (reaction 12a)

$$HC_3H_2^+ + X \rightleftharpoons (C_3H_3X^+)^* \rightarrow c\text{-}C_3H_3^+ + X \tag{12a}$$

$$\downarrow M$$

$$C_3H_3X^+ + M \tag{12b}$$

with $X = C_2H_2$. Further, Moini[54] examined this isomerization in a quadrupole ion trap and observed the collisional association reaction 12b in parallel with reaction 12a at pressures above 10^{-3} torr. To explore this in more detail, McEwan et al. made a combined ICR/SIFT study.[55] In both sets of measurements (ICR and SIFT), $HC_3H_2^+$ could only be produced in a mixture with $c\text{-}C_3H_3^+$ (about 50%), but the $c\text{-}C_3H_3^+$ could be produced as the unique isomer (but with another ion, $C_4H_5^+$, also present at 25%). In these ICR and SIFT studies, $HC_3H_2^+$ was reactive (see Table 1; with the exception of the ICR reaction with C_2H_2 for which channels other than isomerization were slow as shown previously) and $c\text{-}C_3H_3^+$ was unreactive (rate coefficients $<2(-11)$ and $<5(-12)$ cm^3 s^{-1} for the ICR and SIFT studies, respectively) with almost all the neutral species used. For the reactions of the $HC_3H_2^+$ isomer, there was a consistent pattern between the two sets of measurements. For neutral reactants C_2H_2, NH_3, CH_3OH, and CH_3CN, association channels were observed in every case in the SIFT, but not in the ICR (for the latter two reactants, binary channels were also observed at the pressures of both experiments). This is expected for reaction 12b because, at the high pressure in the SIFT, collisional stabilization of the intermediate can occur in parallel with binary channels, but not

Table 1. Summarized Rate Coefficients (cm^3 s^{-1}) and Product Ion Distributions (%) for the Reactions of the Acyclic Isomer, $HC_3H_2^+$, of $C_3H_3^+$ with the Reactant Neutral Indicated[55] a

Reactant Neutral	Product Ions	Low-Pressure ICR		High-Pressure SIFT	
		Percent Products	Rate Coefficient	Percent Products	Rate Coefficient
C_2H_2	$C_5H_5^+$	—	$<2(-11)^b$	100	1.2(−9)
NH_3	NH_4^+	60	3.0(−10)	55	2.0(−9)
	H_2CN^+	40		20	
	$C_3H_3^+ \cdot NH_3$			25	
CH_3OH	CH_2OH^+	100	3.4(−10)	60	1.8(−9)
	$C_3H_3^+ \cdot CH_3OH$	0		40	
CH_3CN	products	100	1.6(−10)	0	3.3(−9)
	$C_3H_3^+ \cdot CH_3CN$	0		100	

Notes: ªThe data clearly illustrate that at the low pressures in the ICR binary channels dominate, but that the intermediate complex is sufficiently long lived to be collisionally stabilized in the higher pressure SIFT experiments. The cyclic isomer, $c\text{-}C_3H_3^+$, was unreactive with these reactant neutrals.

ᵇAn isomerization reaction to yield $c\text{-}C_3H_3^+$ also occurs.[52,53]

at the low pressure of the ICR; reaction 12a. From these measurements, lifetimes against unimolecular decomposition of about 20, 13, and 8 μs were deduced. In the case of the reaction with C_2H_2, calculations of the possible $C_5H_5^+$ intermediate structures are available. Leszczynski et al.[42] and Feng et al.[43] have shown that there are five possible isomeric products of the association, all with the large well depths required for the above observations. Additional theoretical studies have been made by Glukhovtsov et al.[44] For the reaction with NH_3, the potential surface of $C_3H_6N^+$ is accessed (see Section III.G for information on this surface).

B. $CH_3O_2^+$

Eighteen isomeric structures of $CH_3O_2^+$ (both stable and transition states) and their relative stabilities were first identified by Ha and Nguyen with ab initio SCF and 3-21G and 6-31G** Gaussian basis sets.[56] More recently, Cheung and Li[57] with calculation at the MP2/6-31G(d) level established six stable equilibrium structures and their relative energies. They later extended this to eight with calculated structures and relative energies of the transition states between them.[58] The present state of the potential energy surface for this system is depicted in Figure 3. Also included on this figure are the asymptotic limits to the dissociation channels, i.e. $HCO^+ + H_2O$, $H_3O^+ + CO$, and $CH_3^+ + O_2$. Not included is $HCO_2^+ + H_2$ since this exhibits a large activation energy (48 kcal mol^{-1}) leading to the formation of $HC(OH)_2^+$.[58] It can be seen that the surface is in two distinct parts centered on the most stable structure, protonated formic acid. This structure has been well established since the early appearance potential measurement of Munson and Franklin.[59] It can be seen from the potential surface that this form is not readily accessed by ion–molecule associations. As already stated, its formation from $HCO_2^+ + H_2$ has a large activation energy barrier and does not even form the weakly bonded polarization structure $HCO_2^+ \cdot H_2$. A possible route is from $HCO^+ + H_2O$ which forms an association intermediate complex with a barrier of 10 kcal mol^{-1} to $HC(OH)_2^+$ (see Figure 3). The reaction of HCO^+ with H_2O has been studied and found to proceed by rapid proton transfer to give $H_3O^+ + CO$ (the rate coefficient is 2.6(–9) cm^3 s^{-1}).[25] Such exothermic proton transfers are usually gas-kinetic.[36] No such channel is possible for $H_3O^+ + CO$ and a slow association is observed (the rate coefficient is ~4(–30) cm^6 s^{-1} in He) which presumably gives the structure $H_3O^+ \cdot CO$ which is weakly bonded by 15 kcal mol^{-1} according to the calculations. A small three-body association rate coefficient is consistent with this bond strength; the protonated formic acid minimum is inaccessible being separated by a 32 kcal mol^{-1} barrier relative to the $H_3O^+ + CO$ dissociation limit. The magnitude of this rate coefficient is too small for significant amounts of the weakly bonded isomer to be produced in the flow tube and its reactivity could not be studied.

Other isomeric forms can be created in binary reactions such as,

$$O_2^+ + CH_4 \rightarrow CH_3O_2^+ + H \tag{13}$$

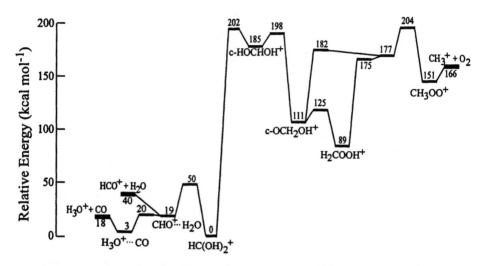

Figure 3. Potential energy surface for the $CH_3O_2^+$ system showing both stable structures and transition states (from the theoretical work of Cheung and Li[57,58]) and some of the lower dissociation limits. The more critical structures are given as well as the energies of the dissociation limits. The association between HCO_2^+ and H_2 is not included since it has a large activation energy barrier (48 kcal mol^{-1}). The values are corrected for zero-point energies.

$$\Delta H^\circ = -83 \text{ kcal mol}^{-1} \text{ for } CH_2OOH^+$$

and,

$$HCOOH^+ + HCOOH \rightarrow CH_3O_2^+ + HCO_2 \tag{14}$$

$$\Delta H^\circ = -38 \text{ kcal mol}^{-1} \text{ for } CH(OH)_2^+$$

which access specific points on the potential surface, the actual point being dependant on the ΔH° of the reaction, the amount of the internal energy in the reactant ions and the degree of collisional quenching that has occurred before the ion is probed. (The energetics of these and the other reactions have been calculated using thermodynamic information in refs. 27–30). This approach of generating $CH_3O_2^+$ by reactions 13 and 14 has been used by Villinger et al.[60] Having created the ions, they investigated the structural form by exciting the ion in the drift tube in collision with Ar. In the case of these two ions, the same fragmentation pattern was observed independent of the interaction energy from 0.2 to 1.2 eV, viz.:

$$CH_3O_2^+ + Ar \rightarrow HCO^+ + H_2O + Ar \quad 50\% \tag{15a}$$

$$\rightarrow H_3O^+ + CO + Ar \quad 50\% \tag{15b}$$

This does not necessarily mean that the two isomers have the same form, but rather that on excitation they pass through at least one intermediate which is the same. Further excitation then results in the observed fragmentation. However, the situation with regard to reaction 15 appears to be straightforward. From Figure 3, if the two reactions, 13 and 14, produce a single isomer and that is $HC(OH)_2^+$, then the excitation would have to take the species over the 50 kcal mol^{-1} barrier to excited $H_2O\cdots CHO^+$, which could then decompose to either directly to HCO^+ or via a further motion on the surface to H_3O^+. For the product ratios to be constant requires that, after the barrier is surmounted, the reaction explores the potential surface reaching the products thermodynamically rather than kinetically. The variation of the overall rate coefficient with center-of-mass energy, however, led to an Arrhenius behavior with a barrier height of 22 kcal mol^{-1}; i.e. much smaller than the barrier for $HC(OH)_2^+$ as determined by theory. It is interesting that no $HCO_2^+ + H_2$ was observed which has a barrier of only 1.8 times larger.

It was noted that the isomeric ion species was unlikely to be the weakly bonded CH_3–O–O$^+$ since switching reactions such as,

$$CH_3^+ \cdot O_2 + N_2 \rightarrow CH_3^+ \cdot N_2 + O_2 \tag{16}$$

were not observed contrary[61] to when $CH_3O_2^+$ was generated from:

$$CH_3^+ + O_2 + He \rightarrow CH_3O_2^+ + He \tag{17}$$

These identifications were questioned by Ha and Nguyen,[56] who, from their theoretical calculations and geometrical considerations, concluded that reactions 13 and 14 would not yield $CH(OH)_2^+$ but rather $CH(O)OH_2^+$. However, the later calculations of Cheng and Li[57,58] showed that this isomer has only a very small barrier (~ 1 kcal mol^{-1}) against fragmentation to $H_3O^+ + CO$, which is exothermic by 1.4 kcal mol^{-1} (see Figure 3). Consistent with the findings of Villinger et al.,[60] Holmes et al.[62] from collision-induced dissociation studies with 8 keV ions, reported that the product of reaction 13 is $HC(OH)_2^+$, and that this and H_2COOH^+ give the same dominant fragment ions HCO^+ and H_3O^+ (as in reaction 15). However, in a later similar CID study, Kirchner et al.[63] concluded that the product of reaction 13 is $H_2CO_2H^+$ in agreement with the reactivity studies of Van Doren et al.[64] (discussed below). The collisional dissociation of the $CH_3O_2^+$ isomers produced in reactions 13 and 14 was recently investigated further by Glosik et al.[65] in a selected ion-flow drift tube at collision energies from thermal to 0.2 eV. In some ways, the results were similar to the earlier drift-tube studies in that the dissociation of the products of these reactions gave HCO^+ and H_3O^+ (in the proportions 60%/40% and 50%/50%, respectively, for reactions 13 and 14) but the activation energy barriers were different being 24 kcal mol^{-1} (similar to before) and 38.4 kcal mol^{-1}, respectively. Thus, they confirmed the finding of Van Doren et al.[64] that the product of reaction 13 is not protonated formic acid. They concluded that the earlier studies of Villinger et al.[60] were contaminated with formic acid which adheres strongly to surfaces. It was further concluded that the reactivity studies by Villinger et al.[66] of $CH_3O_2^+$ with D_2O were similarly flawed. Indeed, Glosik et al. performed collision dissociation studies of deuterated formic acid produced by proton transfer from D_3O^+, and showed that the deuteron was attached to oxygen and not to carbon. Also, isotope exchange of protonated formic acid with D_2O,

$$HC(OH)_2^+ + D_2O \rightarrow CH_2DO_2^+ + HDO \tag{18}$$

showed the incorporation of two deuterium atoms in consecutive reactions, again consistent with the structure $HC(OD)_2^+$.

As mentioned above, the structure of the product of reaction 13 had been investigated experimentally by Van Doren et al.[64] who were concerned that the production of either of the isomers suggested by the earlier experimental and theoretical studies would require the breaking and remaking of four bonds. To obtain structural information, Van Doren et al. made a detailed SIFT study of the reactivity of $CH_3O_2^+$ (produced in reaction 13) with a series of alkanes, alkenes, alkyl chlorides, as well as carbonyl compounds and CS_2, COS, CO_2, H_2S, $(CH_3)_2S$, NH_3 (ND_3), and H_2O (D_2O, $H_2^{18}O$). The reactivity of this ion was compared with the $CH_3O_2^+$, produced by electron impact on HCO_2H in the remote high-pressure ion source followed by reaction 14. The ions were then mass selected and injected into the flow tube. This latter ion was identified as $CH(OH)_2^+$ by bracketing its proton affinity (179 kcal mol^{-1} for formic acid) between that of H_2S (170 kcal mol^{-1}) and methyl formate (189 kcal mol^{-1}). This $CH_3O_2^+$ isomer does not react with propane, n-butane, ethylene, water, or hydrogen sulfide, but does exhibit two H/D exchanges in reaction with D_2O (as also observed by Villinger et al.[66] and by Glosik et al.[65]). In contrast to this, the $CH_3O_2^+$ from reaction 13 did react with all of these species showing that these are different isomeric species. In the D_2O reaction, the H/D exchange channel was not observed (i.e. no $CHD_2O_2^+$), the only primary product being DH_2O^+. In these studies, the dominant reaction mechanism was hydride abstraction or OH^+ transfer to the neutral reactant, with proton transfer also occurring in some cases, Van Doren et al. concluded that this reactivity is consistent with the resonant structures $H_2C=O^+-OH$ and H_2C^+-OOH with a heat of formation, ΔH_f°, of 184 ± 2 kcal mol^{-1}. The ΔH_f° of protonated formic acid is 96 kcal mol^{-1}, which is lower by 88 kcal mol^{-1} and almost exactly the difference of 89.2 kcal mol^{-1} in the Cheng and Li calculations.[57,58] This makes the identification quite secure. The difference is 111.2 kcal mol^{-1} for the structure, c-OCH_2OH^+, which has the next nearest heat of formation. It is, in some senses, surprising that this species is not also produced in the $O_2^+ + CH_4$ reaction since the direct proton transfer is exothermic by 0.8 kcal mol^{-1}. However, if the reaction proceeds via $H_2CO_2H^+$, the reaction exothermicity of 23 kcal mol^{-1} is not sufficient for the 36 kcal mol^{-1} barrier to c-OCH_2OH^+ to be overcome. Note that it may be accessed in the CH_4 reaction if the O_2^+ is internally or kinetically excited.

One of the other isomeric forms of $CH_3O_2^+$ that has been alluded to is CH_3OO^+ which is accessed by the association reaction 17 with no other isomeric forms being accessible according to Figure 3. Studies of the reactivity of this species, CH_3^+/O_2, have been made recently by Fisher and Adams[67] and these are listed in Table 2 together with equivalent reactions of $HC(OH)_2^+$[67] and of CH_2OOH^+ (from Van Doren et al.[64]). Consistent with Table 2, Van Doren et al. found that the $HC(OH)_2^+$ ions were unreactive with propane. The combined studies clearly show that the isomeric forms produced in the three ways are different in both reactivity and products, establishing that they are independent isomeric forms. Note that for CH_3^+/O_2, like the simple reaction 16 with N_2,[61] there is some evidence for the weakly bonded form in the ligand switching reaction with H_2O which could be,

$$CH_3^+ \cdot O_2 + H_2O \rightarrow CH_3^+ \cdot H_2O + O_2 \tag{19}$$

or which could produce a more strongly bonded form of CH_5O^+ (see Section III.C on the isomers of that species). All reactions in the table are exothermic for the CH_3OO^+ isomer.

The $CH_3O_2^+$ surface can also be accessed by other binary reactions. In particular, the proton transfer reactions between HCO_2H and CH_5^+, HCO^+, $C_2H_6^+$, H_3O^+, H_3S^+, CH_3O^+, $CH_3OH_2^+$, and CH_2N^+ will produce protonated formic acid[24-26] with different degrees of excitation equal to a maximum of the differences in the proton affinities, i.e. varying from zero excitation to 47 kcal mol^{-1} for proton transfer from CH_5^+. According to Figure 3, the product is still trapped within the $CH(OH)_2^+$ well until an energy of 50 kcal mol^{-1} is deposited in the $CH(OH)_2^+$, above which the ion

Table 2. Summarized Rate Coefficients (cm^3 s^{-1}) and Product Ion Distributions (%) for the Reactions of $CH_3O_2^+$ with the Reactant Neutral Indicated[a]

Reactant Neutral	$CH(OH)_2^+$	CH_3^+/O_2	$CH_2O_2H^+$
CH_3OH	$CH_3OH_2^+$ 100%	$CH_3OH_2^+$ 100%	—
181.9	1.3(−9)	1.7 (−9)	
C_2H_5I	$C_2H_6I^+$ 100%	$C_2H_6I^+$ 100%	$C_2H_6OI^+$ 100%
176	1.6(−10)	2.4 (−9)	5.9(−10)
H_2O	—	$CH_3OH_2^+$ 100%	H_3O^+ 100%
166.5	<3 (−11)	1.8(−9)	2.1 (−11)
C_3H_8	$C_3H_5O^+$ 100%	$C_3H_7^+$ 100%	$C_3H_7^+$ 100%
150	1.8(−12)	2.0(−9)	4.6(−10)
C_2H_6	—	$HCO^+/C_2H_5^+$ 100%	—
143.6	≤3 (−13)	5.3(−10)	≤(−12)

Note: [a]The data were obtained from isomers generated in a high pressure ion source containing HCO_2H, from the association of CH_3^+ with O_2[67] (CH_3^+/O_2) and in the binary reaction of O_2^+ with CH_4.[64] These isomers are all different and are identified as $CH(OH)_2^+$, CH_3OO^+, and $CH_2O_2H^+$, respectively (see Figure 3). Proton affinities (kcal mol^{-1}) are given below each reactant neutral (that of HCO_2H is 178.8).[28-30]

will dissociate to either $HCO^+ + H_2O$ or $H_3O^+ + CO$; i.e. dissociative proton transfer will be observed (this assumes that the deprotonated molecule cannot form a compound with the neutral fragment making additional energy available). Note that in addition to the above, the potential surface is accessed by:

$$OH^+ + H_2CO \rightarrow H_2CO^+ + OH \quad 40\% \tag{20a}$$

$$\rightarrow CH_2OH^+ + O \quad 60\% \tag{20b}$$

However, it is not known into which part of the potential energy surface these species couple. The reactions of H_2CO^+ with OH and CH_2OH^+ with O-atoms would also access the surface although these are not experimentally very tractable. The surface is also accessed to a limited extent by the gas kinetic proton transfer from HCO^+ to H_2O yielding H_3O^+.

Overall, the potential surface explains the experimental data quite well, having the appropriate stable forms connected to the relevant dissociation limits. Quantitatively, there is a discrepancy between the heights of the barriers for $HC(OH)_2^+$ and $CH_2O_2H^+$ fragmentation to HCO^+ and to H_3O^+ as determined experimentally (24 and 38 kcal mol^{-1})[65] and theoretically (50 and 113 kcal mol^{-1}).[58] Tunneling through the barrier, or the existence of a reaction coordinate which circumvents the barrier, may be a possibility in the case of $CH_2O_2H^+$ fragmentation; however, for $HC(OH)_2^+$, such would not be sufficient since the fragmentation to $HCO^+ + H_2O$ is endothermic by 40 kcal mol^{-1}. Thus, either a collisional pumping mechanism is required or the internal energy from the various other vibrational modes (which would be excited and equilibrated in the collision process) would have to be accessed. If such occurs, then the Arrhenius barrier in the multicollisional fragmentation environment of the drift tube cannot be simply equated to the barrier height. For such information, a lower pressure single collisional fragmentation experiment would be more suitable.

C. CH_5O^+

This isomeric form is of interest from an interstellar point of view since the isomer, $CH_3OH_2^+$, is a possible route, via dissociative electron–ion recombination, to the observed methanol.[14] A proposed reaction[68] leading to this isomer is the radiative association,

$$CH_3^+ + H_2O \rightarrow CH_3OH_2^+ + h\nu \tag{21}$$

which accesses the CH_5O^+ potential surface. Limited theoretical calculations are available for this surface with one stable isomer and one transition state having been identified.[69] Some information on the potential surface, as deduced from thermochemical data and experimental reaction rate data, are also collected in Figure 4.

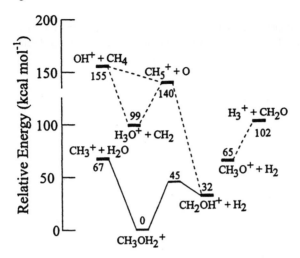

Figure 4. Part of the potential energy surface of CH_5O^+ calculated by Chunxiano et al.[69] (solid lines) and constructed from thermochemical data in the literature.[27–30] The dashed lines indicate that the connected species are coupled by binary reactions, as determined experimentally. It is not meant to imply that there are no barriers between these species.

To investigate this system, Fisher and Adams[70] studied the CH_5O^+ isomer, CH_3^+/H_2O, produced in the collisionally stabilized reaction 22a,

$$CH_3^+ + H_2O \xrightarrow{He} CH_5O^+ + He \quad 12\% \tag{22a}$$

$$\rightarrow CH_3O^+ + H_2 \quad 88\%, \tag{22b}$$

which has a rate coefficient of 3.3(–10) cm³ s⁻¹ at 0.5 torr. This isomer was probed by its reactivity and this was compared with the reactivity of $CH_3OH_2^+$ produced in a high-pressure ion source by the reaction:

$$CH_3OH^+ + CH_3OH \rightarrow CH_3OH_2^+ + CH_3O \tag{23}$$

In these reactivity studies, reactions 22a and b were studied and the rate coefficient and product distribution determined as reported above. This product distribution is at variance with a much earlier study where only an association channel was reported, although with a similar rate coefficient ~1(–26) cm⁶ s⁻¹, equivalent to a binary rate coefficient of ~2(–10) cm³ s⁻¹ at 0.5 torr.[61] The CH_5O^+, produced in this way and by reaction 23, was reacted with a series of molecules with proton affinities varying from 166 to 193 kcal mol⁻¹ and encompassing that of CH_3OH; see Table 3. For the production of CH_5O^+ in the association reaction 22a, sufficient water was

Table 3. Summarized Rate Coefficients (cm^3 s^{-1}) and Product Ion Distributions (%) for the Reactions of CH$_5$O$^+$ with the Reactant Neutrals Indicated[a]

Reactant Neutral	Product Ions	CH$_3$OH$_2^+$		CH$_3^+$/H$_2$O	
		Product Percent	Rate Coefficient	Product Percent	Rate Coefficient
C$_2$H$_5$OCHO	C$_2$H$_5$OCHOH$^+$	100	1.8 (−9)	100	2.0(−9)
193.1					
(CH$_3$)$_2$CHOH	(CH$_3$)$_2$CHOH$_2^+$	100	1.9(−9)	100	2.6(−9)
191.6					
CH$_3$COOH	CH$_3$COOH$_2^+$	92	1.8 (−9)	100	2.1(−9)
190.2	CH$_3$CO$^+$	8		0	
CH$_3$CHO	CH$_3$CHOH$^+$	100	2.4(−9)	95	2.6(−9)
186.6	CH$_5$O$^+$·CH$_3$CHO	0		5	
H$_2$C$_3$H$_2$	C$_3$H$_5^+$	20	2.9(−10)	30	1.9(−10)
186.3	CH$_5$O$^+$·C$_3$H$_4$	80		70	
c-C$_3$H$_6$	C$_3$H$_7^+$	100	3.3(−10)	100	3.8(−10)
179.8					
H$_2$O	CH$_5$O$^+$·H$_2$O	100	1.6(−11)	100	5.4(−12)
166.5					

Note: [a]The reactions were studied by Fisher and Adams in a SIFT at 300 K.[70] CH$_3$OH$_2^+$ was produced from CH$_3$OH in a high-pressure ion source by reaction 23 and CH$_3^+$/H$_2$O by associating CH$_3^+$ from a low-pressure ion source with H$_2$O in the flow tube. Proton affinities (kcal mol^{-1}) are given below each reactant neutral and that of CH$_3$OH is 181.9 kcal mol^{-1}.[28–30]

added to the limit imposed by the saturated vapor pressure. It can be seen from the table that CH$_3$OH$_2^+$ rapidly proton transfers when this is exothermic. For the reaction with c-C$_3$H$_6$, proton transfer is endothermic by 2 kcal mol^{-1} if the cyclic isomer is the product or exothermic by 5.5 kcal mol^{-1} if the product is acyclic. Thus the latter product seems more likely on energetic grounds. This is a further example of where isomeric effects need to be investigated in more detail. That the proton transfer does not occur at the collisional rate is probably a manifestation of the inefficiency of this isomerization. Where proton transfer is definitely endothermic, e.g. as in the case of the H$_2$O reaction, it does not occur; the association product CH$_7$O$_2^+$ is produced. Therefore the CH$_5$O$^+$ produced in reaction 23 is undoubtedly CH$_3$OH$_2^+$. It can also be seen from Table 3 that CH$_3^+$/H$_2$O is very similar in reactivity (in both rate coefficients and product ions) to CH$_3$OH$_2^+$ even for the slower reactions. This isomer is then also identified as CH$_3$OH$_2^+$. Thus, on the CH$_5$O$^+$ potential surface, there must be only a small, if any, potential well for CH$_3^+$·H$_2$O and only a small barrier to isomerization to CH$_3$OH$_2^+$. This is consistent with the calculations and CID experiments of Chunxiao et al.[69]

Other routes to CH$_5$O$^+$ by association are not possible since the reactions,

$$OH^+ + CH_4 \rightarrow H_3O^+ + CH_2 \qquad 87\% \qquad (24a)$$

$$\rightarrow CH_5^+ + O \qquad 13\% \qquad (24b)$$

and,

$$H_3^+ + CH_2O \rightarrow CH_3O^+ + H_2 \qquad (25)$$

are very efficient, with rate coefficients of 1.5(−9) and 6.3(−9) cm^3 s^{-1} respectively.[25] The reverse of reaction 25, where proton transfer is endothermic, is very slow with no detectable association. Thus, CH$_3$O$^+\cdot$H$_2$ must have a very shallow well with a substantial barrier to CH$_3$OH$_2^+$. It is not possible to study the reaction of H$_3$O$^+$ with CH$_2$ (the reverse of reaction 24a), since it would be difficult to produce the CH$_2$ radical. The reaction of CH$_5^+$ with O (the reverse of reaction 24b) is quite rapid giving mainly H$_3$O$^+$ + CH$_2$.[25] CH$_5$O$^+$ is otherwise a very simple ion to produce; it is generated in proton transfer reactions of CH$_3$OH with CH$^+$, HCO$^+$, CH$_2^+$, H$_2$CO$^+$, CH$_3$O$^+$, CH$_4^+$, C$_2$HN$_2^+$, NH$^+$, NH$_2^+$, and H$_3$O$^+$.[24–26] In addition, it is a rearrangement product in the reactions of CH$_3$O$^+$ with H$_2$CO and H$_3^+$ with HCO$_2$CH$_3$.[24] Further studies, in particular theoretical studies of the potential surface are required; however, the experimental data indicate that the potential surface will be relatively simple.

D. C$_2$H$_5$O$^+$

This system is particularly interesting since, in a recent SIFT study by Fairley et al.,[71] it was found that, like reaction 10 which produces C$_3$H$_3^+$ isomers (see Section III.A), the reaction,

$$H_3O^+ + C_2H_2 + He \rightarrow C_2H_5O^+ + He \qquad (26)$$

produces two isomeric forms in approximately equal amounts. This reaction was studied because of its possible implication as a source of the observed interstellar acetaldehyde via the dissociative electron–ion recombination of CH$_3$CHOH$^+$ (the lowest energy isomer). However, earlier studies of the rate coefficient of reaction 26 had suggested that the product was not protonated acetaldehyde, but rather a higher energy isomeric ion.[72]

Detailed structural calculations have been carried out for this system. This is because the neutral isomer, C$_2$H$_5$O, which is implicated in the thermochemistry of ethanol, is of interest in pollution control, atmospheric chemistry, and combustion. Also, there is new information available from photoionization experiments with which to compare theoretical calculations. For details of these comparisons, see Curtiss et al.[73] In the earlier theoretical studies of Nobes et al.,[74] calculations were performed at the MP2 and MP3 levels with basis sets of double ζ plus polarization (6-13G**) with electron correlation. These studies revealed four stable minima for the system: protonated acetaldehyde, CH$_3$–C$^+$H–OH $\leftrightarrow$ CH$_3$–CH=O$^+$H; the methoxymethyl cation, CH$_3$OCH$_2^+$; protonated oxirane, (CH$_2$)$_2$OH$^+$; and vinyloxonium, $CH_2CHOH_2^+$ (protonated vinyl alcohol), with $CH_3CH_2O^+$ and $HOCH_2CH_2^+$ being unstable species. The more recent calculations by Curtiss et al.[73] using G2 theory also found protonated acetaldehyde, $CH_3OCH_2^+$, and $(CH_2)_2OH^+$ as stable species, but found that the planar structure $CH_2CHOH_2^+$ considered by Nobes et al. was only a saddle point of the stable nonlinear isomer. In addition, they found that $CH_3COH_2^+$ and CH_3OHCH^+ were local minima (the latter was previously reported in an experimental and theoretical study by Audier et al.[75] at the MP2/6-311^{++}G*//HF/6-31G* level), and that there is also a weakly bonded complex, $H_3CCO^+·H_2$. To explore more of the surface, Fairley et al.[71] made calculations, also at the G2 level. They located the four stable isomers identified by Nobes et al. and by Curtiss et al. and transition states accessed in the interconversion between these isomers. In addition, the weakly bound complex, $H_3O^+·C_2H_2$ was found. The form of the surface with features from Fairley et al.[71] and Audier et al.[75] is shown in Figure 5. Of the four stable species, all but $CH_2CHOH_2^+$ have been detected experimentally. To investigate this potential surface, Jarrold et al.[76] performed metastable reaction studies and collision-induced dissociation (CID) with the three stable isomers that could be produced experimentally. The metastable spectra of CH_3CHOH^+ and $(CH_2)_2OH^+$ were similar, as were the CID spectra, indicating that these species access the same region of the potential surface. Products of the metastable reaction are $HCO^+ + CH_4$ (major at about 60%), $H_3O^+ + C_2H_2$ (~40%), and $C_2H_3^+ + H_2O$ (minor) in both cases. That these two isomers behaved similarly

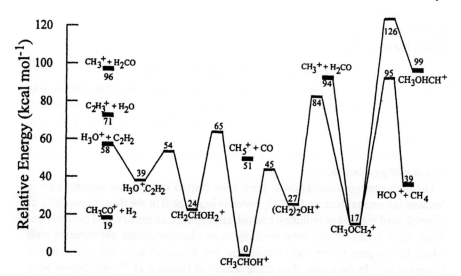

Figure 5. Potential energy surface for the $C_2H_5O^+$ system, showing stable structures, transition states and dissociation limits. The information is predominantly from the theoretical calculations of Fairley et al.[71] with the region of the $HCO^+ + CH_4$ and $CH_3^+ + H_2CO$ dissociation limits from Audier et al.[75]

in the Jarrold et al. studies is not surprising, since the barrier between them (see Figure 5) is lower than the barriers to other isomeric forms and to the $H_3O^+ + C_2H_2$ dissociation limit. That the dissociation limits $H_3O^+ + C_2H_2$ and $C_2H_3^+ + H_2O$ are coupled is reasonable since the exothermic proton transfer reaction,

$$C_2H_3^+ + H_2O \rightarrow H_3O^+ + C_2H_2 \quad \Delta H^\circ = -13.9 \text{ kcal mol}^{-1} \quad (27)$$

is known to proceed rapidly $(1.1(-9) \text{ cm}^3 \text{ s}^{-1})$.[25] That $CH_5^+ + CO$ is not also observed in addition to $HCO^+ + CH_4$ is surprising since these are coupled by the rapid reaction (rate coefficient, $9.9 (-10) \text{ cm}^3 \text{ s}^{-1}$),[25]

$$CH_5^+ + CO \rightarrow HCO^+ + CH_4 \quad \Delta H^\circ = -10.5 \text{ kcal mol}^{-1} \quad (28)$$

and therefore sample the same part of the potential surface (all of these dissociation limits are included on Figure 5). It is, of course, possible that there is a barrier in the exit channel to $CH_5^+ + CO$, as is indicated for $HCO^+ + CH_4$ from the large kinetic energy release in the metastable studies.[76] The metastable spectrum for $CH_3OCH_2^+$ is different from that for the other isomers discussed above since the $HCO^+ + CH_4$ channel is dominant (98%) and the $CH_5^+ + CO$ channel is observed (~2%) as well as $CH_3^+ + H_2CO$, although this is very small. These differences are reasonable since this isomer is more remote on the potential surface from the $H_3O^+ + C_2H_2$ dissociation limit and is separated by a large barrier, making the contribution from this channel small. It is also closer on the surface to the $HCO^+ + CH_4$ dissociation limit, although still separated by a substantial barrier (see Figure 5). The CID spectrum for this isomer is also different with CH_5^+ being much more dominant.

The $CH_3OCH_2^+$ region of the potential surface has recently been probed in detail by Audier et al.[75] who studied metastable dissociation of a series of deuterium and ^{13}C-labeled forms of this isomer. As before, the $HCO^+ + CH_4$ channel dominates with minor channels to CH_5^+ and CH_3^+. These isotope studies showed statistical scrambling of the H/D in the HCO^+ channel with complete equivalence of the C-atoms. Also, the bimodal kinetic energy release distributions that were obtained revealed the production of both HCO^+ and COH^+. Parallel FT-ICR studies of the H/D and ^{13}C-labeled reaction of CH_3^+ with H_2CO again revealed the complete equivalence of the C-atoms, but incomplete scrambling of the H- and D-atoms, in parallel with a binary product channel to $HCO^+ + CH_4$. Such studies provide information on the lifetime of the intermediate complex relative to the times for isotope exchange and suggest that the barrier to the binary products is not as high as theory indicates.

Jarrold et al.[76] also used CID to probe the $C_2H_5O^+$ produced predominantly in the collisional association of H_3O^+ with C_2H_2. The similarity between the CID spectrum and that of CH_3CHOH^+ and $(CH_2)_2OH^+$ indicates that the same parts of the potential surface are being accessed as expected from Figure 5. That vinyloxonium, $CH_2CHOH_2^+$, has not been detected experimentally is surprising since it has

Table 4. Summarized Rate Coefficients (cm^3 s^{-1}) and Product Ion Distributions (%) for the Reactions of $C_2H_5O^+$ Isomers with the Reactant Neutrals Indicated[a]

Reactant Neutral	H_3O^+/C_2H_2	CH_3CHOH^+	$CH_3OCH_2^+$	$(CH_2)_2OH^+$
		184.2 (186.8)	238.2	183.4 (188.0)
C_2H_5Br 171.0	$C_2H_5BrH^+$ $C_2H_5O^+·C_2H_5Br$ 1.8(−9) 55% 5.8(−11) 45%	ns[b]	ns	ns
C_7H_7F 181.1	$C_7H_8F^+$ $C_2H_5O^+·C_7H_7F$ 3.9(−9) 50% 5.4(−10) 50%	$C_2H_5O^+·C_7H_7F$ 4.5(−10)	$C_8H_8F^+$ ~85% $C_8H_9F^+$ ~15% $C_2H_5O^+·C_7H_7F$ <5% 2.6(−10)	$C_9H_{10}F^+$ 90% $C_2H_5O^+·C_7H_7F$ 10% 6.3(−10)
C_6H_6 181.6	$C_6H_7^+$ other products 2.1(−9) 50% 2.0(−10) 50%	ns	$C_7H_7^+$ 100% 5.0(−10)	ns
CH_3OH 182.1	$CH_3OH_2^+$ $C_2H_5O^+·CH_3OH$ 3.9(−9) 40% 4.7(−10) 60%	$C_2H_5O^+·CH_3OH$ 4.4(−10)	ns	$C_3H_7O^+$ 50% $C_2H_5O^+·CH_3OH$ 50% 5.0(−10)

Notes: [a]The reactions were studied in a SIFT at 300 K by Fairley et al.[71] The ions were produced in a high pressure electron impact ion source containing acetaldehyde (for CH_3CHOH^+), dimethoxymethane (for $CH_3OCH_2^+$) or oxirane (for $(CH_2)_2OH^+$). For H_3O^+/C_2H_2 (produced by injection of H_3O^+ and reaction with C_2H_2 in the flow tube; reaction 26), the separate rate coefficients for the two coexisting isomeric forms are given together with their percentage abundancies. Proton affinities (kcal mol^{-1}) are given below each reactant neutral and the proton detachment energies of the isomeric ions (also in kcal mol^{-1}) are below the ion identifications (both the theoretical and experimental (in parentheses) values are given).[28-30]
[b]ns indicates not studied.

a deep well. However, it may couple to the lower energy $CH_3CO^+ + H_2$ dissociation limit and thus spontaneously fragment. That this channel is not observed for the other isomers, even though it is accessible energetically, is evidence for the importance of dynamic motions over the potential surface.

The isomeric forms of $C_2H_5O^+$ have also been probed reactivity in a SIFT by Fairley et al.[71] The rate coefficients and product ions for the reactions studied are summarized in Table 4. By inspection of this table, it can be seen that the three experimentally accessible isomeric forms all react differently showing that they are all independent as would be expected from the potential surface. Thus, this is an example where reactivity can distinguish between isomers, whereas CID did not distinguish between CH_3CHOH^+ and $(CH_2)_2OH^+$. As stated at the beginning of this section, Fairley et al. found that the association product, $C_2H_5O^+$, of reaction 26 was an equal mixture of two isomeric forms. The reactivity of the rapidly reacting H_3O^+/C_2H_2 is different from the stable isomers as can be seen from Table 4. Some degree of identification was possible on the basis of proton affinity. The more reactive component proton transfers at the gas kinetic rate with all of the reactant neutrals in Table 4 (the proton affinities of these are given in this table in addition to the proton detachment energies of the stable isomers as determined by calculation[71] or obtained from compilations[28-30]). This shows that unlike the experimentally observed stable isomers, the proton affinity of H_3O^+/C_2H_2 is $\leq$ 171.1 kcal mol^{-1}. Only protonated vinyl alcohol, $CH_2CHOH_2^+$, has a proton detachment energy (171.5 kcal mol^{-1}) sufficiently small to be a candidate. Unfortunately, perhaps for the reason stated above, it was not possible to produce this isomer in a well-characterized way, so no direct comparison of reactivity was possible. The identification of the less reactive component of H_3O^+/C_2H_2 was more difficult. From the reactivity and product ions, it is obviously not $CH_3OCH_2^+$ or $(CH_2)_2OH^+$. The form CH_3CH-OH^+ is a possibility as is the weakly bonded adduct, $H_3O^+\cdot C_2H_2$, shown as a shallow minimum in Figure 5 ($CH_2CHOH_2^+$ is not a possibility since it is identified as the other H_3O^+/C_2H_2 component). Fairley et al. concluded that both of the identifications remain possibilities. However, the weakly bonded $H_3O^+\cdot C_2H_2$ (calculated bond strength 18.9 kcal mol^{-1}) might be expected to ligand switch with some of the neutral species used, e.g.:

$$H_3O^+\cdot C_2H_2 + CH_3OH \rightarrow H_3O^+\cdot CH_3OH + C_2H_2 \qquad (29)$$

That no such switching was observed favors CH_3CHOH^+. This needs further testing by more reactivity studies. Thus, Jarrold et al.[76] only obtained a partial identification since they did not consider $CH_2CHOH_2^+$ as a possible structure.

Protonated acetaldehyde has been accessed in studies reported in the literature[24,25] by rapid proton transfer reactions of acetaldehyde with HCO^+, $C_2H_3O^+$, and H_3O^+, and by:

$$CH_3CHO^+ + CH_3CHO \rightarrow CH_3CHOH^+ + C_2H_3O \quad 3.0(-9) \text{ cm}^3 \text{ s}^{-1} \qquad (30)$$

Isomers of $C_2H_5O^+$ have also been accessed by the binary reactions,[24,25]

$$CH_2O^+ + CH_4 \rightarrow C_2H_5O^+ + H \quad 15\% \qquad (31a)$$

$$\qquad\qquad 1.1(-10) \text{ cm}^3\text{s}^{-1}$$

$$\rightarrow CH_3O^+ + CH_3 \quad 85\% \qquad (31b)$$

$$CH_5O^+ + H_2CO \rightarrow C_2H_5O^+ + H_2O \quad 2.1(-11) \text{ cm}^3 \text{ s}^{-1} \qquad (32)$$

$$CH_3CHO^+ + (CH_2O)_3 \rightarrow C_2H_5O^+ + C_3H_5O_3 \quad 3.5(-9) \text{ cm}^3 \text{ s}^{-1} \qquad (33)$$

and,

$$C_3H_5^+ + (CH_3)_2O \rightarrow C_2H_5O^+ + C_3H_6 \qquad 5.3(-10) \text{ cm}^3 \text{ s}^{-1} \qquad (34)$$

although the forms of the $C_2H_5O^+$ products have not been identified. In addition, ternary association occurs in the reaction:

$$CH_3^+ + H_2CO + He \rightarrow C_2H_5O^+ + He \qquad 3.5(-26) \text{ cm}^6 \text{ s}^{-1} \qquad (35)$$

However, again the isomeric form of the product ion is not known.[24] For this, Jarrold et al.[76] concluded that, under low-pressure conditions, unimolecular decomposition of the $C_2H_5O^+$ will occur giving $HCO^+ + CH_4$ (as expected from inspection of Figure 5) and that radiatively stabilized association is unlikely.

E. $C_2H_7O^+$

The isomeric forms of this ion have been extensively investigated because of its significance in the production of the C_2H_5OH and $(CH_3)_2O$ detected in interstellar space.[14] These species are believed to be formed by the dissociative recombination of specific forms of $C_2H_7O^+$, i.e.,

$$C_2H_7O^+ + e \rightarrow C_2H_5OH + H \qquad (36a)$$

$$\rightarrow (CH_3)_2O + H \qquad (36b)$$

(e.g. see Herbst,[22] this volume). The reactions producing the $C_2H_7O^+$ in this environment were suggested to be the radiative associations,[15,77,78]

$$CH_3^+ + CH_3OH \rightarrow C_2H_7O^+ + h\nu \qquad (37)$$

and:

$$H_3O^+ + C_2H_4 \rightarrow C_2H_7O^+ + h\nu \qquad (38)$$

See the reviews by McMahon[16] and by Dunbar[17] for a discussion of this mechanism. Note that the reaction channels 36a and 36b are unlikely to be the only channels, since a considerable amount of energy is available in the recombination (~ 130 kcal mol^{-1}, depending on the isomer) and some further fragmentation probably occurs. For example, the recombination of H_3O^+ has recently been shown to yield the channels,[21]

$$H_3O^+ + e \rightarrow H + H_2O \qquad (39a)$$

$$\rightarrow OH + H_2 \qquad (39b)$$

$$\rightarrow OH + 2H \qquad (39c)$$

$$\rightarrow O + H + H_2 \qquad (39d)$$

with the reaction channels 39b to 39d being 95% of the products (a second recent determination suggests that this is somewhat less at 65%[79]). The details of the $C_2H_7O^+$ recombination will only become clear when the specific product distributions are experimentally determined and when the potential energy surfaces of both $C_2H_7O^+$ and C_2H_7O are known. This also requires the ways in which excited C_2H_7O surfaces couple, including the repulsive interactions to be established.

Partly because of the interstellar implications, the potential surface for $C_2H_7O^+$ has received considerable attention. It has been estimated by Jarrold et al.,[76] and partially calculated by Herbst,[77] Swanton et al.,[80] and Audier et al.[81] The calculations of Swanton et al. at the MP4/6-311G** and MP2/6-31G* levels are the most comprehensive of these and they will be further discussed together with the recent detailed MP2/6-31G* calculations of Fairley et al.[82] A composite of the potential energy surface defined by these calculations is given in Figure 6. It can be seen that the two most stable structures, $(CH_3)_2OH^+$ and $C_2H_5OH_2^+$, which are of most interstellar interest, are completely decoupled by a large activation energy barrier. On the $(CH_3)_2OH^+$ side, there is only the stable weakly bonded $CH_3^+ \cdot CH_3OH$ cluster, whereas on the $C_2H_5OH_2^+$ side, there is the proton bound dimer, $C_2H_4 \cdots H^+ \cdots H_2O$ connecting the $C_2H_5^+ \cdot H_2O$ and $H_3O^+ \cdot C_2H_4$ structures which connect to the analogous dissociation limits. Also included, from the potential surface of Swanton et al., are the separate, large activation energy barriers from $C_2H_5OH_2^+$ to $[CH_2OH \cdot CH_4]^+$ and $[CH_3CHOH \cdot H_2]^+$ which lead to CH_2OH^+ and CH_3CHOH^+ respectively.

This potential surface has been probed experimentally in a series of ways. Studies of proton transfer reactions have shown that

$$C_2H_5^+ + H_2O \rightarrow H_3O^+ + C_2H_4 \tag{40}$$

occurs at the gas kinetic rate $(1.9(-9)\ cm^3\ s^{-1})$, illustrating the close coupling between these dissociation limits.[25] This region of the potential surface has been further probed in studies of the dissociation of $C_2H_5OH_2^+$ in multiple collisions with He in a selected ion flow drift tube experiment up to a center-of-mass energy of 3.2 kcal mol^{-1}.[9] The $C_2H_5OH_2^+$ was produced from C_2H_5OH in an ion source by the reaction:

$$C_2H_5OH^+ + C_2H_5OH \rightarrow C_2H_5OH_2^+ + C_2H_5O \tag{41}$$

These studies showed that dissociation to $H_3O^+ + C_2H_4$ was dominant with that to $C_2H_5^+ + H_2O$ being <10%, consistent with the energetics in Figure 6. In this experimental study, it was concluded that internal modes of the $C_2H_5OH_2^+$ ion (rovibrational) were rapidly collisionally equilibrated on the time scale of the experiment (0.1 to 1 ms). The dissociation pathways of metastable excited $C_2H_5OH_2^+$ (generated by proton transfer to C_2H_5OH using CH_4 chemical ionization following electron impact) was investigated by Jarrold et al. in a reverse geometry (ZAB-2F) mass spectrometer.[76] Only metastable reaction to H_3O^+, the slightly

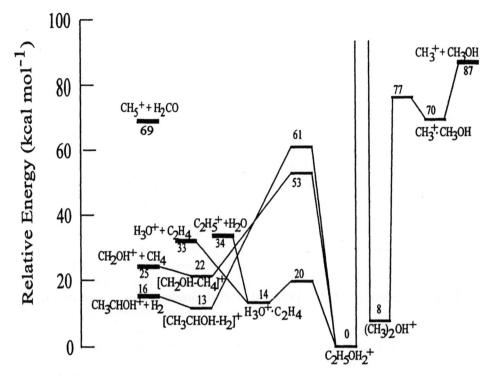

Figure 6. Potential energy surface for the $C_2H_7O^+$ system, showing stable structures, transition states and dissociation limits. The information comes from the theoretical calculations of Fairley et al.[82] and Swanton et al.[80] Note that Audier et al.[81] couple the $CH_2OH^+ + CH_4$ dissociation limit via a barrier to the weakly bonded $CH_3^+ \cdot CH_3OH$ isomer contrary to the conclusions of Swanton et al.

lower energy pathway, was observed, and the metastable peak was narrow, indicating no barrier to the reverse association with C_2H_4, consistent with the theoretical calculations. They also observed that when the CH_4 in the ion source was replaced by CD_4, substantial scrambling (H_2DO^+ ~25%, H_3O^+ ~ 75%) occurred. Although this is less than expected for full statistical scrambling (43% H_2DO^+), it is substantial. This is not surprising since the metastable $C_2H_5OH_2^+$ would have access to both the $C_2H_5^+ \cdot H_2O$ and $H_3O^+ \cdot C_2H_4$ forms via the $C_2H_4 \cdots H^+ \cdots H_2O$ proton bound dimer, with the proton/deuteron oscillating between freely rotating H_2O and C_2H_4. Since the H-atom sites in each of these molecules, and also in H_3O^+, are equivalent, scrambling will be efficient. Jarrold et al. performed similar studies with $(CH_3)_2OH^+$ (generated by proton transfer, again using $(CH_3)_2O$ in the ion source together with CH_4). In this case, the metastable $C_2H_7O^+$ dissociated to CH_3O^+ and CH_4, the production of which from $C_2H_5OH_2^+$ has substantial barriers. A wide metastable peak was also observed, indicating a significant barrier to the reverse association of CH_2OH^+ with CH_4. Use of CD_4 in the ion source gave little isotope scrambling in this case (<1%), as would be expected since the existence of a proton bound dimer of the form $H_2CO\cdots H^+\cdots CH_4$ is not indicated by the theoretical calculations. It is not clear from the calculations how $(CH_3)_2OH^+$ couples to these channels.

Jarrold et al. also performed studies with $Ar/H_2O/C_2H_4$ (99:0.6:0.4) and $He/CH_3Br/CH_3OH$ mixtures in the ion source. They concluded that the $C_2H_7O^+$ "almost certainly arises from the association of H_3O^+ with C_2H_4," in the former case and what was "believed to be from the association of CH_3^+ with CH_3OH" in the latter case. Collisional-induced dissociation (CID) of these isomers and of $C_2H_5OH_2^+$ and $(CH_3)_2OH^+$ was studied under single-collision conditions with He. These data showed that the CID spectrum of H_3O^+/C_2H_4 was similar to that of $C_2H_5OH_2^+$ and that of CH_3^+/CH_3OH had predominantly the $(CH_3)_2OH^+$ form. These observations are again consistent with the potential energy surface shown in Figure 6.

In these studies the quoted production routes were not unique. For example, with $He/CH_3Br/CH_3OH$ mixtures, a significant amount of $(CH_3)_2OH^+$ was directly produced by the reaction:

$$CH_3OH_2^+ + CH_3OH \rightarrow C_2H_7O^+ + H_2O \qquad (42)$$

To further clarify this situation, the $C_2H_7O^+$ isomers were produced in the controlled associations of H_3O^+ with C_2H_4 and CH_3^+ with CH_3OH in a SIFT experiment by Matthews et al.[83,84] Here the precursor ions were injected into flowing He and the reactant neutral added downstream. Thus, there was no possibility of a competing source for $C_2H_7O^+$. These isomeric forms and $C_2H_5OH_2^+$ and $(CH_3)_2OH^+$, generated in reactions 41 and

$$(CH_3)_2O^+ + (CH_3)_2O \rightarrow (CH_3)_2OH^+ + CH_3OCH_2 \qquad (43)$$

and directly injected into the flow tube, were probed by their reactivity with a series of neutral species. The combined data (Table 5) show that the CH_3^+/CH_3OH is very similar in reactivity to $(CH_3)_2OH^+$ (in both rate coefficients and products) and H_3O^+/C_2H_4 is very similar to $C_2H_5OH_2^+$. A more restricted experimental study by Fairley et al.,[82] involving the reactions of H_3O^+/C_2H_4, $C_2H_5OH_2^+$ and $(CH_3)_2OH^+$ with C_7H_7F and CH_2CHCN, has also shown that H_3O^+/C_2H_4 reacts very similarly to $C_2H_5OH_2^+$. Where there are differences between the products of the associated ion reactions and those of the strongly bound isomer with which they are identified, this can be explained as being due to some residual excitation and some associated complex character ($CH_3^+\cdot CH_3OH$ and $H_3O^+\cdot C_2H_4$) in the $C_2H_7O^+$ produced by ternary association in the flow tube. This implies that, for these systems, the He collisions are not sufficiently efficient to totally relax the associated ions. This is contrary to what is usually assumed in the vast literature on collisionally stabilized reactions[24] and on equilibrium in such reactions.[8] Simple quantal calculations have

Table 5. Summarized Rate Coefficients (cm^3 s^{-1}) and Product Ion Distributions (%) for the Reactions of $C_2H_7O^+$ with the Reactant Neutrals Indicated[a]

Reactant Neutral	Product Ions	CH_3^+/CH_3OH		$(CH_3)_2OH^+$; 192.1		H_3O^+/C_2H_4		$C_2H_5OH_2^+$; 188.3	
		Product Percent	Rate Coefficient	Product Percent	Rate Coefficient	Product Percent	Rate Coefficient	Product Percent	Rate Coefficient
C_2H_5OCHO 193.1	$C_2H_7O^+ \cdot C_2H_5OCHO$	100	1.9(−9)	75	2.2(−9)	10	2.3(−9)		2.6(−9)
	$C_2H_5OCHOH^+$			≤25		90		100	
$(CH_3)_2CHOH$ 191.2	$C_2H_7O^+ \cdot (CH_3)_2CHOH$	70	1.5(−9)	100	2.1(−9)	35	1.8(−9)	75	2.1(−9)
	$(CH_3)_2CHOH_2^+$					55			
	$(CH_3)_2CHO^+$					10			
	$CH_3^+(CH_3)_2CHOH$	30						25	
	$(CH_3)_2CHOH_2^+ \cdot C_2H_4$								
CH_3CO_2H 190.2	$C_2H_7O^+ \cdot CH_3CO_2H$	60	1.8(−9)	100	1.2(−9)	45	1.9(−9)	40	2.1(−9)
	$CH_3CO_2H_2^+$	40				55			
	$CH_3^+ CH_3CO_2H$							30	
	$CH_3CO_2H_2^+ \cdot C_2H_4$								
$CH_3C_6H_5$ 189.8	$C_2H_7O^+ \cdot CH_3C_6H_5$		6.9(−10)	≤35	6.9(−10)		1.7(−9)	30	1.7(−9)
	$CH_3C_6H_5H^+$	100		≥65		100		100	
CH_3CHO 186.6	$C_2H_7O^+ \cdot CH_3CHO$	100	1.3(−9)	100	1.0(−9)	95	1.6(−9)	100	1.6(−9)
	CH_3CHOH^+					5			

Reactant (PA, kcal mol⁻¹)	Product ion	%	k	%	k	%	k	%	k
$H_2C_3H_2$ 186.3	$C_2H_7O^+ \cdot C_3H_4$	50	6.8(-12)	95	5.8(-12)	10	3.0(-10)		4.8(-10)
	$C_3H_7^+$	50		≤5		90			
	$C_3H_7O^+$							100	
CH_3OH 181.9	$C_2H_7O^+ \cdot CH_3OH$	100	4.8(-10)	100	4.3(-10)	100	8.1(-10)	100	6.8(-10)
$c\text{-}C_3H_6$ 179.8	$C_2H_7O^+ \cdot C_3H_6$	25	9.9(-12)	100	6.4(-12)	100	2.4(-10)	40	2.4(-10)
	$C_3H_7^+$	35						60	
	$C_4H_9^+$	40							
H_2O 166.5	$C_2H_7O^+ \cdot H_2O$	100	5.4(-11)	100	5.8(-11)	100	8.1(-11)	100	8.1(-11)

Note: ªThe reactions were studied by Matthews and Adams[83] and Matthews et al.[84] in a SIFT at 300 K. $C_2H_5OH_2^+$ and $(CH_3)_2OH^+$ were produced from ethanol and dimethyl ether in a high pressure ion source by reactions 41 and 43, respectively. The associated ions, CH_3^+/CH_3OH and H_3O^+/C_2H_4, were produced by injecting CH_3^+ and H_3O^+ separately and associating them with CH_3OH and C_2H_4, respectively in the flow tube. Proton affinities (kcal mol⁻¹) are given below each reactant neutral and those of $(CH_3)_2O$ and C_2H_5OH are 192.1 and 188.3 kcal mol⁻¹, respectively.[28–30] C_2H_6 is unreactive with all of the isomers (<1(-13) cm³ s⁻¹).

shown[84] that when an ion can readily isomerize between a weakly and a strongly bonded form (e.g. $CH_3^+ \cdot CH_3OH$ and $(CH_3)_2OH^+$, see Figure 6), the fraction of the time spent in the weakly bonded form varies with vibrational level. Indeed, the amount of time spent in the weakly bonded form can be as much as 90% exhibiting an irregular oscillation with vibrational quantum number. Thus, as the degree of vibrational excitation reduces, the lowest level in the shallow part of the potential, i.e. the weakly bonded form, will be reached, so that further relaxation cannot occur when in that form. Then the rate of relaxation will be reduced, depending on the proportion of the time spent in the strongly bonded form. Eventually, a He collision will relax the ion when in the strongly bonded form so that the weakly bonded form will no longer be accessible and further, unimpeded relaxation to the bottom of the strongly bound well will occur. Since the detailed potential surface calculations suggest that, for both $CH_3^+ \cdot CH_3OH$ and $H_3O^+ \cdot C_2H_4$, there are stable shallow wells; a large proportion of the ions would be expected to be stabilized into these wells. That this is not the case implies that either there are no vibrational levels within the well or the barrier is sufficiently shallow and/or sufficiently thin for quantum tunneling to occur. This needs to be investigated further by, for example, using deuterated species to reduce the rate of tunneling.

F. $C_3H_3O^+$

Studies of the isomers of this species were undertaken because of a particularly vexing problem relating to the chemistry in interstellar molecular clouds. This is that propynal, $HC\equiv C-CHO$, has been observed but its isomer propadienone, $H_2C=C=C=O$, has not, although searches have been made in the most molecule-rich clouds, the Taurus Molecular Cloud and Sagittarius B2.[85] A possible route to these isomers is the reaction,[86]

$$C_2H_3^+ + CO \rightarrow C_3H_3O^+ + h\nu \tag{44}$$

(the calculated rate coefficient is $5.3(-11)$ cm^3 s^{-1} [86]; the collisionally stabilized analog has a ternary rate coefficient of $>3(-26)$ cm^6 s^{-1} at 80 K[87]). Following dissociative recombination, this could perhaps lead to the production of either propynal and/or propadiene plus an H-atom (as mentioned earlier, other energetically possible products are also likely to occur in general in such a recombination).

The structures of the $C_3H_3O^+$ isomers were first investigated by Holmes et al.[88] from fragmentation of metastable ions generated from 11 precursor molecules (methyl vinyl ketone, propargyl alcohol and its methyl homologs, acetylenic alcohols, etc.). From these studies, coupled with LCAO-SCF-MO calculations, they obtained heats of formation associated with five structures. Bouchoux et al.[89] made calculations at up to the CIPSI/4-31G*//HF/3-21G level. In this study, they determined the four most stable structures (out of 11 explored) to be: C-protonated propadienone, CH_2CHCO^+ (the acylium ion); the cyclic structure, c-C_3H_2-OH^+; O-protonated propynal, HC_2CHOH^+; and O-protonated propadienone,

CH$_2$C$_2$OH$^+$. They also characterized the transition states accessed in isomerization and explored dissociative channels to C$_2$H$_3^+$ + CO and HCO$^+$ + C$_2$H$_2$. Further detailed calculations were performed by Hopkinson and Lien[90] at the 6-31G*//6-31G* level which showed, consistent with Bouchoux et al., that the most stable structure is C-protonated propadienone, the dissociation of which leads to C$_2$H$_3^+$ + CO (they also studied isomeric forms of C$_3$H$_3^+$, C$_4$H$_5^+$, C$_3$H$_4$N$^+$, and C$_3$H$_2$F$^+$). Very recently, Maclagan et al.[91] have revisited the problem at the G2 level of theory, and calculated the relative energies of 15 isomeric forms and the various dissociation limits. These energies are depicted on the potential surface in Figure 7, together with some data from Bouchoux et al. Maclagan et al. also compared their results with the previous theoretical calculations; in particular, in agreement with Hopkinson and Lien, they concluded that the association of C$_2$H$_3^+$ with CO would lead to protonated propadienone.

To identify the isomeric form of the product of the collisionally stabilized analog of reaction 44 experimentally, Scott et al.[92] studied reactions of C$_3$H$_3$O$^+$ (C$_2$H$_3^+$/CO) produced in that reaction and compared it with that produced directly from propynal in an electron impact ion source or by proton transfer from HCO$^+$ to propynal in the flow tube (these latter two production methods yielded ions with

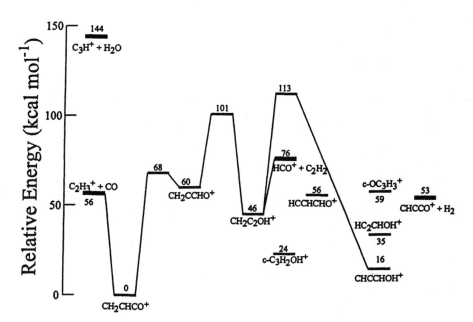

Figure 7. Potential energy surface for the C$_3$H$_3$O$^+$ system, showing stable structures, transition states and dissociation limits. The information originates mainly from the theoretical calculations of Maclagan et al.,[91] although some data from the earlier calculations of Bouchoux et al.[89] are included. Dissociation limits are noted where possible.

identical reactivities). The species were then reacted with a series of neutrals to search for similarities and differences in reactivity. These data are summarized in Table 6, where it can be seen that the two forms of $C_3H_3O^+$ react very differently. The rate coefficients for both of the isomeric ions are close to the collision limit in all cases except for $C_2H_3^+/CO + NH_3$ which is 2 orders of magnitude smaller. The reactions of HC_2CHOH^+ all occur predominantly by proton transfer consistent with the calculated proton affinity of propynal of 180.9 kcal mol^{-1}.[28-30] To bracket the proton affinity experimentally, reactions with C_6H_6 and C_2H_5I (with proton affinities of 181.6 and 176.1 kcal mol^{-1} respectively) were studied. These gave 20% proton transfer with C_6H_6 (the remainder being association) at the collisional rate, whereas the rate coefficient for reaction with C_2H_5I was an order of magnitude smaller with no proton transfer, the only product being association. Since proton transfer occurs whenever it is energetically possible,[36] this brackets the proton affinity of propynal between 181.6 and 176.1 kcal mol^{-1}, again consistent with the theoretical value, so the identification of this isomer is secure. The proton detachment energy of the $C_2H_3^+/CO$ ion can be determined in a similar way since proton

Table 6. Summarized Rate Coefficients (cm^3 s^{-1}) and Product Ion Distributions (%) for the Reactions of $C_3H_3O^+$ Isomers with the Reactant Neutrals Indicated[a]

Reactant Neutral	Ion Products	$C_2H_3^+/CO$		HC_2CHOH^+	
		Product Percent	Rate Coefficient	Product Percent	Rate Coefficient
$C_2H_5NH_2$	$C_2H_5NH_3^+$	85	1.7(−9)	85	1.8(−9)
234.5	$C_3H_3O^+ \cdot C_2H_5NH_2$	15		15	
CH_3NH_2	$CH_3NH_3^+$	75	1.5(−9)	90	2.0(−9)
214.4	$C_3H_3O^+ \cdot CH_3NH_2$	25		10	
C_4H_5N	$C_4H_5NH^+$	0	2.0(−9)	80	2.2(−9)
207.7	$C_3H_3O^+ \cdot C_4H_5N$	100		20	
NH_3	NH_4^+	0	3(−11)	100	1.6(−9)
204.3	$C_3H_3O^+ \cdot NH_3$	100		0	
$(n\text{-}C_4H_9)_2O$	$(n\text{-}C_4H_9)_2OH^+$	0[b]	1.6(−9)	100	2.0(−9)
203.8	$C_3H_3O^+ \cdot (n\text{-}C_4H_9)_2O$	≤25		0	
$(C_2H_5)_2CO$	$(C_2H_5)_2COH^+$	0	1.3(−9)	100	2.5(−9)
201.7	$C_3H_3O^+ \cdot (C_2H_5)_2CO$	100		0	
$C_6H_{10}O$	$C_6H_{10}OH^+$	0	2.9(−9)	100	3.2(−9)
201.7	$C_3H_3O^+ \cdot C_6H_{10}O$	100		0	

Note: [a]The reactions were studied by Scott et al. in a SIFT at 300 K.[92] Protonated propynal, HC_2CHOH^+ was produced from propynal in a high pressure ion source and by direct proton transfer from HCO^+ in the flow tube. The $C_2H_3^+/CO$ isomer was produced directly in the flow tube by injecting $C_2H_3^+$ and associating with CO. Proton affinities (kcal mol^{-1}) are given below each reactant neutral and the calculated value for HC_2CHO is 180.9 kcal mol^{-1}.[28-30]

[b]Other prroducts are $C_4H_9OC_4H_8^+$ and $C_4H_9O^+$.

transfer is observed with CH_3NH_2, but not with C_4H_5N (proton affinities of 214.4 and 207.7 kcal mol^{-1} respectively) bracketing the proton detachment energy between 208 and 214 kcal mol^{-1}. The proton affinity of propadienone with the proton in the C-2 position has been calculated to be 214.4 kcal mol^{-1} [91] (the proton detachment energies of the other isomers are lower than this), identifying the product of the $C_2H_3^+$ association with CO as C-2 protonated propadienone, consistent with the potential surface in Figure 7.

Thus, the route to the propynal observed in interstellar clouds is not established. A survey of the literature shows that a few experimentally studied reactions access isomeric forms of $C_3H_3O^+$. The $(CO)_2^+$ dimer apparently reacts with CH_4 to give $C_3H_3O^+ + OH$, but with a small rate coefficient $(7.6(-11)$ cm^3 s^{-1}).[24] The rapid binary reaction of HCO^+ with C_2H_2 $(1.4(-9)$ cm^3 s^{-1})[24] passes through the potential surface giving >90% $C_2H_3^+ + CO$, thus coupling these dissociation limits. This is important since the calculated potential surface has a substantial barrier between these two limits (see Figure 7), indicating that there must be an alternative reaction coordinate not shown in Figure 7 or that the calculated barrier height is incorrect. A further interesting binary reaction is

$$C_3H^+ + H_2O \rightarrow C_2H_3^+ + CO \quad 40\% \tag{45a}$$

$$\rightarrow HCO^+ + C_2H_2 \quad 55\% \quad 4.5(-10) \text{ cm}^3 \text{ s}^{-1} \tag{45b}$$

$$\rightarrow CHC_2O^+ + H_2 \quad 5\% \tag{45c}$$

which again couples $HCO^+ + C_2H_2$ with $C_2H_3^+ + CO$, with $C_3H^+ + H_2O$ and, to a limited extent, with $CHC_2O^+ + H_2$. This latter part of the potential surface needs to be explored.

G. $C_3H_6N^+$

Some of the isomers in this system have recently been investigated both theoretically and experimentally in a SIFT experiment.[93,94] The primary purpose of this study was to identify the structural forms of the isomers produced in the association channels,

$$H_2CN^+ + C_2H_4 + He \rightarrow C_3H_6N^+ + He \tag{46}$$

and:

$$CH_3^+ + CH_3CN + He \rightarrow C_3H_6N^+ + He \tag{47a}$$

$$(\rightarrow C_2H_5^+ + HCN) \tag{47b}$$

$$(\rightarrow H_2CN^+ + C_2H_4) \tag{47c}$$

This would reveal possible routes to the $C_2H_5CNH^+$ that is believed to be produced in the analogous radiative association reactions that occur in interstellar gas clouds.

The observed C_2H_5CN could then be produced,[14] following electron–ion recombination. The rate coefficients for the collisionally stabilized analogs are large with values of >7 (-27)[72] and 1(-23)[94] cm^6 s^{-1} for reactions 46 and 47a, respectively, at 300 K. The forms of the association products were probed reactively by Wilson et al.[93] in parallel with the $C_3H_6N^+$ isomers generated in an ion source containing either C_2H_5CN or C_2H_5NC (producing presumably $C_2H_5CNH^+$ and $C_2H_5NCH^+$, respectively, in secondary reactions involving the source gas) and by proton transfer from HCO^+ in the flow tube, viz.:

$$HCO^+ + C_2H_5CN(C_2H_5NC) \rightarrow C_2H_5CNH^+(C_2H_5NCH^+) + CO \quad (48)$$

The $C_3H_6N^+$ produced in these ways (reactions 46 and 48) were then further reacted with $H_8C_4O_2$ (1,4 dioxane), $(CH_3)_2CO$ and NH_3 (and for C H/C HCN with CH_3NH_2 and $C_2H_5NH_2$). These data are summarized in Table 7. They show the simple picture of either no reactivity or efficient reaction leading to proton transfer or association. The occurrence of proton transfer is consistent with the known proton affinities of C_2H_5CN or C_2H_5NC and of the reactant neutrals (also given in Table 7).

Some ab initio calculations have been made for the $C_3H_6N^+$ system by Smith et al.[94] using up to MP4SDQ/6-311G**//HF/6-311G** and by Bouchoux et al.[95] using

Table 7. Summarized Rate Coefficients (cm^3 s^{-1}) and Product Ion Distributions (%) for the Reactions of $C_3H_6N^+$ Isomers with the Reactant Neutrals Indicated[a]

Reactant Neutral	$HCNH^+/C_2H_4$	CH_3^+/CH_3CN	$C_2H_5NCH^+$	$C_2H_5CNH^+$
			204	193
$H_8C_4O_2$	adduct	ns[b]	adduct	adduct 80%
194				$H_8C_4O_2H^+$ 20%
	3.7(−10)		~3.5(−10)	1.1(−9)
$(CH_3)_2CO$	adduct	ns	adduct	$(CH_3)_2COH^+$
197	7.2(−10)		7.9(−10)	2.8(−9)
NH_3	NH_4^+ 95%		NH_4^+ 85%	ns
204	adduct 5%		adduct 15%	
	1.0(−9)	<5(−13)	8.5(−10)	
CH_3NH_2	ns	<5(−13)	ns	ns
214				
$C_2H_5NH_2$	ns	adduct	ns	ns
217		2.8(−11)		

Note: [a]The reactions were studied by Wilson et al. in a SIFT at 300 K.[93] $HCNH^+/C_2H_4$ and CH_3^+/CH_3CN were produced in the flow tube by reactions 46 and 47a. $C_2H_5NCH^+$ and $C_2H_5CNH^+$ were produced directly from C_2H_5NC and C_2H_5CN in a high pressure ion source and by proton transfer to these neutrals from HCO^+ in the flow tube. Proton affinities (kcal mol^{-1}) are given below each reactant neutral and the proton detachment energies of the ions are below the ion identification.[28–30]
[b]ns indicates not studied.

MP4SDTQ/6-31G*//HF/3-21G level of theory. These calculations, together with information on known enthalpies of formation of the stable isomers,[93] are given as a potential energy surface in Figure 8. This shows six stable isomeric forms. The experimental data in Table 7 indicate that $HCNH^+/C_2H_4$ reacts very similarly to $C_2H_5NCH^+$ with the only slight difference in the rate coefficient and product ion distribution being in the NH_3 reaction. This strongly suggests that the ions are identical, consistent with the potential surface. From this surface it is obvious that $C_2H_5CNH^+$ could not be accessed even though it has a lower energy, because of the intervening barrier,[96] which is too large to overcome even with the additional energy in the $HCNH^+ + C_2H_4$ reactants. Note that the reactivity of $HCNH^+/C_2H_4$ is very different from that of $C_2H_5CNH^+$, consistent with the potential surface. In a similar manner, the stable form $CH_3CNCH_3^+$ is not accessible from $HCNH^+ + C_2H_4$, because of an even higher barrier.[94] That proton transfer does not occur in the $HCNH^+ + C_2H_4$ reaction is not surprising since it is endothermic by 8.9 kcal mol^{-1} [28–30] (similar to the energy difference determined theoretically; see Figure 8). Indeed, the reverse proton transfer reaction ($C_2H_5^+ + HCN$) proceeds with a rate coefficient of 2.7 (–9) cm^3 s^{-1}, close to the collisional value.[25] This shows that both $HCNH^+ + C_2H_4$ and $C_2H_5^+ + HCN$ are coupled through the same intermediate, presumably the $HCN\cdots H^+\cdots C_2H_4$ complex which is an excited form of $C_2H_5NCH^+$. However, $CH_3CNCH_3^+$ would be accessed from the $CH_3^+ + CH_3CN$ association and, indeed, CH_3^+/CH_3CN reacts very differently with NH_3 than $C_2H_5NCH^+$ and $HCNH^+/C_2H_4$ (the $C_2H_5CNH^+$ reaction was not studied). The CH_3^+/CH_3CN ion must

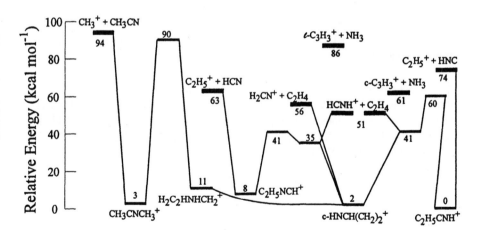

Figure 8. Potential energy surface for the $C_3H_6N^+$ system, showing stable structures, transition states and dissociation limits. The information is from the work of Bouchoux et al.,[95,96] Smith et al.[94] and Wilson et al.[93] An additional part of the potential surface connecting the stable forms, $C_2H_5CNH^+$ and $C_2H_5NCH^+$ (not shown) has been calculated by Bouchoux et al. This a higher energy pathway than that illustrated in the figure.

have a large equivalent proton affinity (>217 kcal mol^{-1}) since it does not proton transfer with either CH_3NH_2 or $C_2H_5NH_2$ (proton affinities of 214 and 217 kcal mol^{-1} respectively); indeed it associates in the latter case. It is unlikely for $C_2H_5CNH^+$ to react in this way since rapid proton transfer is energetically possible and is observed with both $(CH_3)_2CO$ and $H_8C_4O_2$ and would occur at the gas kinetic rate with CH_3NH_2 and $C_2H_5NH_2$.[36] Its proton affinity of 193 kcal mol^{-1} must therefore be much smaller than that of CH_3^+/CH_3CN. Theory predicts that the CH_3^+/CH_3CN isomer will be $CH_3CNCH_3^+$. Available experimental data is consistent with this, although there has been no comparative study between the reactivity of this ion and CH_3^+/CH_3CN.

In parallel with the association leading to CH_3^+/CH_3CN, there are binary channels giving $C_2H_5^+ + HCN$ and $H_2CN^+ + C_2H_4$ (reactions 47b and 47c). This shows that the barrier to these species can be surmounted, which is again consistent with Figure 8. In Section III.A and Table 1, it is shown that the reaction of acyclic $C_3H_3^+$ with NH_3 also accesses the $C_3H_6N^+$ potential surface, although the isomeric form of the association product is not known. The dissociation limits, $C_3H_3^+$ (both cyclic and acyclic) + NH_3 are included on Figure 8 as well as that for $HCNH^+ + C_2H_4$. Isomers of $C_3H_6N^+$ have been shown to be produced in a series of binary reactions ($C_2H_4N^+$ + CH_3OH, $C_2H_4N^+ + CH_3NC$, $C_2H_5OH_2^+ + CH_3CN$, $(CH_3)_2OH^+ + CH_3CN$ and $C_2H_5SH_2^+ + HCN$),[24] although with rate coefficients in the $1(-11)$ cm^3 s^{-1} range and with an isomeric form being suggested in some cases.

H. $C_2H_5O_2^+$

The studies discussed above have shown that many isomeric forms of ions can be generated by association and their reactivity probed. However, this is not straightforward in all cases. The isomeric forms of $C_2H_5O_2^+$ are a clear example for which, unfortunately, potential surface information is not available. This is of interest since it is a possible route, following dissociative electron–ion recombination, to the HCO_2CH_3 and CH_3CO_2H observed in interstellar gas clouds.[14,97] Logical routes to these isomeric ions are the associations: $CH_3^+ + HCO_2H$; $C_2H_5^+ + O_2$; $O_2H^+ + C_2H_4$; $CH_3OH_2^+ + CO$; $HCO^+ + CH_3OH$; $CH_5^+ + CO_2$; $HCO_2^+ + CH_4$; $H_3O_2^+ + C_2H_2$; $C_2H_3^+ + H_2O_2$ and $H_3CO^+ + H_2CO$. In the reaction,

$$HCO_2^+ + CH_4 \rightleftharpoons CH_5^+ + CO_2 \qquad (49)$$

proton transfer occurs in both directions (with rate coefficients of $7.2(-10)$ and $3.3(-11)$ cm^3 s^{-1} respectively[25]) with no association in either direction. For the reaction,

$$HCO^+ + CH_3OH \rightarrow CH_3OH_2^+ + CO \qquad (50)$$

the forward rate coefficient is $2.6(-9)$ cm^3 s^{-1} with no significant reverse reaction[25] ($<1(-13)$ cm^3 s^{-1}). Thus, although the reactant and product species are separately coupled on the potential surface in these two reactions, no stable intermediate is indicated. In the rapid reaction,

$$O_2H^+ + C_2H_4 \rightarrow C_2H_3^+ + O_2 + H_2 \quad 77\% \tag{51a}$$

$$\rightarrow C_2H_5^+ + O_2 \qquad 23\% \tag{51b}$$

(overall rate coefficient of 1.1(−9) cm^3 s^{-1}), the reaction proceeds by dissociative and non-dissociative proton transfer.[25] The reverse of reaction 51b gives

$$C_2H_5^+ + O_2 \rightarrow \text{products} \qquad 95\% \tag{52a}$$

$$\rightarrow C_2H_5O_2^+ + He \qquad 5\% \tag{52b}$$

with a small amount of the association product. The ion products of the other channels could not be determined, presumably because the secondary reactions are more rapid than the slow primary reaction (rate coefficient of 1.6(−13) cm^3 s^{-1}). Also,

$$CH_3^+ + HCO_2H \rightarrow HC(OH)_2^+ + CH_2 \tag{53}$$

is rapid (2.1(−9) cm^3 s^{-1}).[25] The association,

$$H_3CO^+ + H_2CO + He \rightarrow C_2H_5O_2^+ + He \tag{54}$$

does occur and with quite a large ternary rate coefficient (2(−27) cm^6 s^{-1}).[24] The $H_3O_2^+ + C_2H_2$ and $C_2H_3^+ + H_2O_2$ reactions have not yet been studied. In binary reactions, protonated species, presumably protonated acetic acid and methyl formate, are readily produced by rapid proton transfer from H_3^+, HCO^+, and H_3O^+ to those molecules[24,25] and they are well-established stable species. $C_2H_5O_2^+$ is also produced in the fragmentation reactions of H_3O^+ with $CH_3CO_2C_2H_5$ and $C_2H_5CO_2CH_3$.[25]

Thus, there are only a few routes known to access the $C_2H_5O_2^+$ potential surface. Obviously, more work needs to be done to establish the isomeric identities of the species in this system; because of the nature of the reactants, this will not be a simple study.

IV. CONCLUSIONS

The foregoing discussions have shown how valuable ion–molecule reactions are in probing potential energy surfaces of isomers, ABCD$^+$, by accessing the surface with association reactions, transiently with binary reactions where essentially the (ABCD$^+$)* intermediate undergoes unimolecular decomposition, or as a product in binary reactions. In the association reactions, isomers can only be produced if they are accessible from a dissociating limit. Thus, some isomers may be hidden from view in this approach. The same restriction applies where the $(ABCD^+)^*$ is the intermediate, except that in this case the intermediate is not identified but dissociation limits which connect to the same region of the potential surface are. In the latter case, information can often be gleaned about the intermediate from the degree of scrambling using isotopically labeled reactants. The restriction that the $ABCD^+$ isomer be accessible from a dissociation limit does not apply when it is a *product* of a binary reaction. Such species can often be identified reactively if the energetics are known. Collision-induced dissociation under single-collision conditions is also very useful for identifying isomeric species, although if more than one isomer accesses the same dissociation limit, they cannot be distinguished except perhaps by the energy dependence of the fragmentation, which may reveal the relative heights of the different transition states involved. In a somewhat equivalent way, multicollision dissociation in a drift tube or flow drift tube can provide similar information, although not with such a high-energy resolution. Also, for those isomers that are accessible by association, studies of equilibrium can provide bond strengths via van't Hoff plots.

In these studies, theoretical calculations are extremely useful in identifying stable structures and transition states, i.e. specific points on a rudimentary potential surface. They are an essential complement to the experimental measurements. With the present computing capabilities, there are few cases where a significant fraction of the potential surface can be explored and where the dynamics of the reaction process can be modeled. As a first step, it would be valuable to know the shapes of the various potential wells in addition to their depths.

From an experimental viewpoint, the surface can be probed further by studying binary systems that are unreactive at thermal energies by increasing the collision energy in an attempt to open up channels and determine their onsets. This will yield information on activation energy barriers. New channels can also be opened up in reactive binary systems. The study of associated ion reactions as a function of energy in a drift tube will also enable different parts of the potential surface to be probed by varying the internal energy of the ion. All of these different approaches are necessary to get as complete a picture as possible of the potential surface.

There has been a consistent motivation for the work presented in this chapter; the application to molecular synthesis in interstellar gas clouds (see, for example, Herbst,[22] this volume). The species in these regions are detected spectroscopically and are thus automatically isomerically identified. The routes to the observed neutral species consistently involve ion–molecule reactions followed by dissociative electron–ion recombination.[18] The first step in this process is to determine whether an isomeric ion can be formed which is likely to recombine to an observed neutral species. The foregoing discussion has shown that whether this occurs depends on the detailed nature of the potential surface. Certainly, this only occurs in some of the cases studied. Much more understanding will be required before the needs of this application are fulfilled.

Obviously, a great deal more information could be obtained if the isomeric ions could be probed spectroscopically. Vibrational states of the various isomers are not generally well known, but some structural information is available. Thus, the rotational structure of vibrational transitions may provide a better signature for particular isomers. Certainly, insufficient data are available about the potential surfaces of electronically excited states for electronic excitation to be used as a probe, e.g., as in the very sensitive laser induced fluorescence. At present, there are sensitivity limitations in the infrared region of the spectrum, but this may well be an avenue for the future. The study of isomeric systems and their potential surfaces has just begun!

ACKNOWLEDGMENTS

The NSF (Division of Astronomical Sciences) under grant AST-94415485 and the Donors of the PRF (administered by the ACS) under grant 27441-AC6 are both gratefully acknowledged for support of our work reported here. Thanks are due to Eldon E. Ferguson and Murray J. McEwan for providing preprints of their work prior to publication.

REFERENCES

1. Bell, R. P. *The Tunnel Effect in Chemistry*. Chapman & Hall: London, 1980.
2. Radom, L. *Int. J. Mass Spectrom. Ion Proc.* **1992**, *118/119*, 339.
3. Radom, L. *Organ. Mass Spec.* **1991**, *26*, 359.
4. Uggerud, E. *Mass Spec. Revs.* **1992**, *11*, 389.
5. Schreiner, P. R.; Schaefer, H. F.; Schleyer, P. v. R. *Adv. Gas Phase Ion Chem.* **1996**, *2*, 125.
6. Brauman, J. I. In *Kinetics of Ion–Molecule Reactions*; Ausloos, Ed.; Plenum: New York, 1979, p 153.
7. Dodd, J. A.; Brauman, J. I. *J. Phys. Chem.* **1986**, *90*, 3559.
8. Kebarle, P. *Ann. Rev. Phys. Chem.* **1977**, *28*, 445.
9. Smith, S. C.; McEwan, M. J.; Giles, K.; Smith, D.; Adams, N. G. *Int. J. Mass Spectrom. Ion Proc.* **1990**, *96*, 77.
10. Cooks, R. G. In *Collision Spectroscopy*; Cooks, R. G., Ed.; Plenum: New York, 1978, p 357.
11. Marinelli, P. J.; Paulino, J. A.; Sunderlin, L. S.; Wenthold, P. G.; Poutsma, J. C.; Squires, R. R. *Int. J. Mass Spectrom. Ion Proc.* **1994**, *130*, 89.
12. Smith, D.; Adams, N. G. In *Reactions of Small Transient Species*; Fontijn, A.; Clyne, M. A. A., Eds.; Academic Press: London, 1983, p 311.
13. Bowen, R. D.; Williams, D. H.; Schwarz, H. *Angew. Chem. Int. Ed. Engl.* **1979**, *18*, 451.
14. Millar, T. J.; Williams, D. A. In *Dust and Chemistry in Astronomy*; Millar, T. J.; Williams, D. A., Eds.; IOP Publishing: London, 1993, p 1.
15. Smith, D.; Adams, N. G. *Int. Rev. Phys. Chem.* **1981**, *1*, 271.
16. McMahon, T. B. *Adv. Gas Phase Ion Chem.* **1996**, *2*, 41.
17. Dunbar, R. C. *Adv. Gas Phase Ion Chem.* **1996**, *2*, 87.
18. Adams, N. G. *Adv. Gas Phase Ion Chem.* **1992**, *1*, 271.
19. Johnsen, R.; Mitchell, J. B. A. *Adv. Gas Phase Ion Chem.* **1998**, *3*, 49.
20. Bates, D. R. *Adv. At. Mol. Opt. Phys.* **1994**, *34*, 427.
21. Williams, T. L.; Adams, N. G.; Babcock, L. M.; Herd, C. R.; Geoghegan, M. *Mon. Not. Roy. Astr. Soc.* **1996**, *282*, 413.
22. Herbst, E. *Adv. Gas. Phase Ion Chem.* **1998**, *3*, 1.

23. McEwan, M. J. *Adv. Gas. Phase Ion Chem.* **1992**, *1*, 1.
24. Ikezoe, Y.; Matsuoka, S.; Takebe, M.; Viggiano, A. *Gas Phase Ion Molecule Reaction Rate Constants Through 1986.* Maruzen: Tokyo, 1987.
25. Anicich, V. G. *J. Phys. Chem. Ref. Data* **1993**, *22*, 1469.
26. Anicich, V. G. *Astrophys. J. Suppl. Ser.* **1993**, *84*, 215.
27. Rosenstock, H. M.; Draxl, K.; Steiner, B. W.; Herron, J. T. *J. Phys. Chem. Ref. Data* **1977**, *6*, Suppl. No. 1.
28. Lias, S. G.; Liebman, J. F.; Levin, R. D. *J. Phys. Chem. Ref. Data* **1984**, *13*, 695.
29. Lias, S. G.; Bartmess, J. E.; Liebman, J. F.; Holmes, J. L.; Levin, R. D.; Mallard, W. G. *J. Phys. Chem. Ref. Data* **1988**, *17*, Suppl. No. 1.
30. Hunter, E. P.; Lias, S. G. *J. Phys. Chem. Ref. Data.* **1997**. To be published; Web Site: http://webbook.nist.gov/pa-ser.htm.
31. McIver, R. T. Jr. *Rev. Sci. Instrum.* **1978**, *49*, 111.
32. Adams, N. G.; Smith, D. In *Techniques for the Study of Ion-Molecule Reactions*; Farrah, J. M.; Saunders, W. H., Eds.; Wiley: New York, 1988, p 165; Smith, D.; Adams, N. G. *Adv. At. Mol. Phys.* **1988**, *24*, 1.
33. Baranov, V.; Bohme, D. K. *Int. J. Mass Spectrom. Ion Proc.* **1996**, *154*, 71.
34. Adams, N. G.; Smith, D. *Int. J. Mass Spectrom. Ion Phys.* **1976**, *21*, 349.
35. Adams, N. G.; Smith, D. *J. Phys. B* **1976**, *9*, 1439.
36. Bohme, D. K. In *Interactions between Ions and Molecules*; Ausloos, P., Ed.; Plenum: New York, 1975, p 489.
37. Petrie, S.; Freeman, C. G.; Meot-Ner, M.; McEwan, M. J.; Ferguson, E. E. *J. Amer. Chem. Soc.* **1990**, *112*, 7121.
38. Hansel, A.; Glantschnig, M.; Scheiring, Ch.; Lindinger, W.; Ferguson, E. E. *J. Phys. Chem.* **1997**. In press.
39. Maluendes, S. A.; McLean, A. D.; Yamashita, K.; Herbst, E. *J. Chem. Phys.* **1993**, *99*, 2912.
40. Maluendes, S. A.; McLean, A. D.; Herbst, E. *Astrophys. J.* **1993**, *417*, 181.
41. Radom, L.; Hariharan, P. C.; Pople, J. A.; Schleyer, P. v. R. *J. Amer. Chem. Soc.* **1993**, *98*, 10.
42. Leszczynski, J.; Wiseman, F.; Zerner, M. C. *Int. J. Quantum Chem.* **1988**, *S21*, 117.
43. Feng, J.; Leszczynski, J.; Weiner, B.; Zerner, M. C. *J. Amer. Chem. Soc.* **1989**, *111*, 4648.
44. Glukhovtsev, M. N.; Bach, R. D.; Laiter, S. *J. Phys. Chem.* **1996**, *100*, 10952.
45. Smythe, K. C.; Lias, S. G.; Ausloos, P. *Combust. Sci. Technol.* **1982**, *28*, 147.
46. Baykut, G.; Brill, F. W.; Eyler, J. R. *Combust. Sci. Technol.* **1986**, *45*, 233.
47. Smith, D.; Adams, N. G. *Int. J. Mass Spectrom. Ion Proc.* **1987**, *76*, 307.
48. Adams, N. G.; Smith, D. *Astrophys. J.* **1987**, *317*, L25.
49. Matthews, H. E.; Irvine, W. M. *Astrophys. J.* **1985**, *298*, L61.
50. Thaddeus, P.; Vrtilek, J. M.; Gottlieb, C. A. *Astrophys. J.* **1985**, *299*, L63.
51. Cernicharo, J.; Gottlieb, C. A.; Guelin, M.; Killian, T. C.; Paubert, G.; Thaddeus, P.; Vrtilek, J. M. *Astrophys. J.* **1991**, *368*, L39.
52. Ozturk, F.; Baykut, G.; Eyler, J. R. *J. Phys. Chem.* **1987**, *91*, 4360.
53. Wiseman, F. L.; Ozturk, F.; Zerner, M. C.; Eyler, J. R. *Int. J. Chem. Kinet.* **1990**, *22*, 1189.
54. Moini, M. *J. Amer. Soc. Mass Spectrom.* **1992**, *3*, 631.
55. McEwan, M. J.; McConnell, C. L.; Freeman, C. G.; Anicich, V. G. *J. Phys. Chem.* **1994**, *98*, 5068.
56. Ha, T-K; Nguyen, M. T. *J. Phys. Chem.* **1984**, *88*, 4295.
57. Cheung, Y-S.; Li, W-K. *Chem. Phys. Letts.* **1994**, *223*, 383.
58. Cheung, Y-S.; Li, W-K. *J. Mol. Struct. (Theochem.)* **1995**, *333*, 135.
59. Munson, M. S. B.; Franklin, J. L. *J. Phys. Chem.* **1964**, *68*, 3191.
60. Villinger, H.; Saxer, A.; Richter, R.; Lindinger, W. *Chem. Phys. Letts.* **1983**, *96*, 513.
61. Smith, D.; Adams, N. G. *Int. J. Mass Spectrom. Ion Phys.* **1977**, *23*, 123.
62. Holmes, J. L.; Mommers, A. A.; Koster, C. De.; Heerma, W.; Terlouw, J. K. *Chem. Phys. Letts.* **1985**, *115*, 437.

63. Kirchner, N. J.; Van Doren, J. M.; Bowers, M. T. *Int. J. Mass Spectrom. Ion Proc.* **1989**, *92*, 37.
64. Van Doren; J. M.; Barlow, S. E.; DePuy, C. H.; Bierbaum, V. M.; Dotan, I.; Ferguson, E. E. *J. Phys. Chem.* **1986**, *90*, 2772.
65. Glosik, J.; Jordan, A.; Skalsky, V.; Lindinger, W. *Int. J. Mass Spectrom. Ion Proc.* **1993**, *129*, 109.
66. Villinger, H.; Richter, R.; Lindinger, W. *Int. J. Mass Spectrom. Ion Proc.* **1983**, *51*, 25.
67. Fisher, N. D.; Adams, N. G. *Chem. Phys. Letts.* **1998**. Submitted.
68. Smith, D.; Adams, N. G. *Astrophys. J.* **1977**, *217*, 741.
69. Chunxiao, G.; Dongyan, H.; Shuying, L. *Acta Phys.-Chim. Sinica* **1991**, *7*, 420.
70. Fisher, N. D.; Adams, N. G. **1998**. In preparation.
71. Fairley, D. A.; Scott, G. B. I.; Freeman, C. G.; Maclagan, R. G. A. R.; McEwan, M. J. *J. Phys. Chem.* **1996**, *92*, 1305.
72. Herbst, H.; Smith, D.; Adams, N. G.; McIntosh, B. J. *J. Chem. Soc: Farad. Trans. II* **1989**, *85*, 1655.
73. Curtiss, L. A.; Lucas, D. J.; Pople, J. A. *J. Chem. Phys.* **1995**, *102*, 3292.
74. Nobes, R. H.; Rodwell, W. R.; Bouma, W. J.; Radom, L. *J. Amer. Chem. Soc.* **1981**, *103*, 1913.
75. Audier, H. E.; Bouchoux, G.; McMahon, T. B.; Milliet, A.; Vulpius, T. *Org. Mass Spectrom.* **1994**, *29*, 176.
76. Jarrold, M. F.; Kirchner, N. J.; Liu, S.; Bowers, M. T. *J. Phys. Chem.* **1986**, *90*, 78.
77. Herbst, E. *Astrophys. J.* **1987**, *313*, 869.
78. Dalgarno, A. In *Chemistry in Space*; Greenberg, J. M.; Pirronello, V., Eds.; Kluwer: Dordrecht, 1991, p 71.
79. Vejby-Christensen, L.; Andersen, L. H.; Heber, O.; Kella, D.; Pedersen, H. B.; Schmidt, H. T.; Zajman, D. *Astrophys. J.* **1997**, *483*, 531.
80. Swanton, D. J.; Marsden, D. C. J.; Radom, L. *Org. Mass Spectrom.* **1991**, *26*, 227.
81. Audier, H, E.; Koyanagi, G. K.; McMahon, T. B.; Tholmann, D. *J. Phys. Chem.* **1996**, *100*, 8220.
82. Fairley, D. A.; Scott, G. B. I.; Freeman, C. G.; Maclagan, R. G. A. R.; McEwan, M. J. *J. Phys. Chem.* **1997**, *101*, 2848.
83. Matthews, K. K.; Adams, N. G. *Int. J. Mass Spectrom. Ion Proc.* **1997**, *163*, 221.
84. Matthews, K. K.; Adams, N. G.; Fisher, N. D. *J. Phys. Chem.* **1997**, *101*, 2841.
85. Irvine, W. M.; Brown, R. D.; Cragg, D. M.; Friberg, P.; Godfrey, P. D.; Kaifu, N.; Matthews, H. E.; Ohishi, M.; Suzuki, H.; Takeo, H. *Astrophys. J.* **1988**, *335*, L89.
86. Adams, N. G.; Smith, D.; Giles, K.; Herbst, E. *Astron. Astrophys.* **1989**, *220*, 269.
87. Herbst, E.; Smith, D.; Adams, N. G. *Astron. Astrophys.* **1984**, *138*, L13.
88. Holmes, J. L.; Terlouw, J. K.; Burgers, P. C. *Org. Mass Spectrom.* **1980**, *15*, 140.
89. Bouchoux, G.; Hoppilliard, Y.; Flament, J-P. *Org. Mass Spectrom.* **1985**, *20*, 560.
90. Hopkinson, A. C.; Lien, M. H. *J. Amer. Chem. Soc.* **1986**, *108*, 2843.
91. Maclagan, R. G. A. R.; McEwan. M. J.; Scott, G. B. I. *Chem. Phys. Letts.* **1995**, *240*, 185.
92. Scott, G. B. I.; Fairley, D. A.; Freeman, C. G.; Maclagan, R. G. A. R.; McEwan, M. J. *Int. J. Mass Spectrom. Ion Proc.* **1995**, *149/150*, 251.
93. Wilson, P. F.; Freeman, C. G.; McEwan, M. J. *Int. J. Mass Spectrom. Ion Proc.* **1993**, *128*, 83.
94. Smith, S. C.; Wilson, P. F.; Sudkeaw, P.; Maclagan, R. G. A. R.; McEwan, M. J. *J. Chem. Phys.* **1993**, *98*, 1944.
95. Bouchhoux, G.; Flament, J. P.; Hoppilliard, Y.; Tortajada, J.; Flammang, R.; Maquestiau, A. *J. Am. Chem. Soc.* **1989**, *111*, 5560.
96. Bouchoux, G.; Nguyen, M. T.; Longevialle, P. *J. Am. Chem. Soc.* **1992**, *114*, 10000.
97. Mehringer, D. M.; Snyder, L. E.; Miao, Y.; Lovas, F. J. *Astrophys. J.* **1997**, *480*, L71.

DYNAMICS OF GAS-PHASE S$_N$2 NUCLEOPHILIC SUBSTITUTION REACTIONS

William L. Hase, Haobin Wang, and

Gilles H. Peslherbe

Advances in Gas-Phase Ion Chemistry
Volume 3, pages 125–156
Copyright © 1998 by JAI Press Inc.
All rights of reproduction in any form reserved.
ISBN: 0-7623-0204-6

ABSTRACT

Computational and experimental studies of the dynamics of gas-phase S_N2 nucleophilic substitution reactions of the type $X^- + RY \rightarrow XR + Y^-$ are reviewed. The two specific reactive systems considered are $Cl^- + CH_3Cl$ and $Cl^- + CH_3Br$. The computational studies involve classical trajectory simulations on analytic potential energy functions derived from ab initio electronic structure calculations. Ab initio calculations for the $Cl^- + CH_3Cl$ and $Cl^- + CH_3Br$ reactions at different levels of theory are reviewed. The classical trajectory simulations are used to investigate: (1) the dynamics of the $X^- + RY \rightarrow X^-\cdots RY$ association process; (2) the unimolecular and intramolecular dynamics of the $X^-\cdots RY$ and $XR\cdots Y^-$ complexes; (3) a direct mechanism for $X^- + RY \rightarrow XR + Y^-$ substitution; (4) energy partitioning for the $XR + Y^-$ products; and (5) a dynamical model for S_N2 nucleophilic substitution and central barrier recrossing. The dynamics observed in the trajectories are used to interpret the nonstatistical $ClCH_3 + Br^-$ product energies measured following $Cl\cdots CH_3Br$ dissociation and the inability of statistical rate theories to fit the $Cl^- + CH_3Br \rightarrow ClCH_3 + Br^-$ rate constants measured experimentally versus $Cl^- + CH_3Br$ relative translational energy and CH_3Br temperature.

I. INTRODUCTION

Gas-phase S_N2 nucleophilic substitution reactions are particularly interesting because they have attributes of both bimolecular and unimolecular reactions.[1] As discovered from experimental studies by Brauman and coworkers[2] and electronic structure theory calculations,[3] potential energy surfaces for gas-phase S_N2 reactions of the type,

$$X^- + RY \rightarrow XR + Y^- \qquad (1)$$

often have two potential energy minima, which arise from the $X^-\cdots RY$ and $XR\cdots Y^-$ ion–dipole complexes. Two typical S_N2 potential energy curves are depicted in Figure 1. One is for the thermoneutral symmetric reaction,

$$Cl_a^- + CH_3Cl_b \rightarrow Cl_aCH_3 + Cl_b^- \qquad (2)$$

and the other for the exothermic asymmetric reaction:

$$Cl^- + CH_3Br \rightarrow ClCH_3 + Br^- \qquad (3)$$

If trapping in the ion–dipole complexes is important, the S_N2 reaction mechanism may be written as:

$$X^- + RY \rightleftharpoons X^-\cdots RY \rightleftharpoons XR\cdots Y^- \rightarrow XR + Y^- \qquad (4)$$

Calculations have identified three transition states (TS) for an S$_N$2 reaction.[4–6] Two are variational, one of which is located along the X$^-$ + RY association reaction path, and the other along the XR + Y$^-$ association reaction path; i.e. see Figure 1. Variational transition state theory (VTST) calculations show that the third TS is located at the central barrier.[4]

Statistical rate theories have been used to calculate rate constants for gas-phase S$_N$2 reactions.[1,7] For a S$_N$2 reaction like Cl$_a^-$ + CH$_3$Cl$_b$, which has a central barrier higher than the reactant asymptotic limit (see Figure 1), transition state theory (TST) assumes that the crossing of the central barrier is rate-limiting. Thus, the TST expression for the S$_N$2 rate constant is simply,

$$k_{S_N2}(T) = \frac{k_B T}{h} \exp(-\Delta G^{\ddagger}/RT) \tag{5}$$

where $\Delta G^{\ddagger}$ is the difference in free energy between the TS and reactants. If recrossings of the central barrier are unimportant, TST will be valid regardless of the nature of the reaction dynamics for other regions of the potential energy surface. For example, TST would be valid for either a vibrationally adiabatic mechanism without trapping in the ion–dipole complexes or a reaction mechanism with trapping in these complexes.[8]

For highly exothermic S$_N$2 reactions, which have a central barrier significantly lower in energy than that of the reactants, association of the reactants may be the rate controlling step in TST.[1] The S$_N$2 rate constant can then be modeled by a capture theory[9] such as VTST,[10] average dipole orientation (ADO) theory,[11] the statistical adiabatic channel model (SACM),[12] or the trajectory capture model.[13]

Rice–Ramsperger–Kassel–Marcus (RRKM) theory has been used to model the S$_N$2 rate constant when, according to TST, neither central barrier crossing nor reactant association is rate-determining.[1] RRKM theory assumes that for every species crossing the entrance-channel variational transition state, vibrational energy becomes completely randomized so that the subsequent dynamics of the species can be modeled as that of an equilibrated microcanonical ensemble. If RRKM theory predicts XR$\cdots$Y$^-$ → XR + Y$^-$ dissociation to be much faster than XR$\cdots$Y$^-$ → X$^-\cdots$RY isomerization (which is the case for the ClCH$_3\cdots$Br$^-$ complex in Figure 1), the statistical S$_N$2 rate constant, expressed as a function of total energy E and total angular momentum J, may be written as,

$$k_{S_N2}(T) = \int \int \frac{k_{as}(E,J)k_{isom}(E,J)P(E)P(J)dEdJ}{k_{isom}(E,J) + k_{dis}(E,J)} \tag{6}$$

where $P(E)$ and $P(J)$ are the energy and angular momentum probability distributions, $k_{as}(E,J)$ is the microcanonical VTST rate constant for X$^-$ + RY association, and $k_{isom}(E,J)$ and $k_{dis}(E,J)$ are the isomerization and dissociation RRKM rate constants for the X$^-\cdots$RY complex.

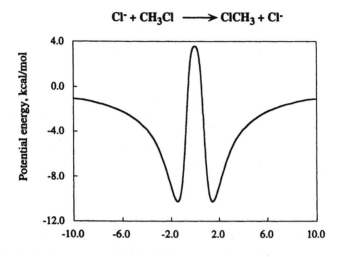

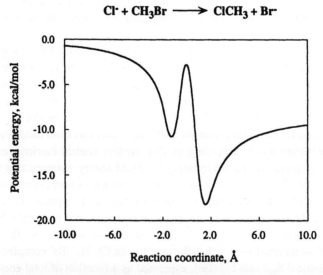

Figure 1. Reaction path potentials for the $Cl_a^- + CH_3Cl_b \rightarrow Cl_aCH_3 + Cl_b^-$ and $Cl^- + CH_3Br \rightarrow ClCH_3 + Br^-$ S_N2 nucleophilic substitutions.

Though statistical models are important, they may not provide a complete picture of the microscopic reaction dynamics. There are several basic questions associated with the microscopic dynamics of gas-phase S_N2 nucleophilic substitution that are important to the development of accurate theoretical models for bimolecular and unimolecular reactions.[1] Collisional association of X^- with RY to form the $X^- \cdots RY$

complex requires energy transfer from relative translational motion to vibrational and/or rotational degrees of freedom of RY. Understanding the dynamics of this process is a fundamental problem in ion–molecule chemistry. If the X$^-$···RY and XR···Y$^-$ potential energy minima have a negligible effect on the S$_N$2 substitution dynamics, the reaction may be akin to the generic displacement reaction A + BC → AB + C. Thus, it is of interest to determine whether the rate of the S$_N$2 reaction can be enhanced by selective excitation of RY's vibrational modes.[14] Also, the two X$^-$···RY and XR···Y$^-$ minima are separated by a central barrier. The dynamics of barrier crossing with the additional possibility of recrossings is a critical issue in reaction rate theory. Comparisons of statistical theory calculations with experiment suggest that for some cases tunneling through the central barrier may be important.[15]

Another important question deals with the intramolecular and unimolecular dynamics of the X$^-$···RY and XR···Y$^-$ complexes. The interaction between the ion and molecule in these complexes is weak, similar to the intermolecular interactions for van der Waals molecules with hydrogen-bonding interactions like the hydrogen fluoride and water dimers.[16] There are only small changes in the structure and vibrational frequencies of the RY and RX molecules when they form the ion–dipole complexes. In the complex, the vibrational frequencies of the intramolecular modes of the molecule are much higher than are the vibrational frequencies of the intermolecular modes, which are formed when the ion and molecule associate. This is illustrated in Table 1, where the vibrational frequencies for CH$_3$Cl and the Cl$^-$···CH$_3$Cl complex are compared. Because of the disparity between the frequencies for the intermolecular and intramolecular modes, intramolecular vibrational energy redistribution (IVR) between these two types of modes may be slow in the ion–dipole complex.[16]

Table 1. Comparison of Vibrational Frequencies for CH$_3$Cl and the Cl$^-$···CH$_3$Cl Complex

| | Frequency (cm^{-1})[a] | |
Mode	CH$_3$Cl	Cl···CH$_3$Cl
A$_1$, C–H str	3171	3225
A$_1$, CH$_3$ def	1464	1396
A$_1$, C–Cl str	788	695
A$_1$, C–Cl str	—	114
E, C–H str	3289	3356
E, CH$_3$ def	1545	1520
E, CH$_3$ rock	1082	1033
E, Cl bend	—	72

Note: [a]The harmonic frequencies are MP2/6-31G** ab initio results from Tucker, S. C.; Truhlar, D. G. *J. Phys. Chem.* **1989**, *93*, 8138.

This chapter surveys studies of the molecular dynamics of the $Cl_a^- + CH_3Cl_b$ and $Cl^- + CH_3Br$ S_N2 nucleophilic substitution reactions; i.e. reactions 2 and 3. Electronic structure calculations of potential energy surface properties and development of analytic potential energy functions for these two reactions are first reviewed. Comparisons are then made between statistical rate theory calculations, based on these surfaces, and experimental measurements of rate constants and product energy partitioning for the $Cl^- + CH_3Br \rightarrow ClCH_3 + Br^-$ reaction. In the next section, classical trajectory calculational studies of the reaction dynamics for both the $Cl_a^- + CH_3Cl_b$ and $Cl^- + CH_3Br$ S_N2 reactions are reviewed, and comparisons are made with experimental studies when possible. The chapter concludes with the outline of a dynamical model for gas-phase S_N2 nucleophilic substitution.

II. POTENTIAL ENERGY SURFACES

A. Ab Initio Calculations of Stationary Point Properties

Different research groups[4,5,17–22] have performed ab initio electronic structure calculations to characterize stationary points on the $Cl_a^- + CH_3Cl_b \rightarrow Cl_aCH_3 + Cl_b^-$ and $Cl^- + CH_3Br \rightarrow ClCH_3 + Br^-$ potential energy surfaces. Energies and structures of stationary points for the former reaction are shown in Table 2, while these properties are listed in Table 3 for the latter reaction. The structure of the molecular moiety in the complexes is only slightly distorted from the isolated molecule's structure. At the highest level of theory the molecular carbon–halogen bond has been lengthened by 0.02–0.06 Å in the complexes.

The most accurate energies and geometries for the $Cl_a^- + CH_3Cl_b$ system are those calculated at the CEPA-1/avtz and G2(+) levels of theory.[18] Without zero-point energies included, the CEPA-1/avtz calculations give a complex well depth of -10.6 kcal/mol and a central barrier height of 2.8 kcal/mol. The G2(+) values for these energies are -10.7 and 3.0 kcal/mol. The most recent experimental value for the 0 K complex well depth is 12.2 ± 2 kcal/mol.[23]

Energies for the $Cl^- + CH_3Br$ system have not been determined at the same high levels of ab initio theory as used for the $Cl_a^- + CH_3Cl_b$ system. Apparently the most accurate ab initio calculation for $Cl^- + CH_3Br$ is that at the G2(+) level of theory which, after neglecting zero-point energy, gives -1.5 kcal/mol for the energy of the $[Cl\cdots CH_3\cdot Br]^-$ central barrier relative to the $Cl^- + CH_3Br$ reactants. The experimental values for the 0 K well depths of the $Cl^-\cdots CH_3Br$ and $ClCH_3\cdots Br^-$ complexes, relative to the reactants and products, respectively, are -10 ± 1 and -8 kcal/mol.[24] The HF and MP2 well depths are similar to these experimental values. Experimental measurements place the 0 K heat of reaction for the $Cl^- + CH_3Br \rightarrow ClCH_3 + Br^-$ reaction at -6 to -9 kcal/mol.[5] The MP2/PTZ+ and G2(+) ab initio calculations give -8.6 and -8.1 kcal/mol, respectively, for the heat of reaction.

Ab initio values of harmonic frequencies for stationary points on the $Cl_a^- + CH_3Cl_b$ potential energy surface are listed in Table 4, while values for these

Table 2. Ab Initio Energies and Structures for Stationary Points on the Cl_a^- + $CH_3Cl_b \rightarrow Cl_aCH_3 + Cl_b^-$ Potential Energy Surface[a]

Property	Ab Initio Method				
	HF/6-31G*[b]	MP2/6-31G**[c]	CEPA-1/avtz[d]	CISD/DZDP// HF/DZDP[e]	G2(+)[f]
Reactant, Cl_a^- + CH_3Cl_b					
r (C–Cl$_a$)	∞	∞	∞	∞	∞
r (C–Cl$_b$)	1.785	1.77	1.805	1.787	1.780
r (C–H)	1.078	1.08	1.085	1.079	1.089
φ(HCCl$_b$)	108.5	109.0	108.0	108.5	108.9
θ(HCH)	110.5	110.1	110.9	110.5	110.2
Energy	0.0	0.0	0.0	0.0	0.0
Complex, $Cl_a^-\cdots CH_3Cl_b$					
r (C–Cl$_a$)	3.266	3.14	3.162	3.384	3.270
r (C–Cl$_b$)	1.828	1.81	1.828	1.823	1.810
r (C–H)	1.073	1.08	1.082	1.074	1.085
φ(HCCl$_b$)	108.0	109.1	108.2	107.8	108.8
θ(HCH)	110.9	110.0	110.7	111.1	110.3
Energy	−10.3	−11.0	−10.6	−9.2	−10.7
Central Barrier, $[Cl_a\cdots CH_3\cdots Cl_b]^-$					
r (C–Cl$_a$)	2.383	2.302	2.323	2.408	2.317
r (C–Cl$_b$)	2.383	2.302	2.323	2.408	2.317
r (C–H)	1.061	1.07	1.071	1.064	1.073
φ(HCCl$_b$)	90.0	90.0	90.0	90.0	90.0
θ(HCH)	120.0	120.0	120.0	120.0	120.0
Energy	3.6	4.6	2.8	6.3	3.0

Notes: [a]Distances in Å, angles in degree, energies in kcal/mol.

[b]Vande Linde, S. R.; Hase, W. L. *J. Phys. Chem.* **1990**, *94*, 2778.

[c]Tucker, S. C.; Truhlar, D.G. *J. Phys. Chem.* **1989**, *93*, 8138.

[d]Seeger, S.; Botschwina, P.; private communication. The avtz basis is similar to a 6-311+G(2df,p)Gaussian type basis.

[e]Vetter, R.; Zülicke, L. *J. Am. Chem. Soc.* **1990**, *112*, 5136. Geometries were optimized at the HF level and energies calculated at the CISD level.

[f]Glukhovtsev, M. N.; Pross, A.; Radom, L. *J. Am. Chem. Soc.* **1995**, *117*, 2024.

frequencies are listed in Table 5 for the Cl^- + CH_3Br potential energy surface. In ascending order the experimental harmonic vibrational frequencies for CH_3Cl are 740, 1038, 1383, 1482, 3074, and 3166 cm^{-1}.[25] For CH_3Br these experimental frequencies are 617, 974, 1333, 1472, 3082, and 3184 cm^{-1}.[25] The highest level ab initio frequencies for CH_3Cl and CH_3Br are the CEPA-1/avtz and MP2/PTZ+ calculations, respectively. On average these ab initio harmonic frequencies, for

Table 3. Ab Initio Energies and Structures for Stationary Points on the
$Cl^- + CH_3Br \rightarrow ClCH_3 + Br^-$
Potential Energy Surface[a]

Property	Ab Initio Method			
	HF/SV4PP/ 6-31G*[b]	MP2/SV4PP/ 6-31G*[c]	MP2/PTZ+[d]	G2(+)[e]
Reactant, $Cl^- + CH_3Br$				
r (C–Cl)	∞	∞	∞	∞
r (C–Br)	1.944	1.939	1.938	1.949
r (C–H)	1.076	1.087	1.083	1.088
ϕ(HCBr)	107.8	107.8	108.1	107.9
θ(HCH)	111.1	111.1	110.9	111.0
Energy	0.0	0.0	0.0	0.0
Complex, $Cl^-\cdots CH_3Br$				
r (C–Cl)	3.216	3.076		3.199
r (C–Br)	1.997	1.994		1.992
r (C–H)	1.071	1.083		1.084
ϕ(HCBr)	106.8	107.1		107.7
θ(HCH)	112.0	111.8		111.2
Energy	−10.7	−12.6		−11.3
Central Barrier, $[Cl\cdots CH_3\cdots Br]^-$				
r (C–Cl)	2.469	2.394	2.322	2.371
r (C–Br)	2.458	2.405	2.392	2.430
r (C–H)	1.062	1.074	1.069	1.073
ϕ(HCBr)	92.2	92.2	91.3	91.4
θ(HCH)	119.8	119.8	119.9	119.9
Energy	−2.9	−5.2	−1.0	−1.5
Complex, $ClCH_3\cdots Br^-$				
r (C–Cl)	1.825	1.832		1.807
r (C–Br)	3.517	3.347		3.457
r (C–H)	1.073	1.084		1.085
ϕ(HCBr)	72.2	72.1		71.1
θ(HCH)	111.1	111.0		110.2
Energy	−21.2	−21.6		−17.5
Product, $ClCH_3 + Br^-$				
r (C–Cl)	1.789	1.796	1.781	1.780
r (C–Br)	∞	∞	∞	∞
r (C–H)	1.077	1.088	1.083	1.089
ϕ(HCCl)	108.1	108.1	108.2	108.9
θ(HCH)	110.8	110.8	110.7	110.2
Energy	−12.6	−11.3	−8.6	−8.1

Notes: [a]Distances in Å, angles in degree, energies in kcal/mol.
[b]Wang, H.; Zhu, L.; Hase, W. L.; *J. Phys. Chem.* **1994**, *98*, 1608.
[c]Wang, H.; Hase, W. L.; unpublished results.
[d]Hu, W.-P.; Truhlar, D. G. *J. Am. Chem. Soc.* **1995**, *117*, 10726.
[e]Glukhovtsev, M. N.; Pross, A.; Radom, L. *J. Am. Chem. Soc.* **1996**, *118*, 6273.

Table 4. Ab Initio Vibrational Frequencies for Stationary Points on the
$Cl_a^- + CH_3Cl_b \rightarrow Cl_aCH_3 + Cl_b^-$
Potential Energy Surface[a]

Property	Ab Initio Method			
	HF/6-31G* [b]	*MP2/6-31G*** [c]	*CEPA-1/avtz* [d]	*HF/DZ* [e]
Reactant, CH$_3$Cl$_b$				
C–Cl$_b$ str (A$_1$)	783	788		679
CH$_3$ rock (E)	1138	1082		1108
CH$_3$ def (A$_1$)	1538	1464		1512
CH$_3$ def (E)	1629	1545		1624
C–H str (A$_1$)	3268	3171		3299
C–H str (E)	3372	3289		3434
Complex, Cl$_a^-$⋯CH$_3$Cl$_b$				
Cl$_a^-$ bend (E)	71	72	70	77
Cl$_a$–C str (A$_1$)	101	114	107	111
C–Cl$_b$ str (A$_1$)	662	695	647	522
CH$_3$ rock (E)	1101	1033	1072	1044
CH$_3$ def (A$_1$)	1479	1396	1342	1421
CH$_3$ def (E)	1605	1520	1588	1590
C–H str (A$_1$)	3323	3225	3130	3361
C–H str (E)	3447	3356	3325	3526
Central Barrier, [Cl$_a$⋯CH$_3$⋯Cl$_b$]$^-$				
Cl$_a$–C–Cl$_b$ bend (E)	204	219		195
Cl$_a$–C–Cl$_b$ str (A$_1$)	214	234		219
CH$_3$ rock (E)	1008	1007		959
out-of-plane bend (A$_2$)	1225	1086		1211
CH$_3$ def (E)	1548	1496		1539
C–H str (A$_1$)	3423	3304		3420
C–H str (E)	3636	3520		3647
reaction coord.	415 i	516 i		324 i

Notes: [a] Frequencies in cm^{-1}.

[b] See footnote (b) in Table 2.

[c] See footnote (c) in Table 2.

[d] See footnote (d) in Table 2.

[e] Vetter, R.; Zülike, L.; *J. Am. Chem. Soc.* **1990**, *112*, 5136. Note that the frequencies were calculated with a different basis set than that for the energy calculations, footnote (e) in Table 2.

Table 5. Ab Initio Vibrational Frequencies for Stationary Points on the $Cl^- + CH_3Br \rightarrow ClCH_3 + Br^-$ Potential Energy Surface[a]

	Ab Initio Method		
Property	HF/SV4PP/6-31G* [b]	MP2/SV4PP/6-31G* [c]	MP2/PTZ+ [d]
Reactant, CH_3Br			
C–Br str (A_1)	642	627	651
CH_3 rock (E)	1066	1015	980
CH_3 def (A_1)	1484	1401	1350
CH_3 def (E)	1620	1537	1492
C–H str (A_1)	3278	3160	3119
C–H str (E)	3392	3283	3235
Complex, $Cl^-\cdots CH_3Br$			
Cl^- bend (E)	71	80	
Cl–C str (A_1)	94	108	
C–Br str (A_1)	518	518	
CH_3 rock (E)	1032	972	
CH_3 def (A_1)	1420	1322	
CH_3 def (E)	1595	1509	
C–H str (A_1)	3334	3207	
C–H str (E)	3471	3349	
Central Barrier, $[Cl\cdots CH_3\cdots Br]^-$			
Cl–C–Br bend (E)	183	187	199
Cl–C–Br str (A_1)	172	193	195
CH_3 rock (E)	974	973	955
out–of–plane bend (A_2)	1203	1069	1029
CH_3 def (E)	1549	1459	1426
C–H str (A_1)	3418	3267	3225
C–H str (E)	3630	3475	3430
reaction coord.	380 i	427 i	490 i
Complex, $ClCH_3\cdots Br^-$			
Br^- bend (E)	65	73	
C–Br str (A_1)	70	84	
Cl–C str (A_1)	674	659	
CH_3 rock (E)	1099	1029	
CH_3 def (A_1)	1483	1381	
CH_3 def (E)	1610	1526	
C–H str (A_1)	3314	3192	
C–H str (E)	3438	3324	

(continued)

Table 5. Ab Initio Vibrational Frequencies for Stationary Points on the
Cl⁻ + CH₃Br → ClCH₃ + Br⁻ Potential Energy Surface[a]

Property	Ab Initio Method		
	HF/SV4PP/6-31G*[b]	MP2/SV4PP/6-31G*[c]	MP2/PTZ+[d]
Product, CH₃Cl			
Cl–C str (A₁)	774	750	766
CH₃ rock (E)	1130	1068	1044
CH₃ def (A₁)	1528	1436	1398
CH₃ def (E)	1627	1546	1504
C–H str (A₁)	3268	3152	3116
C–H str (E)	3374	3271	3227

Notes: [a]Frequencies in cm⁻¹.
[b]Wang, H.; Zhu, L.; Hase, W. L.; *J. Phys. Chem.* **1994**, *98*, 1608.
[c]Wang, H.; Hase, W. L.; unpublished results.
[d]Hu, W. -P.; Truhlar, D. G. *J. Am. Chem. Soc.* **1998**, *117*, 10726.

CH_3Cl and CH_3Br, are 1.06 and 1.01 times larger than the respective experimental harmonic vibrational frequencies. One of the most interesting properties of the vibrational frequencies is the increase in the C–H stretch frequencies in going from the reactants (or products) of the S_N2 reaction to the central barrier.

B. Analytic Potential Energy Functions

A model analytic potential energy function has been developed for S_N2 reactions like those in reaction 1,[4] and here reaction 2 is used to describe this function. The same function is used for reaction 3, except the CH_3Cl_a moiety is replaced by CH_3Br.[5] Two particularly important coordinates for the $Cl_a^- + CH_3Cl_b \rightarrow Cl_aCH_3 + Cl_b^-$ potential energy function are the Cl_a–C and C–Cl_b bond distances denoted by r_a and r_b, respectively. The terms $g_a = r_a - r_b$ and $g_b = r_b - r_a$ measure the extent of reaction. They also conveniently reflect the symmetry of the reactions; i.e. g_a is $+\infty$ for reactants, 0 for the transition state, and $-\infty$ for products. A general model analytic function for a symmetric S_N2 reaction, $Cl_a^- + CH_3Cl_b$, may be written as:

$$V_{total} = V_{Cl}(r_a, g_a) [1 - S_{LR}(g_a)] + V_{Cl}(r_b, g_b) [1 - S_{LR}(g_b)]$$
$$+ V_\phi(r_a, g_a) [1 - S_{LR}(g_a)] + V_\phi(r_b, g_b) [1 - S_{LR}(g_b)]$$
$$+ V_{ClCl}[1 - S_{LR}(g_a)] [1 - S_{LR}(g_b)] + V_\theta(g_a)$$
$$+ V_{HC} + V_{LR}^a S_{LR}(g_a) + V_{LR}^b S_{LR}(g_b) + D_{MC} + D_c \quad (7)$$

In equation 7, the interaction of Cl⁻ with CH_3Cl is divided into two regions: the long-range (electrostatic) region outside the cluster area where g_a (or g_b) is greater than the value of g for the cluster denoted by g_c and the short-range (bond-forming) region where g_a (or g_b) $< g_c$. The long-range potential terms are given by V_{LR}^a and V_{LR}^b. For large separations, these terms approach the ion–dipole potential. The $[Cl–CH_3–Cl]^-$ short-range interaction is described by the V_{Cl} Morse terms for the $Cl_a–C$ and $C–Cl_b$ stretches, the $Cl_a–CH_3$ and $CH_3–Cl_b$ angular deformation terms are denoted by V_ϕ, and the Cl–Cl interaction term by V_{ClCl}. The HCH bending potential of the CH_3 moiety is given by V_θ. V_{HC} represents the potential for the three HC stretches. D_{MC} is the ClC bond energy for CH_3Cl (methyl chloride) and D_c is the ClC bond energy for the cluster.

The long-range potential terms are smoothly connected to the short-range potential terms by the S_{LR} switching functions. For small g_a, $S_{LR}(g_a)$ approaches zero, effectively turning off V_{LR}^a and turning on the short-range potential functions for Cl_a. However, for large g_a where $S_{LR}(g_a) = 1.0$, the total potential function becomes:

$$V_{total} = V_{Cl}(r_b, g_b) + V_\phi(r_b, g_b) + V_\theta(g_a) + V_{HC} + V_{LR}^a + D_{MC} + D_c \qquad (8)$$

Here, the potential function contains terms for the short-range $CH_3–Cl_b$ interactions, long-range $Cl_a^- + CH_3Cl_b$ interactions, V_θ, and V_{HC}.

When g_a and g_b are both less than g_c, the V_{ClCl} term is completely turned on, since both S_{LR} terms equal zero. For this region, the potential becomes:

$$V_{total} = V_{Cl}(r_a, g_a) + V_{Cl}(r_b, g_b) + V_\phi(r_a, g_a) + V_\phi(r_b, g_b)$$
$$+ V_{ClCl} + V_\theta(g_a) + V_{HC} + D_{MC} + D_c \qquad (9)$$

In the intermediate range where Cl_a^- is approaching methyl chloride and $S_{LR}(g_a)$ is changing from 1.0 to 0.0, all terms in the potential function contribute except for V_{LR}^b.

Parameters for the analytic potential energy functions were determined by fitting experimental data and ab initio calculations. In choosing a level of theory for the ab initio calculations, one has to consider both the accuracy required and the expense of the calculations since numerous potential energy points are needed to fit an analytic potential function. After taking both of these considerations into account, the HF/6-31G* and HF/SV4PP/6-31G* levels of ab initio theory were used for the $Cl_a^- + CH_3Cl_b$ and $Cl^- + CH_3Br$ S_N2 reactions, respectively.[4,5] For the latter reaction, the SV4PP basis set developed by Andzelm et al.[26] is used for Cl and Br and the 6-31G* basis set is used for CH_3.

Three potential energy functions, identified as PES1, PES2, and PES3, have been developed for reaction 2 by using different parameters in the above analytic potential energy function. PES1 and PES2, described previously,[4] are identical except for their treatment of the ϕ-bending force constants. The ab initio ϕ force constants are used for PES1, while for PES2 the ab initio force constants are scaled by a factor determined from comparing ab initio and experimental ϕ force constants for CH_3Cl. Both PES1 and PES2 neglect changes in the parameters for the C–H stretching modes along the minimum energy path. For PES3 the C–H "equilibrium" bond length and vibrational frequencies are functions of the reaction coordinate g;[27] i.e. the C–H Morse function is written as,

$$V_{HC}(r, g) = D_{HC}\{1 - e^{-\beta_{HC}(g)[r-r_{HC}(g)]}\}^2 - D_{HC} \tag{10}$$

where r is the H–C bond length and the "equilibrium" H–C bond length, r_{HC}, and Morse parameter, β_{HC}, are,

$$r_{HC}(g) = r_o^{\infty} + (r_o^{\ddagger} - r_o^{\infty})S_r(g) \tag{11}$$

$$\beta_{HC}(g) = \beta_o^{\infty} + (\beta_o^{\ddagger} - \beta_o^{\infty})S_{\beta}(g) \tag{12}$$

In these equations g can be either g_a or g_b, because of the following form of the switching functions, and $D_{HC} = 110.0$ kcal/mol, $r_o^{\ddagger} = 1.0613$ Å, $r_o^{\infty} = 1.0778$ Å, $\beta_o^{\ddagger} = 2.0054$ Å^{-1}, and $\beta_o^{\infty} = 1.867$ Å^{-1}. The switching functions are,

$$S_r(g) = e^{-c_r(g-g_r^{\ddagger})^2} \tag{13}$$

with,

$$c_r = 0.587962907 \text{ Å}^{-1} \quad g_r^{\ddagger} = 0 \text{ Å}$$

Table 6. CH$_3$Br and CH$_3$Cl Harmonic Frequencies[a]

Mode	Ab Initio[b]	Expt.[c]	PES1(Br)[d]	PES2(Br)[d]
		CH$_3$Br		
A$_1$, C–Br str	642	617	620	617
E, CH$_3$ rock	1066	974	1065	927
A$_1$, CH$_3$ def	1484	1333	1497	1374
E, CH$_3$ def	1620	1472	1457	1442
A$_1$, C–H str	3278	3082	3048	3047
E, C–H str	3392	3184	3183	3182
		CH$_3$Cl		
A$_1$, Cl–C str	774	740	739	735
E, CH$_3$ rock	1130	1038	1108	968
A$_1$, CH$_3$ def	1528	1383	1550	1423
E, CH$_3$ def	1627	1482	1460	1440
A$_1$, C–H str	3268	3074	3050	3050
E, C–H str	3374	3166	3181	3181

Notes: [a]Frequencies in cm^{-1}.
[b]Frequencies calculated at the HF/SV4PP/6-31G* level of theory.
[c]Experimental harmonic frequencies, Duncan, J. L.; Allan, A.; McKean, D. C. *Mol. Phys.* **1970**, *18*, 289.
[d]For PES1(Br) the H–C–Cl and H–C–Br bending force constants are the ab initio values, while these ab initio force constants are scaled for PES2(Br) to obtain better agreement with experiment.

and,

$$S_\beta(g) = e^{-c\beta(g-g_\beta^\ddagger)^2} \tag{14}$$

with:

$$c_\beta = 0.503819015 \text{ Å}^{-1} \quad g_\beta^\ddagger = 0 \text{ Å}$$

Table 7. Harmonic Vibrational Frequencies for the Complexes and Central Barrier[a]

Mode	Ab Initio[b]	Scaled[c]	PES1(Br)[d]	PES2(Br)[d]
	Complex, Cl⁻···CH₃Br			
E, Cl⁻ bend	71	64	72	72
A_1, Cl–C str	94	90	91	91
A_1, C–Br str	518	498	500	499
E, CH₃ rock	1032	943	1089	956
A_1, CH₃ def	1420	1276	1467	1344
E, CH₃ def	1595	1450	1452	1434
A_1, C–H str	3334	3134	3147	3147
E, C–H str	3471	3258	3293	3293
	Central Barrier			
E, Cl–C–Br bend	183	165	169	146
A_1, Cl–C–Br str	172	164	161	161
E, CH₃ rock	974	895	1155	995
A_2, out-of-plane bend	1203	1089	1185	1023
E, CH₃ def	1549	1411	1340	1339
A_1, C–H str	3418	3215	3215	3215
E, C–H str	3630	3406	3415	3415
reaction coordinate	380 i		400 i	400 i
	Complex, ClCH₃···Br⁻			
E, Br⁻ bend	65	58	64	64
A_1, C–Br str	70	67	68	68
A_1, Cl–C str	674	644	646	644
E, CH₃ rock	1099	1010	1123	987
A_1, CH₃ def	1483	1342	1548	1418
E, CH₃ def	1610	1467	1462	1439
A_1, C–H str	3314	3117	3101	3101
E, C–H str	3438	3226	3238	3238

Notes: [a] Frequencies in cm⁻¹.

[b] The ab initio calculations performed here at the HF/SV4PP/6-31G* level of theory.

[c] Scale factors determined from the CH₃Br and CH₃Cl ab initio and experimental frequencies were used to scale ab initio frequencies for the complexes and central barrier; see text.

[d] For PES1(Br) the H–C–Cl and H–C–Br bending force constants are the ab initio values, while these ab initio force constants are scaled for PES2(Br) to obtain better agreement with experiment.

For PES3, the ab initio ϕ-bending force constants are scaled as described above for PES2.

Two different analytic potentials, identified as PES1(Br) and PES2(Br), were developed for reaction 3 by varying the scaling of the H–C–Cl and H–C–Br ϕ-bending force constants as described above.[5] For both of these potentials, equations 10–14 are used to represent the C–H stretching potentials. The HF/SV4PP/6-31G* ab initio ϕ-bending force constants are used for PES1(Br). For PES2(Br), these force constants are scaled by factors determined from ratios of experimental and ab initio force constants for CH_3Cl and CH_3Br. The parameters for the PES1(Br) and PES2(Br) potential energy functions have been listed previously. Ab initio, experimental, PES1(Br), and PES2(Br) harmonic frequencies for CH_3Br and CH_3Cl are compared in Table 6. PES1(Br) and PES2(Br) harmonic vibrational frequencies for the complexes and central barrier are compared with ab initio and scaled ab initio values for these frequencies in Table 7.

III. COMPARISONS BETWEEN STATISTICAL THEORY CALCULATIONS AND EXPERIMENT FOR $Cl^- + CH_3Br \rightarrow ClCH_3 + Br^-$

Rate constants and product energy distributions, determined for reaction 3 by statistical theory calculations,[28-30] have been compared with experimental measurements.[28,29,31] Using the central barrier height as an adjustable parameter, not one of the ab initio or analytic potentials, when combined with a statistical rate theory calculations, gives both $k(T)$ and k_H/k_D values for $Cl^- + CH_3Br$ in quantitative agreement with experiment.[30] The difference between statistical theory and experiment is most pronounced for the rate constant versus relative translational energy and CH_3Br temperature, i.e. $k(E_{rel}, T)$.[30]

According to the statistical model,[30] once the $Cl^-\cdots CH_3Br$ complex is formed, the probability of product formation is given by $k_{isom}/(k_{isom} + k_{dis})$. Thus, to obtain the S_N2 rate constant for fixed E_{rel} and T, the $Cl^- + CH_3Br \rightarrow Cl^-\cdots CH_3Br$ capture rate constant is multiplied by the appropriate average of this probability. The resulting expression for $k_{S_N2}(E_{rel}, T)$ is more detailed than that in equation 6 and given by,

$$k_{S_N2}(E_{rel}, T) = k_{cap}(E_{rel}, T) \sum_{\ell=0}^{\ell_{max}} P(\ell) \sum_{j=0}^{\infty} \sum_{j_z=-j}^{j} P(j, j_z) \sum_{n=0}^{\infty} P(n)$$

$$\times \sum_{J=|\ell-j|}^{J=|\ell+j|} P(J) \frac{k_{isom}(E,J)}{k_{dis}(E, J) + k_{isom}(E, J)} \tag{15}$$

where $P(\ell)$ is the normalized probability density of forming the $Cl^-\cdots CH_3Br$ complex by collisions with orbital angular momentum quantum number ℓ. $P(\ell)$ is related to the collision impact parameter b by $P(\ell) \propto bP_r(b)$, where $P_r(b)$ is the probability of complex formation for collisions with b.

$P(j,j_z)$ in equation 15 is the probability density of the CH_3Br total rotational angular momentum quantum number, j, and rotational quantum number, j_z. It is given by,

$$P(j,j_z) = \frac{(2j+1)\exp\{(-\hbar^2/2k_BT)[j(j+1)/I_A + j_z^2(1/I_C - 1/I_A)]\}}{Q_{rot}} \quad (16)$$

where Q_{rot} is the CH_3Br rotational partition function, and I_A and I_C are the two principal moments of inertia with I_C being the one for the symmetric axis. The symmetry number in the partition function is removed so that summing $P(j, j_z)$ overall values of j and j_z yields a normalized probability of unity. This gives the proper reaction path degeneracy factor of one for reaction 3.

$P(\mathbf{n})$ in equation 15 is the probability distribution of vibrational energy for each set of CH_3Br vibrational quantum numbers $\mathbf{n}$ at temperature T, which is given by,

$$P(\mathbf{n}) = \prod_{i=1}^{9} \exp(-n_i h\nu_i/k_BT)[1 - \exp(-h\nu_i/k_BT)] \quad (17)$$

with $\mathbf{n} = \{n_1, n_2, \ldots, n_9\}$. $P(J)$ in equation 15 is the probability distribution of total angular momentum quantum number J, which is given by the combination of orbital angular momentum quantum number l and the CH_3Br rotational angular momentum quantum number j; i.e.:

$$P(J) = \frac{2J+1}{(2l+1)(2j+1)} \quad (18)$$

The energy E in equation 15 can be written as:

$$E = E_{rel} + \sum_{i=1}^{9} n_i h\nu_i + \frac{j(j+1)\hbar^2}{2I_A} + \frac{j_z^2\hbar^2}{2}\left(\frac{1}{I_C} - \frac{1}{I_A}\right) \quad (19)$$

The rate constants $k_{isom}(E,J)$ and $k_{dis}(E,J)$ in equation 15 are given by the RRKM expression,

$$k(E,J) = \frac{N^{\ddagger}(E,J)}{h\rho(E,J)} \quad (20)$$

where $N^{\ddagger}(E,J)$ is the transition state sum of states and $\rho(E,J)$ is the density of states in the $Cl^-\cdots CH_3Br$ complex. Inserting the RRKM expressions for $k_{isom}(E,J)$ and $k_{dis}(E,J)$ into equation 15 gives,

$$k_{S_N2}(E_{rel}, T) = k_{cap}(E_{rel}, T) \sum_{\ell=0}^{\ell_{max}} P(\ell)$$

$$\times \sum_{j=0}^{\infty} \sum_{j_z=-j}^{j} P(j, j_z) \sum_{n=0}^{\infty} P(n) \sum_{J=|\ell-j|}^{|\ell+j|} P(J) \frac{N_{bar}^{\ddagger}(E, J)}{N_{var}^{\ddagger}(E, J) + N_{bar}^{\ddagger}(E, J)} \quad (21)$$

where $N_{bar}^{\ddagger}(E, J)$ is the sum of states for the central barrier transition state and $N_{var}^{\ddagger}(E, J)$ is the sum of states for the $Cl^- \cdots CH_3Br \rightarrow Cl^- + CH_3Br$ microcanonical variational transition state. Equation 21 is similar to those used by Wladkowski et al.[32] and Graul and Bowers[29] to calculate the S_N2 rate constant.

The rate constant for reaction 3 was calculated at fixed E_{rel} in the range of 0.5 to 2.5 kcal/mol and for CH_3Br temperatures of 207, 300, 538, and 564 K. These energy and temperature ranges are the same as those for the experiments.[31] The calculated and experimental results are compared in Figure 2. The calculated values are the same order of magnitude as the experimental ones by Viggiano and coworkers.[31] However, the decrease in the calculated rate constant, with an increase in E_{rel}, is

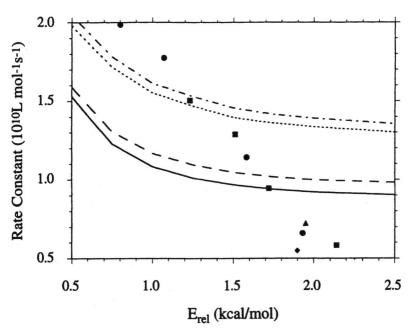

Figure 2. S_N2 rate constant versus $Cl + CH_3Br$ relative translational energy and CH_3Br temperature. The curves are the predictions of statistical theory, equation 21: (———) 207 K, (— —) 300 K, (- - - -) 538 K, and (— · —) 564 K. The points, (●) 207 K, (■) 300 K, (♦) 538 K, and (▲) 564 K, are experimental values[31] (from ref. 30).

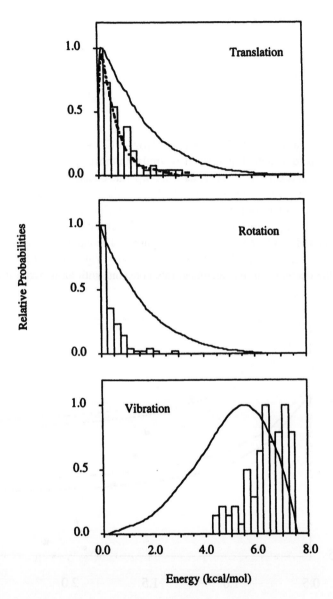

Figure 3. Product energy distributions for the $Cl^-{\cdots}CH_3Br \rightarrow ClCH_3 + Br^-$ reaction: histogram, trajectory result;[6] dashed line, experiment;[29] and solid line, prediction of OTS/PST. The trajectory results are scaled to match the experimental exothermicity.

much less than that observed in the experiments. In addition, at higher E_{rel} the calculated rate constant becomes nearly constant, while the experimental ones continue to decrease. For even higher E_{rel}, the calculated rate constant begins to increase. A striking difference between the calculated and experimental rate constants is the effect of CH_3Br temperature. As the CH_3Br temperature increases, the calculated rate constant increases if E_{rel} is kept constant. This is in contrast with the experimental observation where the rate constant is nearly independent of CH_3Br temperature changes. In a statistical theory calculation similar to that in equation 21, but based on the phase space, average dipole orientation, and RRKM theories, Graul has found that $k(E_{rel},T)$ varies with E_{rel} and T in a manner similar to the calculational results in Figure 2.[33]

In an experimental study, Graul and Bowers[29] prepared the $Cl^-\cdots CH_3Br$ complex with sufficient energy to dissociate to the $Cl^- + CH_3Br$ products. They measured the relative translational energies of these products and, on average, found them to be much smaller than the prediction of orbiting transition state/phase space theory (OTS/PST).[6,29] This is shown in Figure 3, where the CH_3Br vibrational and rotational energy distributions given by OTS/PST are also plotted. The assumptions of OTS/PST are:[34] (1) an orbiting transition state at the centrifugal barrier for dissociation of the $ClCH_3\cdots Br^-$ postreaction complex; (2) statistical energy partitioning at this transition state; and (3) an isotropic long-range potential, so that the potential energy of the centrifugal barrier is transformed to product translational. Equations for performing the OTS/PST calculations are listed in ref. 34. As discussed in the next section, trajectory calculations for $Cl^-\cdots CH_3Br$ dissociation indicate that the available energy is preferentially partitioned to CH_3Br vibration, with an average product E_{rel} much smaller than predicted by OTS/PST.[6]

IV. CLASSICAL TRAJECTORY STUDIES

Classical trajectory calculations, performed on the PES1 and PES1(Br) potential energy surfaces described above, have provided a detailed picture of the microscopic dynamics of the $Cl_a^- + CH_3Cl_b$ and $Cl^- + CH_3Br$ S_N2 nucleophilic substitution reactions.[6,8,35–38] In the sections below, different aspects of these trajectory studies and their relation to experimental results and statistical theories are reviewed.

A. Dynamics of $X^- + RY \rightarrow X^-\cdots RY$ Association

To form the $X^-\cdots RY$ complex during an $X^- + RY$ collision, the relative translational energy of $X^- + RY$ must be transferred to either RY rotation or vibration; otherwise the reactants will simply rebound without forming the complex. In a trajectory simulation of $Cl^- + CH_3Cl$ collisions,[38] it was found that the $Cl^-\cdots CH_3Cl$ complex is formed by a translation to rotation (T → R) energy transfer process, which involves coupling between the orbital angular momentum of the reactants and the rotational angular momentum of CH_3Cl. Complex formation does not involve energy transfer to CH_3Cl vibration. Thus, energy is initially stored only in the three low-frequency intermolecular modes of the $Cl^-\cdots CH_3Cl$ complex (that is, the $Cl^-\cdots C$ stretch and degenerate bending modes). This initially prepared complex is called an *intermolecular complex*[8] because only the intermolecular modes are excited.

The energy transfer mechanism for complex formation can easily be determined from the trajectories by calculating the CH_3Cl internal energy as a function of time. Initially, the internal energy is only rotational, and transfer of this energy to CH_3Cl vibration ($R \rightarrow V$) occurs on a much longer time scale. For $T \rightarrow R$ energy transfer to occur, the reactants must sample the anisotropic part of the interaction potential.

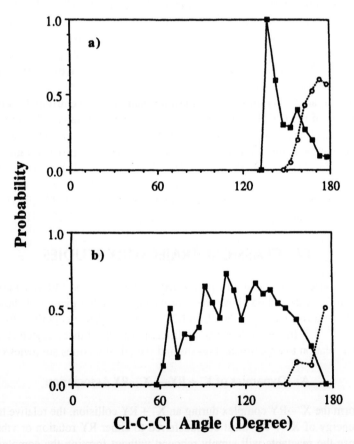

Figure 4. Probability of different trajectory events versus the $Cl_a–C–Cl_b$ angle θ, which is evaluated at the first inner turning point (ITP) for complex formation and at the central barrier for the trajectories which attain this configuration: (■), association probability; (○), probability of attaining the central barrier. $E_{rel} = 0.5$ kcal/mol and $n_{C–Cl_b} = 6$. $T_{rot} = 0$ K in (a) and 300 K in (b) (from ref. 38).

Collisions with a $Cl^-\cdots C-Cl$ angle of approximately π, with the CH_3Cl dipole oriented, do not form the $Cl\cdots CH_3Cl$ complex. Collisions oriented in this manner could yield complexes by $T \rightarrow V$, but this is an unimportant energy transfer mechanism.[38]

Figure 4 illustrates the effect of orientation, i.e. *dynamical stereochemistry*, on the association probability. The angle considered, θ, is the $Cl_a^- - C - Cl_b$ angle and it was evaluated at the first inner turning point (ITP) in the $Cl_a^- + CH_3Cl_b$ center of mass radial motion.[38] Plotted in Figure 4, as a function of θ, is the probability of complex function. It is seen that there is a reduced probability for complex formation as the ion and dipole are aligned with $\theta \approx \pi$. The complexation probability peaks for θ in the range of $100°-120°$.

Because $T \rightarrow V$ energy transfer does not lead to complex formation and complexes are only formed by unoriented collisions, the $Cl^- + CH_3Cl \rightarrow Cl^-\cdots CH_3Cl$ association rate constant calculated from the trajectories is less than that given by an ion–molecule capture model. This is shown in Table 8, where the trajectory association rate constant is compared with the predictions of various capture models.[9] The microcanonical variational transition state theory (μCVTST) rate constants calculated for PES1, with the transitional modes treated as harmonic oscillators (ho) are nearly the same as the statistical adiabatic channel model (SACM),[13] μCVTST,[40] and trajectory capture[14] rate constants based on the ion–dipole/ion–induced dipole potential,

$$V(r, \theta) = -\alpha q^2/2r^4 - q\mu_D\cos\theta/r^2 \qquad (22)$$

where $\alpha = 4.53$ Å^3, $\mu_D = 1.87$ D, and θ is the angle between Cl^- and the CH_3Cl dipole. As expected,[9] the CVTST/ho rate constant is larger than the μCVTST/ho

Table 8. $Cl^- + CH_3Cl \rightarrow Cl^-\cdots CH_3Cl$ Thermal Association Rate Constants[a]

| T (K) | PES1 | | | Two-Body Potential (Eq. 21) | | |
	traj.[b]	ho[c]	ho[d]	SACM[e]	μCVTST[f]	traj. cap.[g]
200	1.33	2.76	1.61	1.78	1.76	1.60
300	1.04	2.27	1.44	1.57	1.51	1.37
500	0.74	1.92	1.35	1.37	1.27	1.14
1000	0.42	1.56	0.93	1.14	1.05	0.94

Notes: [a]The rate constants are given in units of 10^{12} L mol^{-1} s^{-1}.

[b]Determined from classical trajectories; ref. 37.

[c]Canonical variational transition state theory, with transitional modes treated as harmonic oscillators; refs. 5 and 37.

[d]Microcanonical variational transition state theory, with transitional modes treated as harmonic oscillators; ref. 37.

[e]Calculated from Equation 5 of ref. 13.

[f]Calculated from Equations 3.19–3.21 of ref. 40.

[g]Calculated from Equation 3 of ref. 14.

rate constant. This is because in μCVTST transition states are determined at each energy and angular momentum, instead of only determining one transition state at each temperature as is done in CVTST. Also, the CVTST rate constants calculated by treating the transitional modes as hindered rotors are nearly the same as those in Table 8 for a harmonic oscillator treatment of the transitional modes.[37]

The criterion for association in the classical trajectory calculations is the presence of multiple inner turning points between Cl_a^- and the CH_3Cl_b center of mass. Two types of $Cl_a^- \cdots CH_3Cl_b$ complexes, *short-range* and *long-range*, were observed in the trajectories. The two different complexes were identified by the Cl_a^-–C distance

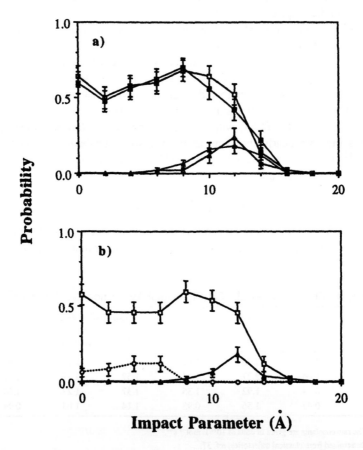

Figure 5. Reaction probabilities vs. impact parameter for E_{rel} = 0.5 kcal/mol and T_{rot} = 300 K. Part (a) is for n_{C-Cl_b} = 0. The solid lines are for E_{vib} = 0 and the dotted lines are for E_{vib} = zpe. The squares are for total complex formation and the triangles for long-range complex formation. Part (b) is for n_{C-Cl_b} = 6 and E_{vib} = 0; with (□, —) for total complex formation, (△, —) for long-range complex formation, and (o, ····) for attaining the central barrier with one or no ITP (from ref. 38).

at the first ITP in the $Cl_a^- + CH_3Cl_b$ center of mass separation and a value for this distance of ~5.0 Å was found as a demarcation between the two types of complex forming trajectories. When CH_3Cl_b is not rotationally excited, i.e. $T_{rot} = 0$ K, no long-range complex formation is observed. Figures 5 and 6 compare the opacity function for long-range complex formation with the opacity function for total (short- and long-range) complex formation for T_{rot} of 300 and 1000 K, respectively. Long-range complex formation is seen to become more important as T_{rot} is increased. At $T_{rot} = 1000$ K, approximately one-half of the complexes are of the long-range type. Moreover, long-range complex formation shifts to smaller impact parameters as T_{rot} is increased.

Computer animation was used to help determine the dynamical origin of the long-range complexes. It was found that for a fixed $Cl_a^- + CH_3Cl_b$ relative translational energy E_{rel} it becomes more difficult for Cl_a^- to orient the CH_3Cl_b dipole and move along the $Cl_a^- + CH_3Cl_b$ attractive minimum energy path as the CH_3Cl_b rotational temperature T_{rot} was increased. For a long-range complex, Cl_a^- begins to move toward CH_3Cl_b when the positive end (i.e. CH_3) rotates towards Cl_a^-. However, just as this attraction begins to take hold, the Cl_b end of CH_3Cl_b rotates around and Cl_a^- is pushed back. This continues for several ITPs in the $Cl_a^- + CH_3Cl_b$ center of mass motion, and then the reactants separate. As E_{rel} is decreased, the CH_3Cl_b rotational temperature required for significant long-range complex formation be-

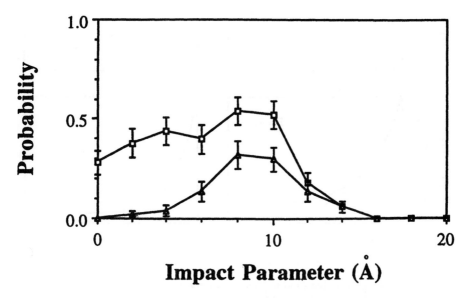

Figure 6. Probability of total ($\square$, —) and long-range ($\triangle$, —) complex formation for E_{rel} = 0.5 kcal/mol, T_{rot} = 1000 K, E_{vib} = 0, and n_{C-Cl_b} = 0 (from ref. 38).

comes smaller. This dynamical picture for long-range complex formation is consistent with the analysis of Lim and Brauman for long-range complex formation in proton transfer reactions.[41,42]

B. Unimolecular and Intramolecular Dynamics of the $Cl^-\cdots CH_3Cl$ Complex

In the trajectory study of $Cl^-\cdots CH_3Cl$ complex formation by $Cl^- + CH_3Cl$ association, the number of complexes with a lifetime t, i.e. $N(t)$, was evaluated for different $Cl^- + CH_3Cl$ initial conditions.[36,37] The resulting plots of $N(t)$ are highly nonexponential and plots of $N(t)/N(0)$ were fit with the biexponential function

$$\frac{N(t)}{N(0)} = \sum_i a_i \exp(-k_i t) \tag{23}$$

with the values for a_i and k_i listed in ref. 37. Plots of $N(t)/N(0)$ are shown in Figure 7 for $Cl^- + CH_3Cl$ collisions with $T_{rot} = 0$ K, $E_{vib} = zpe$, and E_{rel} of 0.5, 1.0, 2.0, and 3.0 kcal/mol.

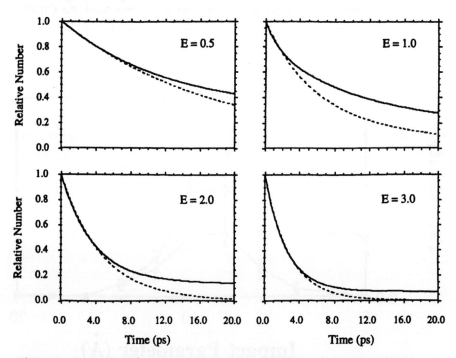

Figure 7. Comparison between trajectory (—) and fitted anharmonic RRKM (- - -) values of the relative number of $Cl^-\cdots CH_3Cl$ complexes versus time, i.e., $N(t)/N(0)$. $E_{rel} = E_\infty$, the energy in excess of the CH_3Cl zero-point level, is given in units of kcal/mol and denoted by E in the figure (from ref. 39).

To analyze the trajectory $N(t)/N(0)$ distributions with a RRKM model in which only the three intermolecular modes of the $Cl^-\cdots CH_3Cl$ complex are active and exchange energy, it is necessary to average over the angular momentum j of the $Cl^-\cdots CH_3Cl$ complexes, i.e.:

$$\frac{N(t)}{N(0)} = \int_0^{j_{max}} \exp\,[-k_{anh}(E,j)t]P(j)dj. \tag{24}$$

The analysis is performed for the calculations with $T_{rot} = 0$ K for the CH_3Cl reactant, so that the angular momentum distribution for the complex $P(j)$ is the distribution of orbital angular momentum for complex formation $P(\ell)$. This latter distribution is given in ref. 37. J_{max}, the quantum number for j_{max}, varies from 282 for $E_{rel} = 0.5$ kcal/mol to 357 for $E_{rel} = 3.0$ kcal/mol. The term $k_{anh}(E,j)$ in equation 24 is written as $k_{anh}(E,j) = k_h(E,j)/f_{anh}(E)$, where $k_h(E,j)$ is the classical RRKM rate constant with the CH_3Cl intramolecular modes inactive and $f_{anh}(E)$ is treated as a fitting factor.

A comparison of the trajectory and RRKM $N(t)/N(0)$ plots is given in Figure 7. The distributions are plotted out to 20 ps, the longest time considered in the trajectory calculations. Quantum inactive and adiabatic models give $N(t)/N(0)$ curves nearly identical to those for the classical inactive model[39] plotted in Figure 7. The plots agree for the initial decay, but at longer times the RRKM model predicts more decay than observed from the trajectories. The anharmonic fitting factor $f_{anh}(E)$ is chosen so that the trajectory and RRKM $N(t)/N(0)$ plots are in agreement for the initial decay, when only the $Cl^-\cdots CH_3Cl$ intermolecular modes are excited. The resulting values of $f_{anh}(E)$ are 4.3, 3.7, 7.5, and 7.9 for $E_{rel} = E_\infty$ of 0.5, 1.0, 2.0, and 3.0 kcal/mol, respectively.

This RRKM model does not fit the long-time tail of the trajectory $N(t)/N(0)$ plots. The long-lived trajectories may be moving on and undergoing transitions between vague tori,[43,44] which are the remnants of stable tori with quasiperiodic motion. Semiclassical theory[45,46] indicates that such vague and stable tori are the classical analogs of quantal compound state (i.e. Feshbach) resonances.

An anharmonic correction for the density of states was also evaluated by solving the phase integral for the $Cl^-\cdots CH_3Cl$ intermolecular complex;[39] i.e.:

$$\rho_{anh}(E,j) = dN_{anh}(E,j)/dE = \int \cdots \int dp\,dq/h^3 \tag{25}$$

For $J = 0$, the resulting values of f_{anh} are ~8, ~6, and ~4.5 for E_∞ of 0.5, 1.0, and 2.0 kcal/mol, respectively. For $J = 220$, the anharmonic correction is independent of E_∞ and is ~5. These calculated f_{anh} values agree with the above fitted values to within 50%.

Recently, McMahon and coworkers[47] have measured a collision-averaged rate constant for $Cl^-\cdots CH_3Cl \rightarrow Cl^- + CH_3Cl$ dissociation at ~310 K. Their experiments are described by the mechanism:

$$Cl^- + CH_3Cl \underset{k_{dis}}{\overset{k_{as}}{\rightleftharpoons}} [Cl^- \cdots CH_3Cl]^* \overset{\omega}{\rightarrow} Cl^- \cdots CH_3Cl \qquad (26)$$

The apparent rate constant for the loss of Cl^- or formation of $Cl^- \cdots CH_3Cl$ is $k_{app} = \omega k_{as}/(k_{dis} + \omega)$, which can be arranged to give:

$$\frac{1}{k_{app}} = \frac{1}{\omega} \frac{k_{dis}}{k_{as}} + \frac{1}{k_{as}} \qquad (27)$$

Thus, the unimolecular rate constant k_{dis} can be found from a plot of k_{app}^{-1} vs. ω^{-1}. Expressing ω as 3.3×10^7 torr^{-1} s^{-1} × pressure (torr) and using the Su–Chesnavich[13] capture rate constant of 2.28×10^{-9} cm^3 molecule^{-1} s^{-1} for k_{as}, a value of 0.083 ps^{-1} was determined for k_{dis} from experiments in the 3–10 torr range.

Following previous work by Slater,[48] Bunker,[49] Forst,[50] and Hase and coworkers,[43,51] the above k_{app} rate constant can be written as,[39]

$$k_{app} = k_{as}\left[1 - \sum_i a_i k_i/(\omega + k_i)\right] \qquad (28)$$

where a_i and k_i are the parameters in equation 23. Expressing ω as 3.3×10^7 torr^{-1} s^{-1} × pressure (torr), as done by McMahon and coworkers, the inverse of k_{app} in equation 28 is plotted versus inverse pressure in torr in Figure 8. The plots are for the trajectory $N(t)/N(0)$ distributions at $T_{rot} = 300$ K, $E_{vib} = $ zpe, and $E_{rel} = 0.5$ and 1.0 kcal/mol. Also plotted in Figure 8 are the experimental measurements of McMahon and coworkers at $T \sim 310$ K.

The average value of E_{rel} at 310 K is 0.9 kcal/mol, so a comparison can be made between the trajectory and experimental results in Figure 8. The trajectory plots for E_{rel} of 0.5 and 1.0 kcal/mol have slopes (i.e. k_{dis}/k_{as}) which are 1.6 and 2.0 times larger, respectively, than the slope of the experimental plot. Thus, if the same value of k_{as} is used to deduce k_{dis} from the trajectory and experimental slopes, the trajectory k_{dis} will be larger than the experimental k_{dis} by the above ratio of slopes. Given the expected accuracy of the trajectory results,[52] the agreement between the trajectories and experiment is quite good.

C. Direct Mechanism for S$_N$2 Nucleophilic Substitution

As discussed above, Cl_a^- and CH_3Cl_b do not associate under thermal conditions when the CH_3Cl_b dipole is oriented with a $Cl_a^- \cdots C-Cl_b$ angle of approximately π. Because these oriented trajectories follow the reaction path, it is of interest to determine whether they can be modified to promote S$_N$2 nucleophilic substitution. What was discovered from the $Cl_a^- + CH_3Cl_b$ trajectory simulations on PES1,[37,38] was that the addition of three or more quanta to the C–Cl stretch mode of the CH_3Cl_b reactant opens up a direct substitution mechanism without trapping in either of the potential energy wells. As shown in Figure 4, this direct reaction only occurs for

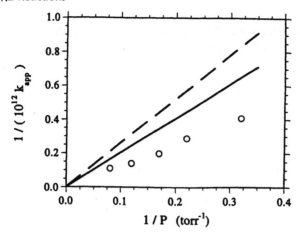

Figure 8. Apparent reaction rate constant k_{app}, equation 27, in cm^3/molecule s, vs. pressure P (torr): (—) and (——), trajectory results at E_{rel} = 0.5 and 1.0 kcal/mol, respectively, with T_{rot} = 300 K and E_{vib} = zpe. The points (o) are the experimental results of McMahon and coworkers[43] at 310 K (from ref. 39).

oriented collisions with a collision angle $\theta \approx \pi$. Thus, the value of θ as the reactants interact, the dynamical stereochemistry,[53] can be used to distinguish between the association and direct substitution mechanisms.

Adding quanta to the C–Cl bond promotes bond extension, so that the central barrier can be reached as Cl⁻ approaches. This dynamical effect is in accord with the role of vibrational energy in A + BC → AB + C triatomic displacement reactions.[15] The plot in Figure 5 of the probability of directly attaining the central barrier versus Cl_a^- + CH_3Cl_b collision impact parameter shows that direct substitution occurs at small impact parameters. In contrast, association extends to larger impact parameters.

The importance of this direct substitution mechanism is strongly dependent on the rotational energy of CH_3Cl_b. If CH_3Cl_b has no rotational energy, most of the collisions have dipole orientation and direct substitution is important. For Cl_a^- + CH_3Cl_b trajectories with a relative translational energy corresponding to 300 K (that is, $3RT/2$), no CH_3Cl_b rotational energy, and six quanta in the C–Cl vibrational mode, direct substitution is about two times more probable than association.[38] However, by simply increasing the CH_3Cl_b rotational temperature to 300 K, one can make the direct substitution/association ratio less than 0.06.

For the Cl_a^- + CH_3Cl_b trajectories on PES1, direct substitution only occurs when the C–Cl stretch normal mode is excited with three or more quanta. For CH_3Cl_b at 300 K, the probability of this vibrational excitation and the rate constant with vibrational excitation is too small to make direct substitution an important contributor to Cl_a^- + CH_3Cl_b → Cl_aCH_3 + Cl_a^- S_N2 nucleophilic substitution on PES1. However, the direct substitution mechanism may become more important if less vibrational excitation is needed to promote it. This may be the case for $Cl^- + CH_3Br$ and $F^- + CH_3Cl$, which have lower central barriers than does the $Cl^- + CH_3Cl$ reaction.

D. Product Energy Partitioning

In an experimental study, Graul and Bowers[29] formed the $Cl^-\cdots CH_3Br$ complex and measured the relative translational energy distribution of $ClCH_3 + Br^-$ formed by dissociation of this complex. To model these experiments, quasiclassical trajectory calculations were performed in which the $Cl^-\cdots CH_3Br$ complex on the PES1(Br) analytic potential function was excited with 2.7 kcal/mol of energy in excess of the $Cl^- + CH_3Br$ asymptotic potential plus the CH_3Br zero-point energy.[6] Dissociation of the complex was studied by adding the excitation energy to each individual intermolecular and intramolecular mode. As discussed above, upon exciting an intermolecular mode the predominate event is dissociation to $Cl^- + CH_3Br$. However, when the intramolecular modes of CH_3Br in the complex are excited, dissociation to form $ClCH_3 + Br^-$ becomes important.

The trajectories were analyzed for product relative translation, E_{rel}, rotation, E_{rot}, and vibration, E_{vib}, energies. Average values for these product energies are statistically the same for each of the intramolecular mode excitations, and the energy partitioning results can be combined to give distributions of product E_{rel}, E_{rot}, and E_{vib} to compare with the Graul/Bowers experiments and OTS/PST. From the trajectories, the product energy is preferentially partitioned to $ClCH_3$ vibration. The $ClCH_3$ average rotational energy is approximately two-thirds of the average translational energy. A comparison of the trajectory average product energies, with those of OTS/PST, shows that OTS/PST predicts a higher relative translational energy and a lower vibrational energy than found from the trajectories.

In their experimental study of $Cl^-\cdots CH_3Br$ dissociation, Graul and Bowers[29] also found that the $ClCH_3 + Br^-$ product relative translational energies are less than the predictions of OTS/PST. However, a direct comparison cannot be made between their work and the trajectory calculations, since the PES1(Br) potential energy surface has a reaction exothermicity including zero-point energies of 12.27 kcal/mol, while the best experimental estimate of the exothermicity is 8–9 kcal/mol. An indirect comparison can be made by scaling the total product energy for the trajectories to match that of the Graul and Bowers experiments, which is estimated[29] to approximately equal the reaction exothermicity. Figure 3 compares the experimental, OTS/PST, and scaled trajectory product energy distributions.

E. Dynamical Model for S_N2 Substitution and Central Barrier Recrossing

A dynamical model for S_N2 nucleophilic substitution that emerges from the trajectory simulations is depicted in Figure 9. The complex formed by a collision between the reactants is an intermolecular complex $C_{inter,R}$. To cross the central barrier, this complex has to undergo a unimolecular transition in which energy is

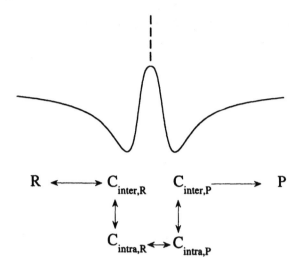

Figure 9. Dynamical model for S$_N$2 nucleophilic substitution. The labels R and P denote the reactant and product sides of the central barrier, respectively.

transferred from the intermolecular modes to the CH$_3$Cl intramolecular modes, which forms the intramolecular complex C$_{intra,R}$. The C$_{inter,R}$ → C$_{intra,R}$ rate does not involve the crossing of a potential energy barrier, but results from a dynamical barrier for energy transfer.[54] The intramolecular complex accesses the central barrier region of the potential energy surface and X$^-$···RY → XR···Y$^-$ isomerization occurs by a C$_{intra,R}$ → C$_{intra,P}$ transition. Products are formed when C$_{intra,P}$ → C$_{inter,P}$ occurs.

The dynamical model described in Figure 9 indicates that the trajectories may recross the central barrier several times if the C$_{intra,R}$ ↔ C$_{intra,P}$ transition is faster than the transitions between the intermolecular and intramolecular complexes. For a S$_N$2 reaction such as Cl$_a^-$ + CH$_3$Cl$_b$, for which TST assumes that crossing of the central barrier is rate-determining, such recrossings would make TST an incomplete model for calculating the S$_N$2 rate constant. It would be necessary to correct the TST rate constant by a factor κ, which is the number of reactive trajectories divided by the number of central barrier crossings in the R → P direction; that is, k_{S_N2} = κ k_{TST} (TST assumes κ = 1).

If crossing the central barrier is not rate-controlling in TST, then trapping in the ion–dipole complex must be incorporated into the statistical model and it is more difficult to represent the effect of central barrier recrossings; correcting TST with the κ factor is not sufficient. The recrossings and presence of both intermolecular and intramolecular complexes are expected to affect the k_{as}, k_{isom}, and k_{dis} rate constants in equation 6. The value for k_{as} should be smaller than that of a capture model, and k_{isom} and k_{dis} should disagree with the predictions of RRKM theory.

V. FUTURE DIRECTIONS

Additional experimental, theoretical, and computational work is needed to acquire a complete understanding of the microscopic dynamics of gas-phase S_N2 nucleophilic substitution reactions. Experimental measurements of the S_N2 reaction rate versus excitation of specific vibrational modes of RY (equation 1) are needed, as are experimental studies of the dissociation and isomerization rates of the $X^-\cdots RY$ complex versus specific excitations of the complex's intermolecular and intramolecular modes. Experimental studies that probe the molecular dynamics of the $[X\cdots R\cdots Y]^-$ central barrier region would also be extremely useful.

The analytic potential energy surfaces, used for the $Cl_a^- + CH_3Cl_b$ and $Cl^- + CH_3Br$ trajectory studies described here, should be viewed as initial models. Future classical and quantum dynamical calculations of S_N2 nucleophilic substitution should be performed on quantitative potential energy functions, derived from high-level ab initio calculations. By necessity, the quantum dynamical calculations will require reduced dimensionality models. However, by comparing the results of these reduced dimensionality classical and quantum dynamical calculations, the accuracy of the classical dynamics can be appraised. It will also be important to compare the classical and quantum reduced dimensionality and classical complete dimensionality dynamical calculations with experiment.

Finally, accurate theoretical kinetic and dynamical models are needed for calculating S_N2 rate constants and product energy distributions. The comparisons described here, between experimental measurements and statistical theory predictions for $Cl^- + CH_3Br$, show that statistical theories may be incomplete theoretical models for S_N2 nucleophilic substitution. Accurate kinetic and dynamical models for S_N2 nucleophilic substitution might be formulated by introducing dynamical attributes into the statistical models or developing models based on only dynamical assumptions.

REFERENCES

1. Hase, W. L. *Science* **1994**, *266*, 998.
2. Moylan, C. R.; Brauman, J. I. In *Advances in Classical Trajectory Methods; Dynamics of Ion-Molecule Complexes*; Hase, W. L., Ed.; JAI Press: Greenwich, CT, 1994, Vol. 2, p 95.
3. Shaik, S. S.; Schlegel, H. B.; Wolfe, S. *Theoretical Aspects of Physical Organic Chemistry: The S_N2 Mechanism*; Wiley: New York, 1992.
4. Vande Linde, S. R.; Hase, W. L. *J. Phys. Chem.* **1990**, *94*, 2778.
5. Wang, H.; Zhu, L.; Hase, W. L. *J. Phys. Chem.* **1994**, *98*, 1608.
6. Wang, H.; Peslherbe, G. H.; Hase, W. L. *J. Am. Chem. Soc.* **1994**, *116*, 9644.
7. Farneth, W. E.; Brauman, J. I. *J. Am. Chem. Soc.* **1976**, *98*, 7891.
8. Cho, Y. J.; Vande Linde, S. R.; Zhu, L.; Hase, W. L. *J. Chem. Phys.* **1992**, *96*, 8275.
9. Hase, W. L.; Wardlaw, D. M. In *Advances in Gas-Phase Photochemistry and Kinetics: Bimolecular Collisions*; Ashfold, M. N. R. and Baggott, J. E., Eds.; Royal Society: London, 1989, p. 171.
10. Hase, W. L. *J. Chem. Phys.* **1976**, *64*, 2442; Truhlar, D. G.; Issacson, A. D.; Garrett, B. C. In *Theory of Chemical Reaction Dynamics*; Baer, M., Ed.; CRC Press: Boca Raton, FL, 1985, Vol. 4.

11. Su, T.; Bowers, M. T. In *Gas Phase Ion Chemistry*; Bowers, M. T., Ed.; Academic Press: New York, 1979, Vol. 1, p. 65.
12. Troe, *J. Chem. Phys. Lett.* **1985**, *122*, 425.
13. Su, T.; Chesnavich, W. J. *J. Chem. Phys.* **1982**, *76*, 5183.
14. Polanyi, J. C. *Science* **1987**, *236*, 680.
15. Seeley, J. V.; Morris, R. A.; Viggiano, A. A.; Wang, H.; Hase, W. L. *J. Am. Chem. Soc.* **1997**, *119*, 577.
16. Miller, R. E. *Science* **1988**, *240*, 447.
17. Tucker, S. C.; Truhlar, D. G. *J. Phys. Chem.* **1989**, *93*, 8138.
18. Seeger, S.; Botschwina, P. Private communication.
19. Vetter, P.; Zülicke *J. Am. Chem. Soc.* **1990**, *112*, 5136.
20. Glukhovtsev, M. N.; Pross, A.; Radom, L. *J. Am. Chem. Soc.* **1995**, *117*, 2024.
21. Hu, W. -P.; Truhlar, D. G. *J. Am. Chem. Soc.* **1995**, *117*, 10726.
22. Glukhovtsev, M. N.; Pross, A.; Radom, L. *J. Am. Chem. Soc.* **1996**, *118*, 6273.
23. Larson, J. W.; McMahon, T. B. *J. Am. Chem. Soc.* **1985**, *107*, 766; McMahon, T. B. In *Advances in Gas Phase Ion Chemistry*; Adams, N. G.; Babcock, L. M., Eds.; JAI Press: Greenwich, CT, 1996, Vol. 2, p. 54.
24. Caldwell, G.; Magnera, T. F.; Kebarle, P. *J. Am. Chem. Soc.* **1984**, *106*, 959.
25. Duncan, J. L.; Allan, A.; McKean, D. C. *Mol. Phys.* **1970**, *18*, 289.
26. Andzelm, J.; Klobukowski, M.; Radzio-Andzelm, E. *J. Comput. Chem.* **1984**, *5*, 146.
27. Mann, D.; Cho, Y. J.; Wang, H.; Hase, W. L. Unpublished results.
28. Viggiano, A. A.; Paschkewitz, J. S.; Morris, R. A.; Paulson, J. F.; Gonzalez-Lafont, A.; Truhlar, D. G. *J. Am. Chem. Soc.* **1991**, *113*, 9404.
29. Graul, S. T.; Bowers, M. T. *J. Am. Chem. Soc.* **1991**, *113*, 9696; Graul, S. T.; Bowers, M. T. *J. Am. Chem. Soc.* **1994**, *116*, 3875.
30. Wang, H.; Hase, W. L. *J. Am. Chem. Soc.* **1995**, *117*, 9347.
31. Viggiano, A. A.; Morris, R. A.; Paschkewitz, J. S.; Paulson, J. F. *J. Am. Chem. Soc.* **1992**, *114*, 10477.
32. Wladkowski, B. D.; Lim, K. F.; Allen, W. D.; Brauman, J. I. *J. Am. Chem. Soc.* **1992**, *114*, 9136.
33. Graul, S. T. Private communication.
34. Chesnavich, W. J.; Bowers, M. T. In *Gas Phase Ion Chemistry*; Bowers, M. T., Ed.; Academic Press: New York, 1979, Vol. 1, p. 119.
35. Vande Linde, S. R.; Hase, W. L. *J. Am. Chem. Soc.* **1989**, *111*, 2349.
36. Vande Linde, S. R.; Hase, W. L. *J. Phys. Chem.* **1990**, *94*, 6148.
37. Vande Linde, S. R.; Hase, W. L. *J. Chem. Phys.* **1990**, *93*, 7962.
38. Hase, W. L.; Cho, Y. J. *J. Chem. Phys.* **1993**, *98*, 8626.
39. Peslherbe, G. H.; Wang, H.; Hase, W. L. *J. Chem. Phys.* **1995**, *102*, 5626.
40. Chesnavich, W. J.; Su, T.; Bowers, M. T. *J. Chem. Phys.* **1980**, *72*, 2641.
41. Lim, K. F.; Brauman, J. I. *J. Chem. Phys.* **1992**, *97*, 6322.
42. Lim, K. F.; Kier, R. I. *J. Chem. Phys.* **1992**, *97*, 1072.
43. Hase, W. L.; Duchovic, R. J.; Swamy, K. N.; Wolf, R. J. *J. Chem. Phys.* **1984**, *80*, 714.
44. Reinhardt, W. P. *J. Phys. Chem.* **1982**, *86*, 2158. Shirts, R. B.; Reinhardt, W. P. *J. Chem. Phys.* **1982**, *77*, 5204.
45. Marcus, R. A. *Faraday Discus. Chem. Soc.* **1973**, *9*, 55.
46. Miller, W. H. *Adv. Chem. Phys.* **1974**, *25*, 69.
47. Li, C.; Ross, P.; Szulejko, J. E.; McMahon, T. B. *J. Am. Chem. Soc.* **1996**, *118*, 9360.
48. Slater, N. B. *Theory of Unimolecular Reactions*; Cornell University Press: Ithaca, NY, 1973.
49. Bunker, D. L. *J. Chem. Phys.* **1964**, *40*, 1946.
50. Forst, W. *Theory of Unimolecular Reactions*; Academic Press: New York, 1973.
51. Lu, D. -h.; Hase, W. L. *J. Phys. Chem.* **1989**, *93*, 1681.
52. Hase, W. L. *J. Phys. Chem.* **1986**, *90*, 365.

53. Benjamin, I.; Liu, A.; Wilson, K. R.; Levine, R. D. *J. Phys. Chem.* **1990**, *94*, 3937.
54. Hase, W. L., Ed. *Intramolecular and Nonlinear Dynamics; Advances in Classical Trajectory Methods*; JAI Press: Greenwich, CT, 1992, Vol. 1.

BARRIERS TO INTERNAL ROTATION IN SUBSTITUTED TOLUENES AND THEIR CATIONS:

EFFECTS OF ELECTRONIC EXCITATION AND IONIZATION

Erik C. Richard, Kueih-Tzu Lu,

Robert A. Walker, and James C. Weisshaar

Advances in Gas-Phase Ion Chemistry
Volume 3, pages 157–183
Copyright © 1998 by JAI Press Inc.
All rights of reproduction in any form reserved.
ISBN: 0-7623-0204-6

ABSTRACT

Threshold photoionization spectroscopy of substituted toluene cations with 1 cm^{-1} resolution combined with ab initio electronic structure calculations has provided new insight into the nature of barriers to methyl group internal rotation. We review recent progress in this area, including an overview of the ZEKE-PFI technique (*zero kinetic energy* threshold photoelectron spectroscopy with *pulsed field ionization* detection). From microwave and optical spectroscopies, we now know the heights of methyl rotor barriers for a variety of substituted toluenes in three electronic states: S_0, S_1, and D_0 (the cation ground state). In all three electronic states, comparison of the torsional potentials for *o*-fluorotoluene, *o*-chlorotoluene, and 2-fluoro-6-chlorotoluene shows nearly quantitative additivity of the contributions of the two *ortho* substituents to the overall barrier in the disubstituted case. The substantial data base of barrier heights shows several qualitative trends which demand theoretical explanation. With the help of ab initio electronic structure calculations, we suggest that three different effects contribute to the observed barriers. First, an *ortho*-halogen substituent creates a substantial preference for the pseudo-*trans* conformation, primarily due to steric repulsion. Second, a difference in bond order between the two ring CC bonds adjacent to the methyl rotor favors placement of one CH bond in the plane of the ring *cis* to the ring CC bond of higher order. This is due to attractive, donor–acceptor interactions analogous to those that arise in the acyclic case of ethane. Third, π ionization often leads to dramatic changes in local ring geometry. The pattern of substitution controls the choice of π orbital from which the electron is removed and also the orbital *orientation*, which in turn governs the conformational preference of the methyl group in the ground-state cation.

I. INTRODUCTION: THE METHYL ROTOR PROBLEM

Methyl rotors pose relatively simple, fundamental questions about the nature of noncovalent interactions within molecules. The discovery in the late 1930s[1] of the 1025 cm^{-1} potential energy barrier to internal rotation in ethane was surprising, since no covalent chemical bonds are formed or broken as methyl rotates. By now it is clear that the methyl torsional potential depends sensitively on the local chemical environment. The barrier is 690 cm^{-1} in propene,[2] comparable to ethane, but only 5 cm^{-1} in toluene.[3] In addition, the pioneering work of Ito and coworkers[4] in the 1980s showed that threefold barriers in substituted toluenes can vary dramatically from S_0 (ground electronic state) to S_1 (first excited singlet electronic state). In addition to presenting subtle structural questions about the relative stability of different molecular conformations, the presence of a methyl rotor can dramatically affect photochemical pathways and quantum yields and also the rate of intramolecular vibrational energy redistribution.[5] Methyl rotors in different chemical environments contribute very differently to the density of states that enters statistical models of chemical reaction rates. Chemists need simple, reliable models based on rigorous quantum mechanics that can unify the seemingly diverse behavior of methyl rotor conformational preference and barrier height.

Microwave spectroscopy can obtain barriers to internal methyl rotation in S_0 from tunneling splittings when the barriers are not too large.[2,3] With the advent of pulsed tunable dye lasers, optical spectroscopy in free jets resolved pure torsional bands and obtained barriers for the S_1 excited electronic state in a variety of substituted toluenes.[4,6,7–9] More recently, the new threshold photoionization technique of pulsed field ionization (PFI) has extended these data to the corresponding ground state cations.[10,11,12,13] In some cases, Franck–Condon factors reveal *changes* on electronic excitation in the methyl group conformation which is preferred. The minimum energy conformation in S_0 and D_0 (ground electronic state of the molecular cation) can often best be determined by ab initio calculations, as described below. We now have a rich data base encompassing different local chemical structures and different electronic states. Each new example raises the subtle structural question of which methyl conformation lies lowest in energy and why.

In this chapter, we first present a brief overview of the experimental techniques that we and others have used to study torsional motion in S_1 and D_0 (Section II). These are resonant two-photon ionization (R2PI) for S_1-S_0 spectroscopy and pulsed-field ionization (commonly known as ZEKE-PFI) for D_0-S_1 spectroscopy. In Section III, we summarize what is known about sixfold methyl rotor barriers in S_0, S_1, and D_0, including a brief description of how the absolute conformational preference can be inferred from spectral intensities. Section IV describes the threefold example of *o*-cholorotoluene in some detail and summarizes what is known about threefold barriers more generally. The sequence of molecules *o*-fluorotoluene, *o*-chlorotoluene, and 2-fluoro-6-chlorotoluene shows the effects of *ortho*-fluoro and *ortho*-chloro substituents on the rotor potential. These are approximately additive in S_0, S_1, and D_0. Finally, in Section V, we present our ideas about the underlying causes of these diverse barrier heights and conformational preferences, based on analysis of the optimized geometries and electronic wavefunctions from ab initio calculations.

II. EXPERIMENTAL TECHNIQUES

A. Resonant Two-Photon Ionization (R2PI)

The use of resonant two-photon ionization (R2PI) and time-of-flight mass spectrometry (TOF-MS) to obtain mass-specific S_1-S_0 absorption spectra is well known. We form an internally cold, skimmed beam of the substituted toluene of choice, typically seeded at its room temperature vapor pressure in a pulsed expansion of Ar gas at ~2 atm stagnation pressure. A frequency-doubled, pulsed tunable dye laser in the near UV intersects the skimmed beam in the extraction zone of the TOF-MS. Two photons of the same frequency ω_1 are absorbed to form molecular cations whenever ω_1 matches an S_1-S_0 resonance; this is known as *one-color* R2PI. The spectrum consists of the current to an ion detector gated at the mass of interest versus ionizing frequency ω_1. The resolution is 0.2 cm^{-1}, limited by the dye lasers. S_1-S_0 vibronic bands typical exhibit rotational contours of about 2 cm^{-1} FWHM, indicating that the internal temperature of the molecules is roughly 5 K under our expansion conditions. Further details of the technique can be found in the literature,[10-16] and examples of spectra will follow.

Care must be taken not to mistake bands arising from 1:1 or 1:2 van der Waals complexes (e.g. toluene·Ar_x, $x = 1$ and 2) for monomer bands. Fragmentation of either the neutral excited state or the cation can cause the 1:1 and even the 1:2 bands to appear in R2PI spectra detected at the monomer cation mass. The van der Waals bands usually appear red-shifted by a constant value from corresponding strong monomer bands. The safest way to identify van der Waals bands is to carry out R2PI scans gated on successive toluene·Ar_x^+ masses to see which bands disappear at each step.

B. Zero Kinetic Energy Threshold Photoelectron Spectroscopy and Pulsed Field Ionization (ZEKE-PFI)

ZEKE-PFI is a threshold photoionization spectroscopy capable of providing cation spectra with resolution of 1 cm^{-1} or better, adequate for resolving the cation torsional bands of interest here. The technique was developed in the 1980s by Müller-Dethlefs, Schlag, and coworkers.[14] We present only a brief overview here. The interested reader might well begin with the review article by Müller-Dethlefs and Schlag.[15] An excellent recent monograph[16] is devoted entirely to applications of the new family of high-resolution photoionization and photoelectron techniques.

In standard photoelectron spectroscopy, the ionizing photon energy is *fixed*. Cation energy levels are revealed by kinetic energy analysis of the resulting photoelectrons. The state of the art in electron energy analysis is about 4 meV = 32 cm^{-1} using hemispherical analyzers and array detectors.[17] Time-of-flight techniques are capable of 1 meV = 8 cm^{-1} resolution at low kinetic energy.[18] Calibration of the electron energy scale can be a time-consuming, repetitive task in both techniques.

In the ZEKE-PFI technique, the ionizing laser frequency is *scanned* through the manifold of cation states (Figure 1). A ZEKE-PFI signal is detected only when the

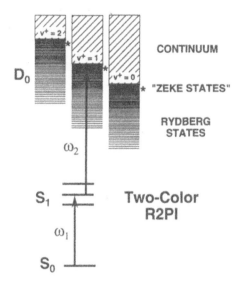

Figure 1. Schematic of a two-color resonant two-photon ionization experiment; "ZEKE states" lie just below each cation threshold, as indicated by asterisks.

scanning laser frequency lies at (or just below) a cation threshold. While ZEKE-PFI spectra have been obtained using one-photon ionization by tunable VUV lasers,[19] the more common experiment uses two-color R2PI ($\omega_1 + \omega_2$) to ionize the molecule. In that case, ω_1 is tuned to a particular S_1-S_0 band already observed in the one-color R2PI spectrum. With ω_1 fixed, ω_2 is scanned through the manifold of cation states. The device we use for pulsed field ionization is shown schematically in Figure 2. The molecular beam and the two dye lasers intersect between a pair of grounded, magnetically shielded stainless steel electron extraction plates. For spectroscopic experiments, both dye lasers are often pumped by harmonics of a single Nd:YAG laser with 5–10 ns pulse width, since the dye laser pulse energies required are modest. At a time delay of 2.0 μs following the laser pulses, a fast voltage step is applied to the lower plate, ionizing long-lived, high-n Rydberg states if they have been excited at ω_2. Extracted electrons travel in a 0.5 V/cm field over 8 mm and then accelerate rapidly to a terminal kinetic energy of 2.6 eV. Gated integration of the delayed electron pulse (60 ns gate width) from a chevron microchannel plate detector 50 cm from the interaction zone produces the PFI spectrum.

The key to the good resolution of ZEKE-PFI is in its discrimination against electrons with ≥ 1 cm^{-1} kinetic energy. This is due to the delayed extraction and time-gated detection. A "bubble" of 1 cm^{-1} electrons expands to a radius of 1.6 cm during the 2.0 μs delay. Such kinetic electrons either miss the detector or arrive

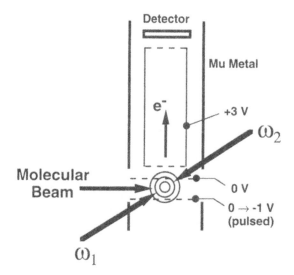

Figure 2. Schematic of apparatus for ZEKE-PFI spectroscopy, including magnetically shielded electron flight path and microchannel plate detector. Not to scale; flight path is 50 cm long; interaction zone is 1 cm long.

outside the signal gate, since they have significantly different terminal velocities than electrons extracted from the center of the interaction region.

The inventors of the technique believed that they were detecting true continuum electrons with near-zero (sub-cm^{-1}) kinetic energy,[14] but this has proven not to be the case. Our own experiment probably does not detect any genuine continuum electrons, since stray fields in the apparatus as small as 1 mV/cm will remove them from the interaction region during the 2 µs delay. Rather, we and others detect electrons arising from PFI[19] of high-n Rydberg states (Figure 1) which lie about 3 cm^{-1} below each true cation threshold and survive in any stray electric fields for at least 2.0 µs. In the simplest view, such Rydberg states are molecular analogues of hydrogen-atom excited states. They have principal quantum number $n \sim 200$, a broad mixture of orbital angular momenta ℓ, and anomalously long lifetimes. The experimental signature of PFI is an asymmetric broadening of bands towards lower energy for larger extraction fields. This explains the hyphenated acronym ZEKE-PFI. The technique was originally called ZEKE but, in fact, it is pulsed field ionization as typically carried out.

The ZEKE-PFI practitioner relies on the existence of a narrow band of long-lived, high-n Rydberg states ("ZEKE states")[15] lying just below each true cation threshold. If the shift of this band of states relative to the true threshold is essentially constant, then spectroscopic information extracted from *differences* in band frequencies will faithfully reflect the true cation energy levels. This assumption seems to hold to an accuracy of 1 cm^{-1} or better in the many molecules studied, as judged by the success of standard spectroscopic models in fitting ZEKE-PFI spectra. Cation vibrational, rotational, and torsional frequency intervals can thus be measured to a similar accuracy. The absolute shift in the lowest energy ZEKE-PFI band relative to the true adiabatic ionization potential is about 3 cm^{-1} under typical conditions. To account for uncertainties in the ionization mechanism, we usually quote adiabatic ionization potentials to an absolute accuracy of ±5 cm^{-1}, although higher accuracy can be achieved by careful work in which the ionizing field strength is varied systematically.

The narrowest PFI bands in the present study are 3 cm^{-1} FWHM, using a 0.5 V/cm extraction field with the lasers attenuated to minimize effects of space charge. We measure band positions at the intensity maxima. These are reproducible to better than ±1 cm^{-1}. The bandwidth is limited by the rotational contour and also by the ionization process. A major advantage of ZEKE-PFI over more traditional photoelectron techniques is that the energy calibration is that of the tunable dye lasers, which are quite stable from day to day. In contrast, both electrostatic analyzers and time-of-flight photoelectron spectrometers require frequent calibration.

The electronic structure of molecules above the first ionization threshold is complex and subtle.[20] Simply put, each cation state supports multiple manifolds of Rydberg series converging to that state. Each Rydberg series might be described in terms of the internal quantum numbers of its cation core (the electronic, vibrational, rotational, and torsional quantum numbers e^+, v^+, J^+, and m^+). Each Rydberg state within a series might be further described by three quantum numbers for the Rydberg electron; these are n (the principal quantum number), ℓ (the orbital angular momentum), and m_ℓ (the projection of the orbital angular momentum onto a space-fixed axis). Rydberg states lying above the adiabatic ionization potential are metastable. They consist of an internally excited ion core to which the Rydberg electron is very weakly bound. The core excitation may be electronic, vibrational, rotational, or some combination of these. Such Rydberg states have sufficient energy to decay by *autoionization*,[20] which can be viewed as a scattering process in which the ion core transfers energy to the electron. The core loses internal energy and becomes a molecular cation, while the electron becomes free. Rydberg states often decay by dissociation to excited state neutral fragments as well.

The physical nature of the ZEKE states has been the subject of intense experimental and theoretical investigation in the past several years. In the well-studied case of NO,[14,21] we know from the 3 cm^{-1} red shift of the ZEKE-PFI threshold band relative to the true adiabatic ionization potential (extrapolated from highly accurate measurements of Rydberg series) that the ZEKE states have principal quantum number $n \sim 200$ and lifetime of 2 μs or longer. Recent work has found ZEKE states with lifetimes as long as 20 μs.[22]

Such long lifetimes are puzzling in view of the well-known n^3 scaling of Rydberg state lifetimes observed from linewidths of resolvable series and predicted by theory. The n^3 scaling derives from a fundamental property of Rydberg states. The normalization factor in the radial part of the Rydberg orbital scales as $n^{-3/2}$. Thus the probability of finding the Rydberg electron in a small sphere about the cation core scales as n^{-3}. Since either autoionization or dissociation involves interaction of the Rydberg electron with the core, the lifetime lengthens with increasing principal quantum number as n^3. In a scattering picture, autoionization results from energy transfer from the ion core to the weakly bound Rydberg electron. The core loses energy as the electron scatters into the continuum. The frequency with which a classical Rydberg electron visits the core scales as the same n^{-3}.

The problem with the observed lifetimes of the ZEKE states is that they are orders of magnitude *too long* compared with estimates obtained by scaling of measured lifetimes of Rydberg states having $n \sim 20$–30 according to the n^3 law. For $n \sim 200$, it appears the lifetimes scale roughly as n^5. Theoretical ideas due to Chupka[23] and recent experiments[24] appear to be converging on the explanation of the breakdown of n^3 scaling at high n. Briefly, electric–dipole optical excitation from a low-n, low-ℓ, low-m_ℓ valence orbital follows $\Delta\ell = \pm 1$ selection rules. Thus the Rydberg states carrying oscillator strength near a cation threshold necessarily have $n \sim 200$, low ℓ, and low m_ℓ just after excitation by the laser. However, n, ℓ, and m_ℓ are good quantum numbers only for a truly spherical potential. In the usual laboratory environment of a ZEKE-PFI experiment, the Stark effect from the small *homogeneous* electric field at the molecule (typically a stray field in the apparatus) mixes states of different ℓ according to the selection rule $\Delta\ell = \pm 1$. Furthermore, the *inhomogeneous* electric field (which apparently arises primarily from the space charge formed by the laser ionization process) mixes states of different m_ℓ. In the limit of complete mixing, each optically active n, low-ℓ, low-m_ℓ state is coupled to *many* states whose quantum numbers ℓ and m_ℓ will be much larger. These background states *do not autoionize or dissociate* on the ZEKE timescale, since the large centrifugal barrier for $\ell \geq 3$ prevents them from visiting the core. This *dilution* of the initially excited low-ℓ character among many long-lived states leads to a corresponding *lengthening* of the lifetime of the ZEKE states beyond the n^3 scaling that applies to low-ℓ, low-m_ℓ states. The lifetime is lengthened in proportion to the number of states with which the optically active state mixes. Since the complete degeneracy of a hydrogenic state is n^2, in the limit of complete ℓ, m_ℓ mixing the lifetimes should scale as n^5, much as observed.

III. SIXFOLD-SYMMETRIC METHYL AND SILYL ROTORS

Sixfold barriers to internal rotation occur in molecules such as toluene and *p*-fluorotoluene whose molecular frame has C_{2v} symmetry about the rotor axis. The simplest spectroscopic model of internal methyl rotation assumes a rigid, threefold symmetric methyl rotor attached to a rigid molecular frame with the C_2 axis coincident with the rotor top axis.[25] The effective one-dimensional sixfold torsional potential takes the traditional form,

$$V(\alpha) = (V_6/2)(1 - \cos 6\alpha), \tag{1}$$

where α is the dihedral angle of the internal rotor relative to the molecular frame and $|V_6|$ is the barrier height. The sign of V_6 fixes the conformation of minimum potential energy. With $\alpha = 0$ defined as the eclipsed conformation (Figure 3a), *positive* V_6 gives the *eclipsed* potential minimum while *negative* V_6 gives the *staggered* minimum ($\alpha = 90°$).

Since sixfold potentials are usually very small, a good starting point for understanding the states of the hindered sixfold rotor is the set of *free rotor* states. These functions of α are the eigenstates of the one-dimensional Hamiltonian including only the kinetic energy of internal rotation, i.e. $V_6 = 0$. They take the form $\phi_m(\alpha) = e^{+im\alpha}$, where $m = 0, \pm1, \pm2, \ldots$. The free rotor energy levels are $E_m = Fm^2$, where F is an effective rotational constant on the order of 5 cm^{-1} for a methyl rotor attached to a heavy frame. A sixfold potential mixes and splits the free rotor wavefunctions in a pattern depending on symmetry. For example, the degenerate $m = \pm3$ levels $\phi_{+3} = e^{+3i\phi}$ and $\phi_{-3} = e^{-3i\phi}$ levels of the free rotor are completely mixed by the smallest hindering potential to form the symmetric and anti-symmetric linear combinations $|3+\rangle = \cos 3\alpha$ and $|3-\rangle = \sin 3\alpha$. These have, respectively, a_1'' and a_2'' torsional symmetry under G_{12}, the appropriate molecular symmetry group for this nonrigid molecule.[26]

A key spectroscopic point is that the *sign* of V_6 (and thus the phase of the potential) determines the energetic order of the two $|m| = 3$ torsional states of the hindered rotor (Figure 4). For negative V_6, $|3+\rangle$ lies *above* $|3-\rangle$ and the minimum is staggered, as occurs in S_0 and S_1 for all molecules studied thus far. For positive V_6, $|3+\rangle$ lies below $|3-\rangle$ and the minimum is eclipsed, as occurs in D_0 of almost all molecules studied thus far. In either case, the splitting between $|3+\rangle$ and $|3-\rangle$ gives $|V_6|/2$. One way to determine the absolute conformation of the energy minimum in a given electronic state is to determine the energy ordering of the $|3+\rangle$ and $|3-\rangle$ states.

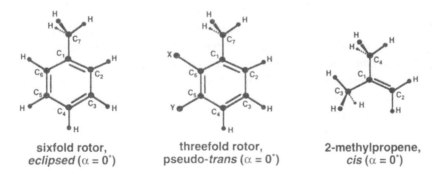

sixfold rotor,
eclipsed ($\alpha = 0°$)

threefold rotor,
pseudo-*trans* ($\alpha = 0°$)

2-methylpropene,
cis ($\alpha = 0°$)

Figure 3a,b,c. Sixfold and threefold rotors; X and Y are halogens in this work. For the sixfold case, we define $\alpha = 0°$ with one CH bond in the plane of the ring (eclipsed). Staggered geometry ($\alpha = 90°$) then has one CH bond perpendicular to the plane of the ring. For the threefold case shown in the center, we define $\alpha = 0°$ with one CH bond "pseudo-trans" to the substituents X and Y.

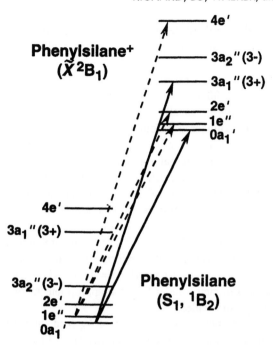

Figure 4. Schematic of torsional energy levels in S_1 electronic state of phenylsilane and D_0 electronic state of phenylsilane$^+$. The torsional state symmetry labels arise under the molecular symmetry group G_{12}.

Microwave spectroscopy can determine the magnitude of V_6 in S_0 but not the sign, since the potential well is too small to localize even the $m = 0$ wavefunction. $S_1 \leftarrow S_0$ absorption spectra of cold molecules with 1 cm^{-1} resolution can reveal the magnitude of V_6 in S_1, a technique pioneered by Ito and coworkers.[4] Pratt and coworkers[7] and Miller and coworkers[8] have made major contributions to the *high-resolution* optical spectroscopy of rotor-containing molecules.

Figure 5 shows a collection of S_1-S_0 R2PI spectra near the origin. The weak bands at low frequency are pure torsional transitions. We can extract the barrier height and the absolute phase of the torsional potential in S_1 from the frequencies and intensities of these bands. The bands labeled m_1^2, m_0^{3+}, and m_1^4 are forbidden in the sense that they do not preserve torsional symmetry. In the usual approximation that the electronic transition dipole moment is independent of torsion–vibrational coordinates, band intensities are proportional to an electronic factor times a torsion–vibrational overlap factor (Franck–Condon factor). These forbidden bands have Franck–Condon factors $|\langle m'|m''\rangle|^2$ that are zero by symmetry. Nevertheless, they are easily observed in jet-cooled spectra. They are comparably intense in many spectra, about 1–5% of the intensity of the allowed origin band.

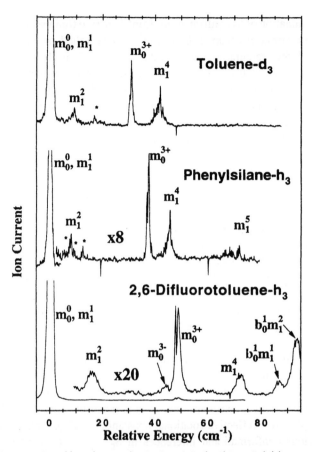

Figure 5. Pure torsional bands near the S_1-S_0 origin for three sixfold rotors as labeled. Band label m_y^x denotes a pure torsional transition from $m = y$ in S_0 to $m = x$ in S_1. The label b_0^1 denotes an increase from zero to one quantum of a low-frequency in-plane bending vibration.

The spectra are assigned by fitting observed band positions to the eigenvalues of a one-dimensional model torsional Hamiltonian in S_1, with the S_0 barrier taken from microwave spectra in most cases. Typically only one of the two possible bands, m_0^{3-} and m_0^{3+}, is observed with appreciable intensity. The observed band always lies *above* the mean of the calculated m_0^{3-} and m_0^{3+} band positions, but its identity as m_0^{3-} or m_0^{3+} remains ambiguous. If the observed band can be properly assigned, the sign of V_6 and the most stable conformer in S_1 follow (Figure 4). All earlier workers have *assumed* the m_0^{3+} assignment, yet efforts to calculate the intensity pattern of the forbidden bands based on torsion-overall rotation coupling within S_1 or torsion-electronic coupling between S_1 and S_2 (second excited singlet state) have failed badly. Determination of S_0 conformations from dispersed fluorescence spectra[9] and of cation conformations from PFI spectra[10,11] have relied on the assumed m_0^{3+} assignment. If the assignment is correct, it implies that S_1 is always *staggered*.

In collaboration with E.L. Sibert, we have learned to interpret these Franck–Condon forbidden, pure torsional band intensities in S_1-S_0 absorption spectra *quantitatively* and thus place the key m_0^{3+} assignment on firm ground.[27] The forbidden bands follow the selection rule $\Delta m = \pm 3$, so we need a perturbation of the form $V_{el} \cos 3\alpha$. Working in an adiabatic representation with the S_0 and S_1 electronic states denoted by $\psi_0(q;\alpha)$ and $\psi_1(q;\alpha)$ and the torsional states by m'' and m', the electric dipole transition moment is,

$$\langle m''|[\langle \psi_1(q;\alpha) |\mu| \psi_0(q;\alpha)\rangle_q] \mid m'\rangle_\alpha, \tag{2}$$

where q represents electronic coordinates, α is the torsional coordinate, and μ is the dipole moment operator. The inner integral over q gives the transition dipole moment $\mu_{10}(\alpha)$; the outer integral is over the torsional coordinate α. The electronic integral thus depends on α. The key step involves approximating each of the three components of the electronic integral $\mu_{10}(\alpha)$ as a *Fourier* series in the lowest order terms such as $\cos 3\alpha$, $\cos 6\alpha$, and $\sin 6\alpha$ permitted by the symmetry species of each vector component under the appropriate molecular symmetry group G_{12}. Details are given elsewhere.[27] The usual Taylor series expansion in powers of α is physically inappropriate because the rotor wavefunctions are delocalized over all α.

Fourier expansion of $\mu_{10}(\alpha)$ allows us to calculate relative intensities of the three forbidden bands m_1^2, m_0^{3+}, and m_1^4. These are in quantitative agreement with experiment. The agreement is excellent over a wide range of ratios of the key model parameters V_6 and F (effective rotational constant), which are taken from experiment. Previous conformational inferences in S_0, S_1, D_0 (ground state cation), including our own, were in fact correct. They now rest on solid ground.

In addition, we have measured new sixfold barrier heights and conformational preferences in S_1 phenylsilane[11] and 2,6-difluorotoluene.[28] For the cation ground state D_0, we have used the 1 cm^{-1} resolution of the ZEKE-PFI technique to determine V_6 for the first time. In Figure 6, we show PFI spectra of D_0 phenylsilane obtained through different S_1 intermediate states.[11] Assignment of the spectra is greatly assisted by the rigorous selection rule preserving nuclear spin symmetry, either a or e under G_{12}. When ω_1 cleanly excites an a-symmetry (or e-symmetry) state of S_1, we observe transitions only to a-symmetry (or e-symmetry) states of the cation. Assuming the same simple rigid-frame, rigid-methyl-rotor model Hamiltonian, we obtain the parameters F (reduced rotational constant for rotation of methyl relative to the frame) and V_6. Among toluene$^+$-h_3, toluene$^+$-d_3, phenylsilane$^+$, and 2,6-difluorotoluene$^+$, V_6 lies in a narrow range from $+12$ to $+19$ cm^{-1}. For p-fluorotoluene$^+$, $|V_6|$ is less than 5 cm^{-1}. The most intriguing feature of these data is that they indicate *eclipsed* minima in most ground-state cations, since $|3+\rangle$ lies below $|3-\rangle$.

In Table 4 of ref. 28, we summarize sixfold barrier heights and conformational preferences in the S_0, S_1, and D_0 states of toluene, *p*-fluorotoluene, 2,6-difluorotoluene, phenylsilane, and *p*-toluidine. In all these sixfold molecules, the barrier height is quite small in all electronic states, less than 5 cm^{-1} in S_0, less than 45 cm^{-1} in S_1, and less than 20 cm^{-1} in D_0. We will explain the uniformly small magnitude of sixfold barriers in Section V. The substantial variation among molecules and among states is perhaps too subtle to explain simply, but some clear trends emerge. Thus far the preferred conformation is always *staggered* in S_1 and almost always *eclipsed* in D_0. Substitution of silyl for methyl or introduction of two sterically bulky *ortho*-fluoro substituents alters only the quantitative details. We know the conformation in S_0 for only one molecule, *p*-fluorotoluene, from Parmenter's dispersed fluorescence work.[9] It is staggered, like the other neutral (S_1) states.

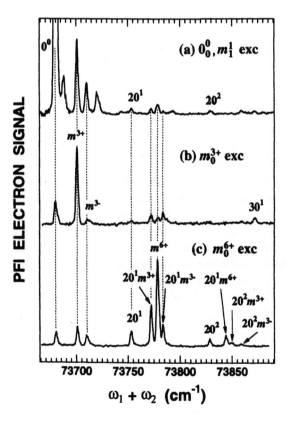

Figure 6. ZEKE-PFI spectra of phenylsilane$^+$ (D_0) taken with ω_1 fixed on three S_1-S_0 bands as indicated and ω_2 scanned through the low-energy manifold of cation states.

IV. THREEFOLD-SYMMETRIC METHYL ROTORS

A. Overview

Chemical substitution in a pattern that breaks the C_{2v} symmetry of the molecular frame introduces a *threefold* component to the one-dimensional model torsional potential:

$$V(\alpha) = (V_3/2)(1 - \cos 3\alpha) + (V_6/2)(1 - \cos 6\alpha). \tag{3}$$

Ethane falls into this category with $V_3 = 1025$ cm^{-1}. In *ortho-* or *meta*-substituted toluenes, we define $\alpha = 0$ as the "pseudo-*trans*" conformation (Figure 3b); $\alpha = 180°$ is then the "pseudo-*cis*" conformation. Typically when symmetry allows a threefold component of the potential, $|V_3|$ is much larger than $|V_6|$. The sign of V_3 then determines the minimum energy conformation. For V_3 positive, the minimum is pseudo-*trans*; for V_3 negative, the minimum is pseudo-*cis*. For threefold symmetric cases, the torsional states can be classified under the molecular symmetry group G_6, whose species designations a_1, a_2, and e lack primes or double primes. In the limit of a large threefold barrier, pairs of torsional states (one nondegenerate a-symmetry state plus one doubly degenerate e-symmetry level) converge in energy and the spectrum approaches that of a threefold-degenerate vibrational mode. Again the conservation of nuclear spin symmetry ($a \leftrightarrow a$ and $e \leftrightarrow e$) is strict in both S_1-S_0 R2PI and D_0-S_1 ZEKE-PFI spectra.

Ito and coworkers[4,12,13] have used both R2PI and more recently PFI spectroscopies to show that in substituted toluenes, the threefold barrier heights and minimum energy conformations depend sensitively not only on the location and nature of substituents but also on the electronic state S_0, S_1, or D_0. Franck–Condon factors in either the S_1-S_0 or D_0-S_1 spectra indicate clearly whether or not the preferred conformation changes on electronic excitation or π-ionization. The absolute conformation in S_0 is available for *o*-fluorotoluene from careful microwave work.[29] More often, ab initio calculations[30] yield accurate barrier heights in S_0 and thus make an unambiguous prediction of the S_0 conformation. The spectral intensities then yield absolute conformations in S_1 and D_0. Earlier work in several groups has measured the effects of methyl, amino, and fluorine substituents.[9]

Figure 7 displays the data of Ito[4,13] and others[29] in a 3×3 matrix of torsional potentials for *o*-fluorotoluene, *m*-fluorotoluene, and *p*-fluorotoluene in the three electronic states S_0, S_1, and D_0. The matrix reveals patterns that hold for other *ortho*, *meta*, and *para* substituents as well.[9] In S_0, *ortho* substitution creates a large barrier, while *meta* substitution does not. A sensible interpretation invokes steric repulsion between methyl and the *ortho* substituent. The *meta*-substituted cases have very small barriers like toluene itself, apparently for lack of steric effects. However, in *ortho*-substituted toluenes, V_3 *decreases* sharply on $\pi^* \leftarrow \pi$ excitation from S_0 to S_1, while in *meta*-substituted toluenes, V_3 *increases* substantially from S_0 to S_1. This suggests that steric interactions are not the complete story. Most intriguing of all,

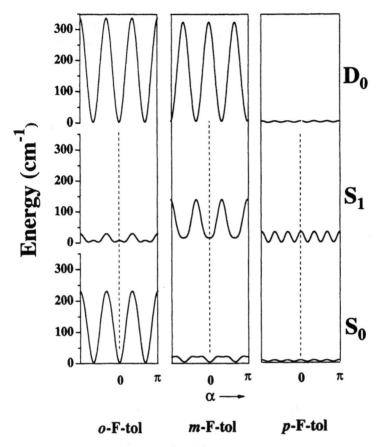

Figure 7. Torsional potentials for *ortho-*, *meta-*, and *para-*fluorotoluene in S_0, S_1, and D_0. Data from ref. 4.

in *m*-fluorotoluene$^+$ the barrier height is 318 cm^{-1}; the preferred conformer places one CH bond in the plane of the ring on the side nearest F (pseudo-*cis*). Apparently these results cannot be understood in terms of steric repulsion alone. Rather, they clearly signal the importance of methyl rotor interactions with the ring. We are beginning to understand this pattern, as explained in Section V.

B. Additivity of *ortho*-Substituent Effects

To provide further experimental clues to the nature of substituent effects, we decided to test the additivity of *ortho*-substituent effects. Specifically, we set out to compare methyl torsional potentials in S_0, S_1, and D_0 for the sequence of molecules *o*-fluorotoluene (studied by Ito and coworkers), *o*-chlorotoluene, and 2-fluoro-6-chlorotoluene.

Figure 8 shows the S_1-S_0 R2PI spectrum of o-chlorotoluene in the first several hundred cm^{-1} above the origin.[31] ZEKE-PFI spectra obtained with ω_1 tuned to various S_1 intermediate states are shown in Figure 9. We have obtained comparable spectra for 2-fluoro-6-chlorotoluene as well.[32] $ortho$-Chlorotoluene has 38 vibrational modes which are classified as either a_1 (in-plane) or a_2 (out-of-plane) under G_6 (the molecular symmetry group suitable for sixfold cases), plus torsion itself. We number the a_1 modes from 1 to 26 in order of decreasing calculated harmonic frequency in S_0; similarly, the a_2 modes are numbered 27 to 38. Torsion is labeled m. The band assignments in Figures 8 and 9 use the standard vibronic notation showing which quantum numbers change in the transition. Thus the band labeled $38_0^1 m_0^{3a_2}$ in the S_1-S_0 spectrum (Figure 8) is a combination band in which the vibrationless, torsionless S_0 level ($v_{38} = 0$, $m = 0$) is excited to the S_1 level having one quantum of v_{38} and three quanta of torsion (in the $3a_2$ level, not $3a_1$). For the D_0-S_1 PFI spectra (Figure 9), the subscript describes quanta in S_1 (the lower state of the transition) while the superscript describes those in D_0.

Assignment of these spectra involves solution of a highly redundant, interlocking puzzle. We seek assignments consistent with all available information. We can calculate the torsional energies and wavefunctions for various one-dimensional

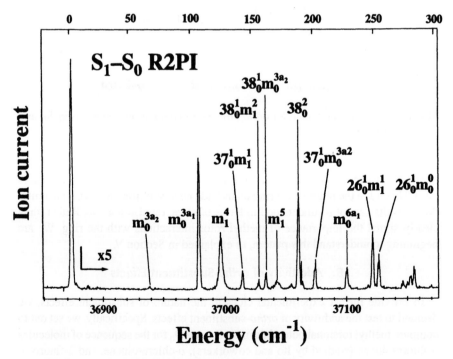

Figure 8. S_1-S_0 absorption spectrum of o-chlorotoluene detected by R2PI.

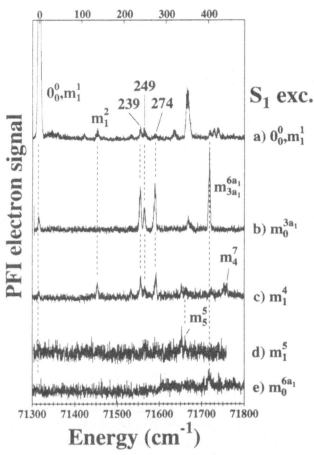

Figure 9. ZEKE-PFI spectra of *o*-chlorotoluene$^+$ (D$_0$) through various S$_1$ intermediate states *a* through *e* as labeled. Energy scale is $\omega_1 + \omega_2$, the total photon energy.

model torsional Hamiltonians in both S$_1$ and D$_0$ to try to fit the observed band frequencies. For the pure torsional transitions, we expect that Franck–Condon factors will predict sensible S$_1$-S$_0$ and D$_0$-S$_1$ band intensities, which provide a further check on the model torsional potentials. Ab initio calculations on S$_0$ and D$_0$ provide accurate barrier heights and reliable predictions of the equilibrium conformation that guide construction of the torsional models.[30] The double resonance nature of PFI spectroscopy is a powerful tool that allows us to correlate S$_1$ levels with cation levels. When the D$_0$ ← S$_1$ geometry change is minimal, we expect the most intense transitions to preserve torsional and vibrational quantum numbers. Harmonic frequencies and normal modes from ab initio calculations on the S$_0$ and D$_0$ states guide vibrational assignments. Some modes exhibit significant isotope effects from ^{35}Cl to ^{37}Cl in mass-gated R2PI spectra. These shifts can be correlated with the ab initio frequencies. Symmetry considerations predict nuclear spin conservation in both S_1-S_0 and D_0-S_1 spectra. In addition, essentially all transitions preserve the overall *torsion–vibration* symmetry, consistent with the Franck–Condon principle. Evidently the strongest couplings between zeroth-order torsion–vibrational states involve the potential energy; we seldom find the need to invoke Coriolis coupling, which could give rise to bands that do not preserve torsion–vibrational symmetry.

Assignment of such spectra and fitting of model potential parameters to the observed band frequencies yields the magnitudes of V_3 and V_6 in the two electronic states involved, either S_1 and S_0, or D_0 and S_1. Franck–Condon modeling of relative band intensities then yields the *relative* phase of the potentials in the two electronic states. We typically fix the *absolute* phase of the potentials from ab initio electronic structure calculations on both the S_0 and D_0 states. This provides an overall consistency check as well.

Our primary interest here is in the model torsional potentials derived from analysis of the spectra. Figure 10 displays our results for *o*-chlorotoluene[31] and 2-fluoro-6-chlorotoluene[32] as well as the data of Ito and coworkers[4,13] for *o*-fluoro-toluene. For the S_0 states of *o*-chlorotoluene and 2-fluoro-6-chlorotoluene, the potentials are calculated from ab initio electronic structure theory[30] as described in Section V, since no microwave data are available. Modest levels of theory give quite good accuracy; in fact, for molecules with substantial barriers, ab initio calculations now provide the easiest means of determining the conformation of minimum energy. The S_1-S_0 R2PI spectral intensities determine the phase of the S_1 potential relative to that of S_0. For the substantial barriers of interest, whenever the S_1 and S_0 phases are the same, the origin bands m_0^0 and m_1^1 dominate the spectrum, as in all cases shown in Figure 10. If the phase changed, the intensity would be distributed broadly over several Franck–Condon allowed bands. Similarly, the D_0-S_1 PFI spectra unambiguously determine the phase of the D_0 potential relative to S_1. Combining the absolute S_0 conformation (from either ab initio calculations, or microwave spectroscopy, or both) with the spectral intensity patterns then determines the absolute minimum energy conformation of all three molecules in all three states, as shown in the 3×3 matrix of potentials in Figure 10.

Comparing the potentials across each row, we can test the idea of additivity of *ortho*-substituent effects for 2-fluoro, 6-chloro, and then 2-fluoro-6-chloro substitution. The definition of $\alpha = 0$ changes across each row to permit easy visual addition of the potentials in the physically appropriate manner to assess the degree of additivity of *ortho*-chlorine and *ortho*-fluorine substituent effects. For *o*-fluorotoluene and 2-fluoro-6-chlorotoluene, $\alpha = 0$ denotes the (minimum-energy) pseudo-*trans* conformation; for *o*-chlorotoluene, $\alpha = 0$ denotes the pseudo-*cis* conformation. The notion of additivity has considerable merit in all three electronic states.

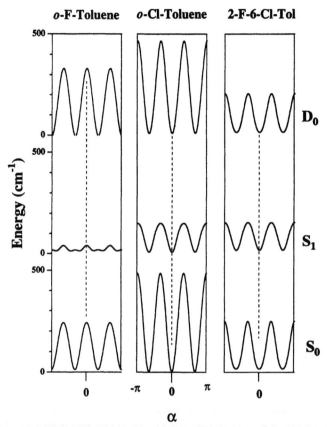

Figure 10. Threefold torsional potentials for *o*-fluorotoluene, *o*-chlorotoluene, and 2-chloro-6-fluorotoluene in the S_0, S_1, and D_0 electronic states. Dashed lines indicate the same reference configuration of the methyl group in all three molecules.

Comparing the potentials in each column of Figure 10, we see the effects of S_1 excitation ($\pi^* \leftarrow \pi$) and of π ionization on the methyl rotor potential. In S_0 *o*-fluorotoluene and *o*-chlorotoluene, the minimum energy conformation is pseudo-*trans*, consistent with the idea that halogen–methyl steric repulsion dominates the potential. The larger barrier for *o*-chlorotoluene would then arise from greater repulsion by the larger chlorine substituent. Electronic excitation to S_1 decreases the barrier by a similar amount in both *o*-fluorotoluene and *o*-cholorotoluene, as if the effects of $\pi^* \leftarrow \pi$ excitation in the ring partially *cancel* the steric effects seen clearly in S_0. Intriguingly, π ionization to D_0 essentially restores the barrier to its S_0 value in both molecules. The disubstituted molecule 2-fluoro-6-chlorotoluene shows very different behavior on $\pi^* \leftarrow \pi$ excitation and π ionization, in that effects on the barrier height are modest. We seek qualitative explanations of these electronic effects in the next section.

V. UNDERSTANDING BARRIERS TO INTERNAL ROTATION

In this section, we present a unified picture of the different electronic effects that combine to determine methyl rotor potentials in the S_0, S_1, and D_0 electronic states of different substituted toluenes. Our approach is based on analysis of ab initio wavefunctions using the *natural bond orbitals* (NBOs)[33] of Weinhold and coworkers. We will attempt to decompose the methyl torsional potential into two dominant contributions. The first is *repulsive steric interactions*, which are important only when an *ortho* substituent is present. The second is *attractive donor–acceptor interactions* between CH bond pairs and empty antibonding orbitals vicinal to the CH bonds. In the NBO basis, these attractive interactions dominate the barrier in ethane ($1025\ cm^{-1}$) and in 2-methylpropene ($1010\ cm^{-1}$); see Figure 3. By analogy, donor–acceptor attractions are important in toluenes whenever there is a substantial difference in bond order between the two ring CC bonds adjacent to the C–CH$_3$ bond. Viewed the other way around, we can use the measured methyl rotor potential as a sensitive probe of *local ring geometry*.

A. Vibrationally Adiabatic Torsional Potentials

We have used standard ab initio electronic structure theory to calculate the methyl rotor potential for a variety of substituted toluenes in the S_0 and D_0 states at various levels of theory. The first excited state, S_1, presents computational difficulties, since it often has the same electronic symmetry as the ground state and may well require a multiconfiguration wavefunction. By fixing the dihedral angle α of the rotor at various values relative to the aromatic frame and *optimizing the rest of the geometry* in each case, we obtain a kind of *vibrationally adiabatic*, one-dimensional potential $V(\alpha)$ for comparison with the spectroscopic model potentials derived from experiment. Detailed study of the optimized geometries and how they change with α provides important clues to the electronic interactions that cause the barriers. In favorable cases, we have attempted a quantitative resolution of the calculated barrier into separate contributions from ethane-like, donor–acceptor interactions and from steric repulsion. Details of these calculations are presented elsewhere;[30] we only summarize the most important results here.

For most molecules studied, modest Hartree–Fock calculations yield remarkably accurate barriers that allow confident prediction of the lowest energy conformer in the S_0 and D_0 states. The simplest level of theory that predicts barriers in good agreement with experiment is HF/6-31G* for the closed-shell S_0 state (Hartree–Fock theory) and UHF/6-31G* for the open-shell D_0 state (unrestricted Hartree–Fock theory). The 6-31G* basis set has double-zeta quality, with split valence plus d-type polarization on heavy atoms. This is quite modest by current standards. Nevertheless, such calculations reproduce experimental barrier heights within ±10%.

B. Steric Repulsion and Donor–Acceptor Attraction

With accurate calculated barriers in hand, we return to the question of the underlying causes of methyl barriers in substituted toluenes. For simpler acyclic cases such as ethane and methanol, ab initio quantum mechanics yields the correct ground state conformer and remarkably accurate barrier heights as well.[34-36] Analysis of the wavefunctions in terms of natural bond orbitals (NBOs)[33] explains barriers to internal rotation in terms of attractive donor–acceptor (*hyperconjugative*) interactions between doubly occupied σ_{CH}-bond orbitals or lone pairs and unoccupied *vicinal* antibonding orbitals.

The NBOs are an optimized set of localized bonding, antibonding, and lone pair orbitals. The bond orbitals typically have occupancies of $1.98e$ in molecules with one simple Lewis structure. In C_2H_6, Weinhold and coworkers[36] have shown that the barrier height $V_3 = 1025$ cm^{-1} can be understood in the NBO terms in the form of $\sigma_{CH} \rightarrow \sigma^*_{CH}$ donor–acceptor interactions. When the corresponding off-diagonal Fock matrix elements are set to zero, the barrier essentially *disappears*. Six donor–acceptor interactions become most favorable in the lowest energy, staggered conformation of C_2H_6. Steric interactions play only a secondary role.

At the next level of complexity, the molecule 2-methylpropene (Figure 3) illustrates clearly what happens when methyl is placed between a localized single bond and a localized double bond, without the complication of multiple resonance structures. The calculated barrier is 1010 cm^{-1}, very similar to ethane, with a strong preference for the conformer with one CH bond in the plane of the carbon frame *cis* to the double CC bond. To see the analogy between staggered ethane and the *cis* conformer of 2-methylpropene, we can envision the CC double bond as a pair of *banana bonds* lying above and below the plane of the carbon frame. One CH bond of the rotor then lies "staggered" between the two banana bonds, much as it lies staggered between two CH bonds in ethane. Thus the 2-methylpropene barrier is *ethane-like*. We could analyze it in terms of σ, σ^*, π, and π^* donor–acceptor interactions akin to those in acyclic analogues.

Turning now to the substituted toluenes, we know from experiment and the ab initio calculations that when the local geometry near methyl has only threefold symmetry, substantial barriers invariably occur. We identify three key effects whose importance differs from case to case. First, a substantial difference in bond order between the local ring CC bonds, denoted a and b (see Figure 13), contributes a threefold potential term favoring the conformation with one CH bond *cis* to the shorter CC bond, analogous to the simple case of 2-methylpropene. For a series of molecules that should suffer essentially no steric effects, we used the natural resonance theory (NRT)[37] to calculate the total bond orders of the ring CC bonds nearest methyl, bonds a and b. The NRT fits the one-electron density matrix as a weighted sum of one-electron density matrices, each associated with a particular resonance structure (and a corresponding set of NBOs):

$$\Gamma^{(1)} = \Sigma_i w_i \Gamma_i^{(1)} \tag{4}$$

The mean bond order computed from the resulting *resonance weights* w_i is called the *natural bond order*. We find a good linear correlation between V_3 and the *difference* in natural bond orders, O_b-O_a (Figure 11). The best-fit slope is 950 cm^{-1}/bond.

The methyl group responds to the *difference* in the three-dimensional electron density distribution about the two nearest ring CC bonds, and the natural bond orders most simply quantifies the key difference in a unified manner across many molecules. At one extreme, 2-methylpropene has essentially localized single and double bonds (O_a-O_b ≈ 1) and a 1010 cm^{-1} barrier. At the other extreme, when the geometry of the ring has good *local* C_{2v} symmetry, as in the S_0 state of toluene, *m*-fluorotoluene, *p*-fluorotoluene, 3,5-difluorotoluene, and 2,6-difluorotoluene, O_a ≈ O_b and the barrier is invariably very small, even for nominal threefold cases. We interpret this equality of bond orders as indicative of essentially equal contributions of the two dominant resonance structures at all α.

Second, *ortho*-halogen substitution places a lone pair ($2p$ on F or $3p$ on Cl) in close proximity to the methyl CH bonds. This leads to steric *repulsion* between the lone pair and methyl and also to *attractive* donor–acceptor interactions between the halogen lone pair and σ_{CH}^* antibonds. These interactions are absent in the simpler

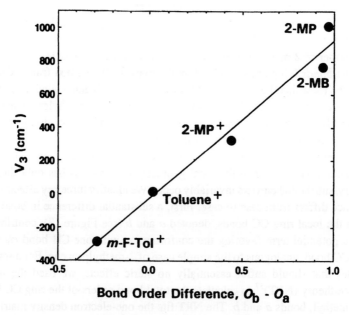

Figure 11. Ab initio potential parameter V_3 vs. the difference in natural bond order, O_b − O_a, between the two ring CC bonds adjacent to the point of methyl substitution.

case of ethane. We find that the donor–acceptor interactions $n_F \rightarrow \sigma^*_{CH}$ mildly favor the pseudo-*cis* geometry ($\alpha = 180°$), as if the halogen and CH bond engage in a weak hydrogen bond. In *o*-fluorotoluene and *o*-chlorotoluene, we have attempted an approximate quantitative resolution of the barrier into additive contributions from donor–acceptor and steric interactions. We combined the established technique of zeroing Fock matrix elements in the NBO representation[33] with the new technique of natural steric analysis (NSA).[30,38] The NSA estimates the steric energy as the increase in total system energy that occurs on the final orthogonalization of the NBOs, in analogy to the repulsion between two He atoms. Our results clearly indicate that in the S_0 state, steric repulsion and donor–acceptor attraction *counteract* each other in *ortho*-substituted toluenes, with the steric interactions dominant.

Finally, π ionization can affect local CC bond order differences and thus methyl rotor barriers in an interesting way. In toluene cations, π ionization creates a pattern of long- and short-ring CC bonds (see Figure 12). Ionization may occur from either of two nearly degenerate π orbitals, which would have symmetry of either b_1 or a_2 under the point-group C_{2v}. *meta* Substitution by the weak π donors fluorine and methyl (as in *m*-fluorotoluene$^+$) *orients* the π-molecular orbitals so that they have approximate C_{2v} symmetry about the pseudo-C_2 axis bisecting the two substituents, as shown in Figure 13. The actual contours of the LUMO of the cation, as revealed in UHF calculations, look remarkably similar to these oriented Hückel orbitals. In forming the ground state of *m*-fluorotoluene$^+$, removal of the a_2 electron *maximizes* positive charge density at the points of substitution, where it can best be stabilized by π donation. This lengthens the two bonds parallel to the pseudo-C_2 axis. Accordingly, the calculated geometry of the *m*-fluorotoluene$^+$ ring (Figure 12) shows two long CC bonds and four short CC bonds oriented exactly as described. Almost as a side effect, this pattern of ring CC π-bond orders causes a substantial difference (O_a-O_b) in local CC bond orders, which produces a large threefold barrier in accord with the correlation of Figure 11! The preferred conformation places one CH bond *cis* to the ring CC bond of higher order, in analogy to the case of 2-methylpropene.

In *o*-fluorotoluene$^+$, the UHF geometries again show the "two-long, four-short" pattern of ring CC bonds (Figure 12), but now the long bonds lie perpendicular to the pseudo-C_2 axis that *bisects* the points of substitution (Figure 13). Similar reasoning explains the orientation of the pattern. Now the local difference in CC π-bond orders favors the pseudo-*trans* conformer, reinforcing the steric interaction with the fluorine lone pair. The qualitative result is a strong preference for the pseudo-*trans* conformer in *o*-fluorotoluene$^+$.

In view of the clear correlation of local ring geometry with methyl rotor barrier height in the S_0 and D_0 states, the strong effects of $S_1 \leftarrow S_0$ excitation on rotor potentials seem to indicate substantial distortion of the ring away from hexagonal symmetry in the S_1 state as well. There is little clear evidence of this from molecular spectroscopy. We have speculated that such a π-*molecular orbital orientation effect* in the S_1 state (similar to that in the cation) might explain the observed characteristic

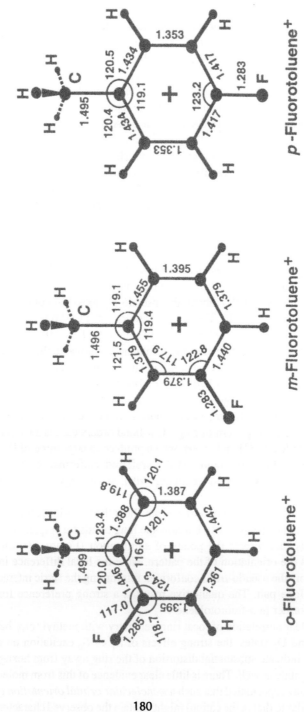

Figure 12. Ab initio optimized cation geometries at UHF/6-31G* level of theory, with α constrained to 90°. Bond lengths in Å and bond angles in degrees.

180

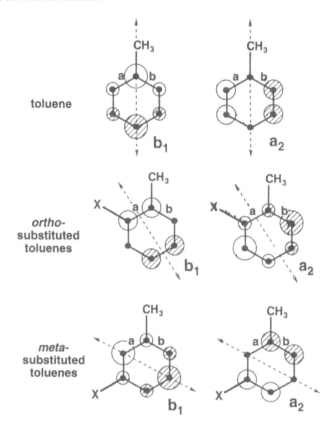

Figure 13. The two highest occupied Hückel molecular orbitals for substituted toluenes assuming pseudo-C_2 symmetry about the axes shown.

changes in barrier height on S_1-S_0 excitation of *ortho-* and *meta*-substituted toluenes. However, we have not yet pursued the multiconfiguration calculations that might show this effect clearly.

Finally, we return to 2-fluoro-6-chlorotoluene (Figure 10), for which $\pi^* \leftarrow \pi$ excitation and π ionization cause only minor changes in the rotor potential. In this disubstituted case, the appropriate pseudo-C_2 axis would coincide with the methyl rotor axis itself. We expect the Hückel orbitals to align with the rotor axis, which would make $O_a \simeq O_b$ in all three electronic states, much as in sixfold cases. Accordingly, there is little effect of electronic excitation on the barrier height, which is presumably dominated by steric repulsion in all three electronic states. In terms of π-electron effects, 2-fluoro-6-chlorotoluene behaves much like a sixfold case.

VI. CONCLUSION

We have seen how the advent of ZEKE-PFI spectroscopy, with its 1 cm^{-1} resolution for cation spectra, has allowed experimental determination of methyl torsional potentials in large molecular cations for the first time. The strong effect of π ionization in m-fluorotoluene$^+$ was sufficiently puzzling to inspire exploration of the electronic structure by ab initio methods. The pattern of long- and short-ring CC bonds in the cation equilibrium geometry then led quite naturally to the notion that weak π-donor substituents direct the orientation of Hückel molecular orbitals along pseudo-C_2 symmetry axes. Our progress provides a good example of how the *combination* of spectroscopic experiments and electronic structure theory can unravel the underlying causes of intramolecular forces in rather complex molecules.

At least two areas require further exploration. The first is the nature of S_1, the first excited singlet state. Excitation to the S_1 state has a strong effect on the methyl rotor potential, but we do not understand why. We know remarkably little about the geometry of the S_1 state from either theory or experiment. In the simple Hückel picture, for C_{2v} geometry there are *two* excitations of comparable energy leading to an excited state of the same symmetry as S_1. This suggests that we will need to carry out multiconfiguration calculations to get a sensible description of the S_1 state geometry and thus of the methyl rotor potential.

The effect of *strong* π-donor substituents on geometry and methyl rotor potentials deserves theoretical and experimental attention. In particular, earlier work shows dramatic effects of amino (-NH$_2$) and hydroxyl (-OH) groups.[9] The barriers in the S_0 state and the changes in barrier height on excitation to the S_1 state are unusually large compared to those of the weak π donors F and Cl. Fujii and coworkers[39] are presently studying the corresponding cations by ZEKE-PFI. Hydroxyl and especially amino π orbitals probably mix substantially with the ring π orbitals, so we must extend our thinking beyond the subtle orientation effects that work for fluorine and chlorine. Ab initio calculations on these systems should again be revealing.

ACKNOWLEDGMENTS

The theoretical aspects of this work were carried out in collaboration with Prof. Frank Weinhold, Prof. Ned Sibert, and Dr. Jay Badenhoop. We have benefitted enormously from the skill and enthusiasm of these colleagues. JCW thanks the U.S. Department of Energy, Division of Chemical Sciences, Office of Basic Energy Research (No. DE-FG02-92ER14306) for generous support of this research. The National Science Foundation provided equipment used in the experimental work. The IBM Corporation generously donated two RS6000 workstations under the IBM-SUR program.

REFERENCES

1. Kemp, J. D.; Pitzer, K. S. *J. Chem. Phys.* **1936**, *4*, 749. Smith, L. G. *J. Chem. Phys.* **1949**, *17*, 139. Wilson, E. B., Jr. *Adv. Chem. Phys.* **1959**, *2*, 367.

2. Durig, J. R.; Guirgis, G. A.; Bell, S. *J. Phys. Chem.* **1989**, *93*, 3487.

3. Rudolph, H.; Dreizler, A.; Jaeschke, W.; Wendling, P. Z. *Naturforsch.* **1967**, *22A*, 940.

4. Murakami, J.; Ito, M.; Kaya, K. *Chem. Phys. Lett.* **1981**, *80*, 203. Okuyama, K.; Mikami, N.; Ito, M. *J. Phys. Chem.* **1985**, *89*, 5617. Mizuno, H.; Okuyama, K.; Takayuki, E.; Ito, M. *J. Phys. Chem.* **1987**, *91*, 5589.

5. Trinkers, P. J.; Parmenter, C. S.; Moss, D. B. *J. Chem. Phys.* **1994**, *100*, 1028; and references therein.

6. Breen, P. J.; Warren, J. A.; Bernstein, E. R.; Seeman, J. I. *J. Chem. Phys.* **1987**, *87*, 1917.

7. Spangler, L. H.; Pratt, D. W. In Hollas, J. M.; Phillips, D., Eds.; *Jet Spectroscopy and Molecular Dynamics;* Chapman and Hall: London, 1994.

8. Lin, T.-Y.; Tan, X.-Q.; Cerny, T. M.; Williamson, J. M.; Cullin, D. W.; Miller, T. A. *Chem. Phys.* **1992**, *167*, 203, and references therein.

9. Zhao, A.-Q.; Parmenter, C. S.; Moss, D. B.; Bradley, A. J.; Knight, A. E. W.; Owens, K. G. *J. Chem. Phys.* **1992**, *96*, 6362. See Tables VI and VII in this reference for summaries of known threefold and sixfold barriers.

10. Lu, K.-T.; Eiden, G. C.; Weisshaar, J. C. *J. Phys. Chem.* **1992**, *96*, 9742.

11. Lu, K.-T.; Weisshaar, J. C. *J. Chem. Phys.* **1993**, *99*, 4247.

12. Takazawa, K.; Fujii, M.; Ebata, T.; Ito, M. *Chem. Phys. Lett.* **1992**, *189*, 592.

13. Takazawa, K.; Fujii, M.; Ito, M. *J. Chem. Phys.* **1993**, *99*, 3205.

14. Müller-Dethlefs, K.; Sander, M.; Schlag, E. W. *Chem. Phys. Lett.* **1984**, *112*, 291.

15. Müller-Dethlefs, K.; Schlag, E. W. *Ann. Rev. Phys. Chem.* **1991**, *42*, 109.

16. Powis, I.; Baer, T.; Ng, C.-Y. (Eds.). *High Resolution Laser Photoionization and Photoelectron Studies*; John Wiley: Chicester, 1995.

17. See, for example: Ho, J.; Polak, M. L.; Lineberger, W. C. *J. Chem. Phys.* **1992**, *96*, 144.

18. Sanders, L.; Hanton, S. D.; Weisshaar, J. C. *J. Chem. Phys.* **1990**, *92*, 3485.

19. Tonkyn, R. G.; Winniczek, J. W.; White, M. G. *Chem. Phys. Lett.* **1989**, *164*, 137.

20. Stebbings, R. F.; Dunning, F. B. (Eds.) *Rydberg States of Atoms and Molecules*; Cambridge University: New York, 1983.

21. Biernacki, D. Th.; Colson, S. D.; Eyler, E. E. *J. Chem. Phys.* **1988**, *88*, 2099.

22. Pratt, S. T. *J. Chem. Phys.* **1993**, *98*, 9241.

23. Chupka, W. A. *J. Chem. Phys.* **1993**, *98*, 4520.

24. Vrakking, M. J. J.; Lee, Y. T. *J. Chem. Phys.* **1995**, *102*, 8818. Vrakking, M. J. J.; Fischer, I.; Villeneuve, D. M.; Stolow, A. *J. Chem. Phys.* **1995**, *103*, 1.

25. Wilson, E. B.; Lin, C. C.; Lide, D. R. *J. Chem. Phys.* **1955**, *23*, 136. Gordy, W.; Cook, R. L. *Microwave Molecular Spectra*, 3rd edn,; Wiley-Interscience: New York, 1984.

26. Longuet-Higgins, H. C. *Mol. Phys.* **1963**, *6*, 445; Bunker, P. R. *Molecular Symmetry and Spectroscopy*; Academic: New York, 1979.

27. Walker, R. A.; Richard, E.; Lu, K.-T.; Sibert III, E. L.; Weisshaar, J. C. *J. Chem. Phys.* **1995**, *102*, 8718.

28. Walker, R. A.; Richard, E. C.; Lu, K.-T.; Weisshaar, J. C. *J. Phys. Chem.* **1995**, *99*, 12422.

29. Schwoch, D.; Rudolph, H. D. *J. Mol. Spectrosc.* **1975**, *57*, 47.

30. Lu, K. T.; Weinhold, F.; Weisshaar, J. C. *J. Chem. Phys.* **1995**, *102*, 6787.

31. Walker, R. A.; Richard, E. C.; Weisshaar, J. C. *J. Chem. Phys.* **1996**, *104*, 4451.

32. Richard, E. C.; Walker, R. A.; Weisshaar, J. C. *J. Phys. Chem.* **1996**, *100*, 7333.

33. Reed, A. E.; Curtiss, L. A.; Weinhold, F. *Chem. Rev.* **1988**, *88*, 899.

34. Pitzer, R. M. *Accts. Chem. Res.* **1983**, *16*, 207.

35. Dorigo, A. E.; Pratt, D. W.; Houk, K. N. *J. Am. Chem. Soc.* **1987**, *109*, 6591. Goodman, L.; Leszczynski, J.; Kundu, T. *J. Am. Chem. Soc.* **1993**, *115*, 11991; and references therein.

36. Corcoran, C. T.; Weinhold, F. *J. Chem. Phys.* **1980**, *72*, 2866.

37. Glendening, E. D.; Hrabal *Abs. Amer. Chem. Soc.* **1997**, 131-Phys. Part 2, 213.

38. Badenhoop, J. K.; Weinhold, F. *J. Chem. Phys.* **1997**, *107*, 5406.

39. Takazawa, K.; Fujii, M. Work in progress.

THE INFLUENCE OF SOLVATION ON ION–MOLECULE REACTIONS

A. W. Castleman, Jr.

Advances in Gas-Phase Ion Chemistry
Volume 3, pages 185–253
Copyright © 1998 by JAI Press Inc.
All rights of reproduction in any form reserved.
ISBN: 0-7623-0204-6

ABSTRACT

Cluster research is a prominent area of investigation in the field of chemical physics, motivated in part by realization of its value in bridging an understanding of phenomena in the gas and condensed phases. The study of cluster ions at varying degrees of solvation is providing new insights into the influence of solvation on the course of ion–molecule reactions. This review displays the breadth of the subject and shows some of the important advances that have been made through the use of various new experimental techniques. Particular consideration is given to aspects related to reaction dynamics and kinetics, also showing how thermochemical and structural information can be derived.

I. INTRODUCTION

One of the scientifically challenging problems in the field of chemical physics is determining the influence which solvation has on the reactions and properties of ionic species.[1,2] There is particular interest in contrasting differences in the behavior and reactivity of ions in the gaseous compared to the condensed phase. In this context, studies of cluster ions and of the "ensuing solvation" that takes place upon ionizing a moiety within a cluster yield information on basic mechanisms of ion reactions within a cluster, and also give considerable insight into those that can occur in the bulk liquid state.[3-22] In terms of basic knowledge, the results of studies of the interaction of ions and molecules provide information on the forces involved, and gas-phase studies of interactions within a cluster contribute to knowledge about the structure and bonding of complexes having analogies to those existing in solutions.

The entire course of a chemical reaction following either a photophysical or ionizing event depends on the mechanisms of energy transfer and dissipation away from the primary absorption site. Neighboring solvent or solute molecules can influence this by collisional deactivation (removal of energy), through effects in which dissociating molecules are kept in relatively close proximity for comparatively long periods of time due to the presence of the solvent (caging), and in other ways where the solvent influences the energetics of the reaction coordinate. Through the use of supersonic molecular beams, it is now possible to produce and tailor the composition of virtually any system of interest. In conjunction with laser spectroscopy, it is possible to selectively solvate a given chromophore (site of photon absorption/site of ionization) and investigate changes in reactivity between the gas and the condensed phase by selectively shifting the degree of solvent aggregation, i.e. the number and in some cases the location of solvent molecules attached to, or bound about, the site of absorption of the electromagnetic radiation. Findings from such studies, together with results obtained on cluster ion kinetics using conventional flow tube reactor studies, are yielding extensive new insights into the dynamics of ion–molecule reactions of solvated systems. Advances in this area potentially impact an enormous range of scientific and applied fields, two in particular being radiological physics and atmospheric science.

This review deals largely with work from my own laboratory. It attempts to show the reader some of the recent developments in the field and the breadth of the scientific questions which are being addressed through investigations of the kinetics and dynamics of ion–molecule reactions as mediated through the presence of bound solvent molecules.

II. EXPERIMENTAL TECHNIQUES

Developments in experimental and theoretical techniques have paved the way for the many advances which have been made in understanding the dynamics and properties of ionic clusters. For instance, the flowing afterglow, stationary afterglow, and ion cyclotron resonance methods have been widely utilized to investigate association reactions responsible for cluster ion formation, as well as their reactivities. The aforementioned techniques also have been employed to derive thermochemical properties, although the most extensive sets of thermodynamic data have been obtained via the high-pressure mass spectrometer technique.[1,23,24] Through the use of lasers, studies of cluster ion dissociation spectroscopy and processes involved in the dynamics of photodissociation are now being unraveled at the molecular level.[25-32] Concomitant investigations of cluster ion unimolecular and collision-induced dissociation are further contributing to an understanding of dynamical processes involved in energy transfer and reactivity.[33-43] Finally, ab initio calculations are contributing to an understanding of the structure and bonding of both strongly and weakly bound cluster ions.[44]

A. Fast Flow Reactions for Studying Reaction Kinetics Under Thermal Conditions

The flowing afterglow technique (FA) developed by Ferguson, Fehsenfeld, and Schmeltekopf[45] and other related flow reactors such as the selected ion flow tube (SIFT)[46] have provided a wealth of data on general ion–molecule reactions,[47,48] with some attention to ion clusters. A typical fast-flow apparatus such as employed in our laboratory is shown in Figure 1; the flow tube is about 1 m long and 8 cm in diameter. In typical experiments, flow velocities are on the order of 10^2 ms^{-1} and pressures in the reaction region are maintained around 0.5 torr. While most of the gas is pumped away by mechanical pumps, a small fraction is sampled through an orifice, allowing the ions to be mass identified and counted. Reactant gases are added uniformly into the flow, so kinetic data (or the approach to equilibrium) can be determined by varying the concentration of reactant added, the position of addition into the reactor, or the bulk flow velocity. In comparison to conventional high-pressure mass spectrometry techniques, the flow tube method affords more versatility in making kinetic measurements and identifying mechanisms. On the other hand, high-pressure mass spectrometry enables measurements at higher pressures and is generally more amenable to conducting equilibrium measurements.

The ions or cluster ions are thermalized by collisions with an inert carrier gas (usually helium), although often argon or even nitrogen is employed. Neutral reactant gas is added through a reactant gas inlet at an appropriate location downstream in the flow tube, and allowed to react with the injected ions. Rate coefficients, k, are determined by establishing pseudo-first-order reaction conditions in which the reactant ion concentration is small compared to the reactant neutral concentration. Bimolecular rate coefficients, k, are obtained from the slope of the natural logarithm of the measured signal intensity, I, of the reactant ion versus the flow rate Q_B of reactant gas:[45,48-50]

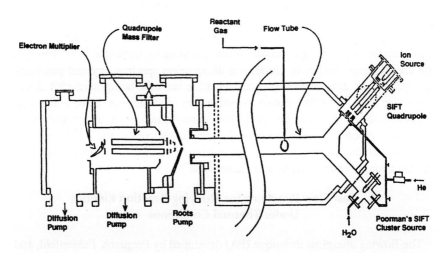

Figure 1. Fast-flow reaction apparatus. Ions or ion clusters are introduced into the flow tube from various sources and reactions proceed after they encounter the reactants added through a ring injector located at a selected position in the flow tube. The disappearance of the reactant ions and formation of products is monitored with the quadrupole mass spectrometer/electron multiplier shown. Taken with permission from ref. 19.

$$\ln\left(\frac{I}{I_o}\right) = -\left(\frac{kzQ_B}{V_iQ_c}\right)\frac{P}{k_BT} \tag{1}$$

Here I_o is the reactant ion intensity at Q_B equal to zero (no reactant gas flow), z the reaction distance (from reactant gas inlet to sampling orifice), Q_c the carrier gas flow rate, P the average pressure (or number density) in the flow tube, V_i the measured ion velocity, k_B the Boltzmann constant, and T the absolute temperature. V_i can be determined by applying a pulsed potential on the reactant gas inlet and measuring the arrival time at the electron multiplier of the resulting disturbance in the ion intensity (see Figure 1). Where appropriate, termolecular rate coefficients[50] are determined from the slope of the apparent bimolecular rate coefficient plotted versus the pressure P. In addition to the determination of rate coefficients, product identification and measurement of branching ratios can also be accomplished.[46,51,52]

B. Molecular Beam–Photoionization–Time-of-Flight Mass Spectrometry

Studying Dynamics and Bonding

The time-of-flight (TOF) mass spectrometer technique is experiencing a resurgence in popularity due to the advent of pulsed lasers, which supply efficient and short durations of ionization in a small volume, and the availability of fast-timing electronic circuitry. In a typical time-of-flight mass spectrometer, either a two element or alternately a single-element accelerating field may be used in the region of ionization. This is followed by a field-free drift region, whereafter the ions are detected. Using the conventional TOF method, dissociation which occurs with rates in the neighborhood of 10^5–10^8 s^{-1} can be investigated by either of two methods. One involves analyzing the peak shape of ions created in a dual field accelerating Wiley–McLaren,[53] in which cases a knee is observed[54] in the arrival spectrum. This occurs due to the fact that ions spend far more time in the first low-field region where ionization is initiated, than in the second high-field region where the bulk of the acceleration occurs. An alternate method is to operate under single-field conditions and deduce rates from the shape of the late arriving tail of the peak.[55]

One of the most useful methods of studying dissociation employs a reflecting electrical field (reflectron). Although originally designed to enhance the resolution of the TOF method,[56] we recognized that a reflectron also can be employed to investigate dissociation in the field-free drift region, so that slower dissociation processes may be observed by separating the parents and daughters.[33] Such experiments are performed by subjecting the cluster beam to multiphoton ionization in the acceleration lens of the TOF, oftentimes using a tunable dye laser with various optical components that provide further frequency selection capabilities. The ions are accelerated to several keV, whereafter they enter the first field-free region and then are electrically reflected and detected in a manner such as the one depicted in Figure 2. With appropriate potentials applied to the reflectron grids, nondissociating

(a)

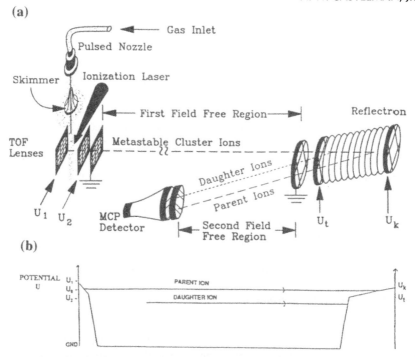

(b)

Figure 2. (a) Reflectron TOF mass spectrometer. (b) Depicts the electrostatic potentials. With a judicious selection of potential, the daughter ions arising from metastable decay arrive at the detector prior to the parent ions which have higher kinetic energy. MCP denotes a microchannel plate charged particle detector. (a) Taken with permission from ref. 22; (b) Taken with permission from ref. 19.

parent ions can be separated from those that dissociate while within the field-free region. A unique identification of these daughter ions can be accomplished by the time separation and by an energy analysis with the reflectron. The separation of the parent and daughter ions is possible as a result of the loss in kinetic energy with essentially no change in velocity of the cluster ion packet upon dissociation, whereby the parent species with greater kinetic energy have a longer path to the detector than do the daughter (dissociation) products (see Figure 3).

Supersonic expansion techniques including both continuous sources as well as pulsed jets are commonly used to produce beams of neutral clusters. In both cases, cooling of the beam is accomplished through the conversion of the random thermal energy of a high-pressure source gas into a directed beam velocity.[57] The latent heat of condensation released during the clustering process leads to internal vibrational and rotational heating of the aggregate, and clusters often do not attain internal temperatures as low as unclustered species. But, cooling collisions with an inert gas serve to reduce the internal temperature of the cluster and enable ones to be produced that have sufficiently long lifetimes to be interrogated in an experiment.

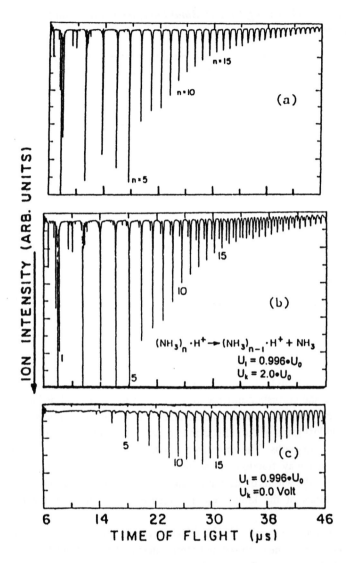

Figure 3. (a) Conventional TOF mass spectrum of $H^+(NH_3)_n$ clusters from the multiphoton ionization of ammonia clusters at 266 nm. (b) Spectrum Taken using the reflectron shown in Figure 2(a). The potentials U_t and U_k are chosen such that all ions are reflected but with the daughter ions arriving at the particle detector earlier than their corresponding parent ions. (c) U_k is lowered to eliminate the parent ions which pass through the reflectron. Only the daughter ions are reflected to the detector. Taken with permission from ref. 2.

Technique for Deducing Cluster Bond Energies from Unimolecular Dissociation Studies

The general versatility of the evaporative ensemble approach for deducing bond energies was first demonstrated[35,43] for ammonia cluster ions, though it is a technique of general applicability. Following ionization of neutral clusters, internal ion–molecule reactions lead to specific ion clusters. In the case of ammonia, the mass spectra are dominated by protonated ammonia clusters. These species experience rapid fragmentation and dissociation in the TOF acceleration region (a time window of typically no more than 1 μs), and thereafter undergo subsequent evaporative cooling through dissociation which extends to longer times. A general cluster ion evaporative dissociation process can be expressed as:

$$IL_n \rightarrow IL_{n-x} + xL \qquad (2)$$

Here, I designates the ion core (NH_4^+ in the case of ammonia) and L the clustering ligand (e.g. NH_3). The intensity and width of the metastable ion peaks carry information on the internal energy of the parent cluster ions.

In the measurement of decay fractions of dissociating cluster ions, the parent and daughter ions are decelerated in the first region (between ground and U_t in Figure 2a and reflected in the second field (between U_t and U_k) of the reflectron. Since the daughter ions have an energy of $U_d = (M_d/M_p)U_o$ as a result of metastable decomposition (U_o is the birth potential, see Figure 2b), and M_d and M_p are the masses of the daughter and parent, respectively), they do not penetrate into the reflection field as deeply as the corresponding parent ions. A critical aspect of deducing accurate kinetic energy release and rate measurements is to vary potential settings on the second and last grids of the reflectron to cause parent and daughter ions to follow the same flight paths. The integrated intensities of the peaks are then used to compute the decay fraction of the original parent cluster. Through studies of differences in the widths of the parents and daughter signals, it is possible to deduce accurate values of the kinetic energy release associated with the dissociation process.

The evaporative ensemble[35,43,58–61(a)] approach assumes that each cluster ion has undergone at least one evaporation before entering the first field-free region of the time-of-flight mass spectrometer. The symbol t_o is the flight time that the parent ion spends from the point of ionization to the last TOF lens (see Figure 2a), and t is the flight time that the parent ion spends from the last TOF lens to the first grid of the reflectron unit. At time t, the remaining population of dissociating cluster ions is given by $P = P_o - D$, where P_o is the population of parent ions at time t_o and D is the population of daughter ions at time t. Based on the evaporative ensemble, the normalized population of daughter ions at time t is given by,

$$D = (C_n/\gamma^2) \ln \{t/[t_o + (t - t_o) \exp(-\gamma^2/C_n)]\} \qquad (3a)$$

where C_n is the heat capacity of the cluster ion (in units of Boltzmann constant k_B), and γ is the Gspann parameter. Electron diffraction experiments on clusters containing many thousands of atoms suggest that γ is ~25, usually independent of cluster size.[35,59,61(a)] However, for small clusters, the Gspann parameter must be modified as follows:

$$\gamma'^2 = \gamma^2/[1 - (\gamma/2C_n)^2] \tag{3b}$$

Replacing γ with the modified Gspann parameter γ' in equation 3a leads to:

$$D = (C_n/\gamma'^2) \ln \{t/[t_0 + (t - t_0) \exp(-\gamma'^2/C_n)]\} \tag{3c}$$

For systems comprised of nonlinear molecules, the heat capacity of the cluster ion of size n is taken to be $6(n - 1)$ (in units of the Boltzmann constant) by considering (only) the cluster modes. The binding energy of a molecule in a cluster ion of size n can be calculated from the equation,

$$\Delta E_n = \gamma <E_r> / [1 - (\gamma/2C_n)] \tag{4}$$

where $<E_r>$ is the average kinetic energy release upon metastable dissociation, a quantity which can be determined experimentally as discussed later in this chapter. Klots has also shown[61(b)] that the daughter ion population can be related to the binding energies of clusters of sizes n and $n + 1$ via,

$$D = 1 - (\alpha W_n)^{-1} \ln\{1 + [\exp(\alpha W_n) - 1]t_0/t\} \tag{5}$$

where:

$$\alpha W_n = \gamma^2(W_n/\Delta E_n)/\{C_n[1 - \gamma/C_n + (\gamma/C_n)^2/12 \ldots]^2\} \tag{6a}$$

$$W_n/\Delta E_n = 1 + [(dE/d\Delta E_n)_k^{-1}](\Delta E_n - \Delta E_{n+1})/\Delta E_n \tag{6b}$$

$$(dE/d\Delta E_n)_k = (C_n/\gamma)[1 - \gamma/2C_n + (\gamma/C_n)^2/12 \ldots]. \tag{6c}$$

By fitting the calculated decay fractions to the measured ones, the values of $(\Delta E_n - \Delta E_{n+1})/\Delta E_n$ are readily determined.

Ultrafast Time Resolved Spectroscopy

The advent of ultrafast pump-probe laser techniques[62] and their marriage with the TOF method also enables study of internal ion–molecule reactions in clusters.[21,63–69] The apparatus used in our experiments is a reflectron TOF mass spectrometer coupled with a femtosecond laser system. An overview of the laser system is shown in Figure 4. Femtosecond laser pulses are generated by a colliding pulse mode-locked (CPM) ring dye laser. The cavity consists of a gain jet, a

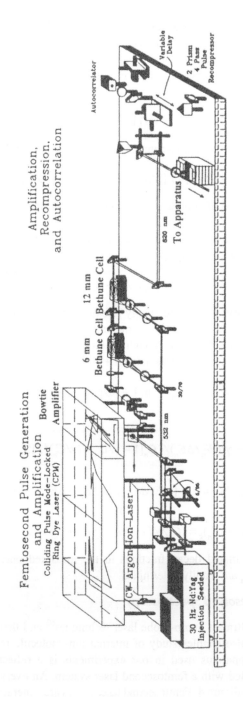

Figure 4. A schematic of the colliding pulse mode-locked femtosecond laser system. Taken with permission from ref. 65.

194

saturable absorber jet, and four recompression prisms. The gain dye is pumped with 5 W, all lines, from an Innova 300 argon ion laser. In order to generate short pulses, passive mode-locking is performed. Four recompression prisms are used to compensate for group velocity dispersion. The output wavelength, pulse width, and energy are ~624 nm, ~100 fs, and ~200 pJ, respectively.

In order to supply the large photon flux needed for multiphoton ionization (MPI), the laser pulses are amplified through additional stages. Each of these is pumped with the second harmonic (532 nm) from an injection seeded GCR-5-30Hz Nd:YAG laser, which is synchronized with the femtosecond laser. The first stage of amplification is a bowtie amplifier. The beam makes six passes through the dye cell giving a total amplification to ~10 µJ per pulse. A white light water cell is employed to generate light in the continuum. After continuum generation, the appropriate wavelength is selected with a 10 nm bandwidth interference filter. The second and third stages of amplification are performed with 6 mm bore prism dye cells, termed Bethune cells. After amplification, the beam is split into pump and probe beams. The probe beam is sent through a delay stage which can be varied from 0.1 µm to 1 nm. Thereafter, the beams are recombined using another 45° high reflector. A Michelson interferometric arrangement is used to set the time delay between the pump and probe beams. Sequential pump energies per pulse typically acquired for the amplification stages are ~33 µJ, ~100 µJ, and ~250 µJ.

After recombination, the laser beams are focused into the interaction region with a 50 cm lens, where they intersect the molecular beam containing the neutral ammonia clusters which are produced via supersonic expansion through a pulsed valve (Figure 2a). The ions formed in the multiphoton ionization process are accelerated in a standard Wiley–McLaren double-electric field arrangement. The ions are directed through the first field free region, which is ~1.5 m long, toward a reflectron. Ions are then reflected, whereupon they travel through a second field-free region which is ~0.5 m long. They are thereafter detected by a chevron microchannel plate detector. The signals received by the detector are directed into a digital oscilloscope coupled to a personal computer.

III. REACTIONS INFLUENCED BY SOLVATION

There are several classes of reactions in which solvation influences reactivity.[2,17,19] These include: (a) solvation effects on the nature of the core ion reaction site, (b) solvation effects on exothermicity (or exoergicity), (c) site specific solvation blocking, and (d) solvation influences on the energy barrier to reaction. Investigations of the influence of clustering on the sites of energy absorption and the ensuing reaction mechanisms are particularly pertinent in further elucidating these classes.

A. Elucidating the Influence of Solvation on the Site of Energy Absorption and Ensuing Ion–Molecule Reactions via Ultrafast Laser Pump-Probe Techniques

The Formation of Protonated Ammonia Clusters: A Paradigm

Through the use of pump-probe techniques pioneered by Zewail and coworkers,[62] it is becoming possible to identify the detailed mechanisms of reactions at the molecular level and follow the actual course of a reaction. The study of ammonia clusters has provided an example of what can be accomplished using these techniques.

The ionization of ammonia clusters (i.e. multiphoton ionization,[33,35,43,70,71] single photon ionization,[72-74] electron impact ionization,[75] etc.) mainly leads to formation of protonated clusters. For some years there has been a debate about the mechanism of formation of protonated clusters under resonance-enhanced multiphoton ionization conditions, especially regarding the possible alternative sequences of absorption, dissociation, and ionization. Two alternative mechanisms[63,64,76,77] have been proposed: absorption–ionization–dissociation (AID) and absorption–dissociation–ionization (ADI) mechanisms; see Figure 5.

Based on the AID mechanism, the neutral clusters are ionized through multiphoton ionization and the protonated clusters are formed through the intracluster ion–molecule reaction between NH_3^+ and the neighboring NH_3 species as expressed in the following:

$$(NH_3)_n + rh\nu \rightarrow (NH_3)_n^* \tag{7}$$

$$(NH_3)_n^* + r'h\nu \rightarrow (NH_3)_n^+ \tag{8}$$

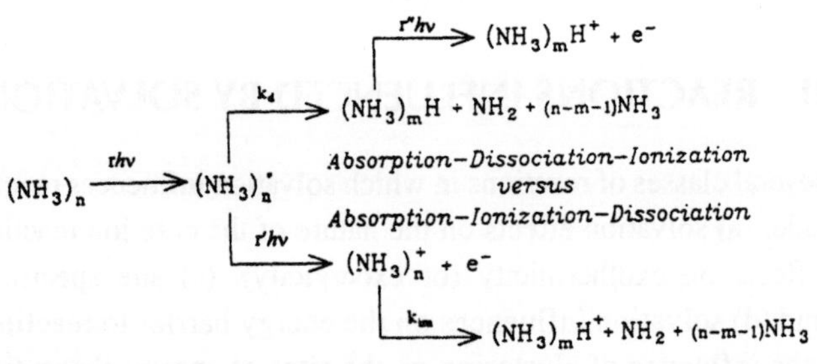

Reaction Schemes of Ammonia Clusters

Figure 5. Possible reaction mechanisms leading to the formation of protonated ammonia clusters from neutral clusters. Taken with permission from ref. 65.

$$(NH_3)_n^+ \rightarrow (NH_3)_m H^+ + NH_2 + (n - m - 1)NH_3 \qquad (9)$$

On the other hand, the alternative ADI mechanism views the processes as occurring where the excited neutral species are also first created through excitation processes. Thereafter, a subsequent intracluster neutral–neutral reaction leads to formation of hydrogenated clusters, $(NH_3)_n H$. Following excitation (reaction 7), ionization of the radical species then results in the observed protonated clusters ions as depicted as follows:

$$(NH_3)_n^* \rightarrow (NH_3)_m H + NH_2 + (n - m - 1)NH_3 \qquad (10)$$

$$(NH_3)_m H \rightarrow NH_4(NH_3)_{m-1} \qquad (11)$$

$$NH_4(NH_3)_{m-1} + r''h\nu_2 \rightarrow NH_4^+(NH_3)_{m-1} + e^- \qquad (12)$$

Since protonated clusters are also produced under electron impact and single photon ionization conditions, the AID mechanism must be operative to some extent. However, theoretical calculations[78,79] supported by the experimental finding that hydrogenated ammonia clusters do have lifetimes of at least a few microseconds following neutralization of the protonated cluster ions,[80] led to the proposal of the ADI mechanism. Nanosecond pump-probe experiments have been carried out[76] to investigate the ionization and dissociation mechanisms of ammonia clusters. By observing the ion signals in a reflectron TOF mass spectrometer as a function of the delay time between the two laser pulses, it was found that the lifetime of the

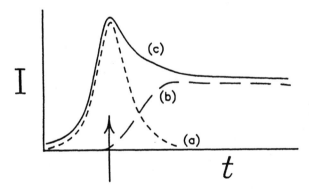

Figure 6. Possible intensity profiles for two mechanisms for the formation of ion clusters: (a) Ionization through AID mechanism. (b) Ionization through ADI mechanism. The signal would persist for long times due to the lifetime of the NH4 in the cluster, and its ensuing ionization. (c) Ionization through both AID and ADI mechanisms. Taken with permission from ref. 65.

intermediate to the formation of $(NH_3)_n H^+$, $n > 1$, is at least 500 times longer than that to the formation of NH_4^+ and $(NH_3)_2^+$. However, due to the broad nanosecond laser pulses employed in the experiments, the issue of ionization and formation mechanisms of protonated ammonia cluster ions was not resolved until a recent series of femtosecond pump-probe experiments were conducted in our laboratory[63,64,67–69] and elsewhere.[81]

The possible profiles of intensity versus the delay between the pump and probe photons for the several potential processes operative in the mechanism of ionization are shown in Figures 6a, b, and c. Several different laser schemes have been employed to investigate the reaction dynamics pertaining to the formation of protonated ammonia clusters through the Ã and C̃' states (the latter possibly including the B̃ state as well). The ionization schemes employed in the present study are shown in Figure 7.

Consider the dynamics of ionization of clusters through the C̃' state of an ammonia molecule, where it should be noted that it is also possible that the excitation leads to some population of the B̃ states due to the broad spectral bandwidth of the femtosecond laser pulses. The measurements indicate lifetimes

Pump-Probe Schemes

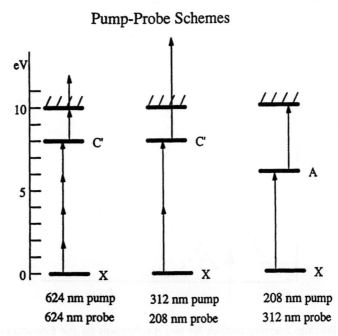

Figure 7. Several of the ionization schemes employed in various pump-probe experiments of ammonia clusters; the energy levels correspond to those of the ammonia monomer. The upper hatched region denotes the ionization limit. Taken with permission from ref. 68.

of less than ~150 femtoseconds for the species $(NH_3)_n^+$ and $(NH_3)_nH^+$ ($n = 1$–4). Moreover, the data reveal virtually no tail in the distribution corresponding to curve 6c, which is a composite and would then indicate contribution from 6b if it were operative. These short lifetimes only can be explained by the AID mechanism. In considering the ADI mechanism, the neutral species $(NH_3)_nH$, if present, would be formed by the predissociation of an ammonia moiety; it would be a long-lived species with a microsecond lifetime.[80] Direct ionization would certainly lead to a sharp intensity peak at $t = 0$, and some initial drop in signal intensity would be expected to be observed irrespective of the mechanism. Predissociation of ammonia would cause the signal intensity to display at least an initially diminishing trend with time. The failure to observe any ionization attributable to that of NH_4 incorporated in the cluster via the predissociation of NH_3 to NH_2 and H, and subsequent reaction of H with NH_3, eliminated ADI as the major formation mechanism of protonated clusters in the $\tilde{C}'$ state.

The results for the $\tilde{A}$ state show that a different mechanism is operative. A series of femtosecond pump-probe experiments were performed at wavelengths corresponding to the Rydberg states $\tilde{A}$ ($v = 0, 1, 2$) of ammonia molecules.[64–66,68,69] The wavelengths used to access these vibrational levels were 214 nm, 211 nm, and 208 nm for the pump laser and 321 nm, 316.5 nm, and 312 nm for the probe laser, respectively.

Figure 8 shows a typical pump-probe spectrum of protonated clusters, $(NH_3)_2H^+$ and $(NH_3)_5H^+$, through a vibrational level corresponding to $v = 2$ of the $\tilde{A}$ state of the monomer. The pump-probe spectra through other vibrational levels, $v = 0$ and 1, show similar response. All of these spectra have some features in common, namely they display a large increase in intensity at $t = 0$ (maximum temporal overlap between the pump and probe pulses), and thereafter a subsequent rapid intensity drop. However, the various spectra do display some noticeable differences with regard to the shape of the fall-off region following the initial substantial peak. Except possibly for $v = 0$, when the vibrational energy of the $\tilde{A}$ states increases, the long-time intensity level of all cluster ions increases. More importantly, the difference between $(NH_3)_2H^+$ and $(NH_3)_5H^+$ becomes evident at higher vibration levels, i.e. $v = 1$ and $v = 2$.

By contrast to the $\tilde{C}'$ state results, a substantially asymmetric transient curve is found for experiments conducted through the $\tilde{A}$ state, which have both a size (solvation) and vibrational level dependence. Experiments have been conducted over a wide range of cluster sizes extending to $n = 25$. For the $\tilde{A}$ state, the pump-probe experiments show two distinct features with respect to the pump-probe delays. A fast decay process is evident, but features showing a leveling off to a non-zero value of ion intensity are also observed (see Figures 8a and b, for example). These observations indicate a competition between the AID and the ADI mechanisms in the case of the $\tilde{A}$ state. The leveling off to a non-zero value has been observed to persist for more than one nanosecond, consistent with expectations based on a contribution from the process represented by Figure 6b.

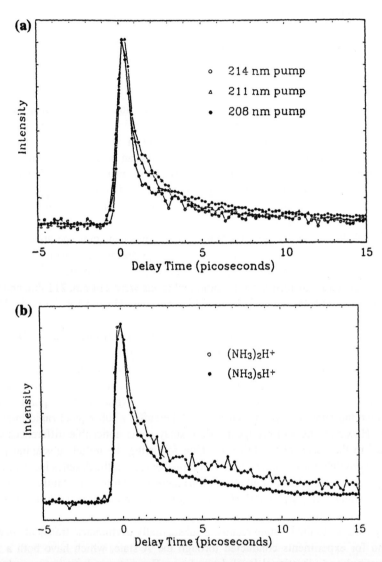

Figure 8. (a) Pump-probe spectra of $(NH_3)_2NH^+$ through the $\tilde{A}$ ($v = 0,1,2$; corresponding to 214, 211, 208 nm, respectively) states; the data reveal the influence of the
vibrational level probed in the experiments. (b) Pump-probe spectrum of $(NH_3)_2H^+$
and $(NH_3)_5H^+$ with pump pulses at 208 nm and probe pulses at 312 nm; $\tilde{A}$ ($v = 2$) of
the ammonia molecule. The role of cluster size is evident. The delay time is the interval
between the pump and probe laser. (a) Taken with permission from ref. 65; (b) Taken
with permission from ref. 68.

The time response features observed for the $\tilde{A}$ state are consistent with the following dynamical processes.

1. The neutral clusters are excited to the $\tilde{A}$ state through absorption of the first photon:

$$(NH_3)_n + h\nu_1 \rightarrow (NH_3)_n^* \tag{13}$$

   Thereupon, the excited clusters undergo intracluster reactions (below).

2. Predissociation of the excited ammonia moiety:

$$(NH_3)_n^* \rightarrow (NH_3)_{n-2} \cdot H_3N \cdot (H \cdots NH_2) \tag{14}$$

3. The intermediate species can lead to the loss of H or NH_2, or reaction of the H to form NH_4.

4. Ionization of the species $(NH_3)_{n-2} \cdot NH_4$ leads to formation of protonated cluster ions:

$$(NH_3)_{n-2} \cdot NH_4 + h\nu_2 \rightarrow (NH_3)_{n-2} \cdot NH_4^+ + e^- \tag{15}$$

5. Competition with direct ionization via an AID process as for the $\tilde{C}'$ state.

The results, which show a rapid decay and leveling off to a non-zero value of intensity, suggest that two processes are operating simultaneously in the $\tilde{A}$ state. It is known[82] that ammonia clusters rapidly predissociate into NH_2 + H and the rapid decay that is observed in the pump-probe spectra suggests that a similar predissociation process is taking place in the clusters. It is also known[80] that the radicals, $(NH_3)_n NH_4$, have long lifetimes (greater than 1 μs), and evidence suggests that formation of these radicals is taking place through intracluster reactions between H and NH_3. This is seen in the leveling off to a non-zero value which persist for longer than 1 ns. Unlike the $\tilde{C}'$ state which almost totally follows the AID mechanism, it is evident that the $\tilde{A}$ state competes between both the AID and ADI mechanisms. The AID, which is the dominant process, is seen when the pump and probe pulses are overlapped ($t = 0$), while the ADI occurs when the probe photon is absorbed at long time delays.

Extensive modeling of the data[69,81] has been undertaken from which rates of several key processes have been ascertained. One of the main reasons for the difference in behavior for the two systems is readily apparent from Figure 9. In the case of excitation through the $\tilde{C}'$ state, any predissociation mechanism which leads to the formation of NH_2H, NH_2, and H gives rise to fast H atoms which are not retained in the cluster and do not contribute to the formation of NH_4, which could then be subsequently ionized. On the other hand, in the case of the $\tilde{A}$ state the reaction channel leading to the formation of NH_2H, NH_2^*, and H has a small

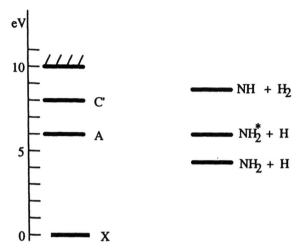

Figure 9. Energy levels of an ammonia molecule showing the states investigated in the pump-probe experiments. Also shown are the threshold energies where H atoms can be produced. The upper hatched region denotes the ionization limit. Taken with permission from , *NATO ASI Series on Large Clusters of Atoms and Molecules*; Kluwer Academic: Dordrecht, 1996, pp 371–404.

exoergicity enabling some fraction of the hydrogen atoms to form NH_4, which is subsequently ionized and detected as the protonated species.

The power of these techniques in elucidating the detailed mechanisms of ion–molecule reactions is well demonstrated by this example. Undoubtedly, we can expect to see increasing use of the ultrafast pump-probe technique in the field of ion–molecule reaction dynamics.

Coulomb Explosion and the Production of High Charge State Ions

Most studies of the influence of solvation on ion–molecule reactions has focused on those involving singly charged ions. However, it is known that multiply charged clusters can be formed which may undergo the processes of fission and Coulomb explosion. These phenomena have been observed in non-hydrogen-bonded clusters (such as those comprised of CO_2[85] and for various metals such as the alkali metal ones[86]), in hydrogen-bonded systems such as water and ammonia clusters,[87] and in rare gas clusters.[88-90] With few exceptions, the data have been confined to observations of stable, multiply charged cluster species. At sizes less than a critical value, fission leads to the loss of the multiple charges from a cluster and hence the production of a singly charged size. For clusters of critical size and above, their cohesive energy is sufficient to overcome the repulsion due to Coulomb forces, and in comparatively large clusters, multiply charged species are "stable" and are readily observed.

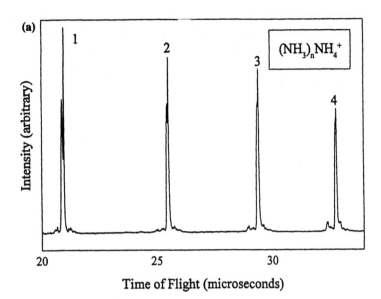

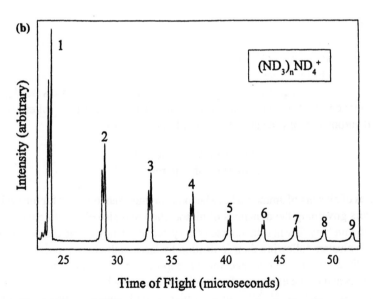

Figure 10. (a) Time-of-flight mass spectrum of ammonia clusters. The characteristic split peaks of a Coulomb explosion process are clearly seen for clusters up to $(NH_3)_3NH_4^+$. A small shoulder is observed for $(NH_3)_4NH_4^+$. The voltage gradient across the first two TOF lenses is 400 V/cm. (b) Time-of-flight mass spectrum of deuterated ammonia clusters. The characteristic split peaks of a Coulomb explosion process are clearly seen for all clusters shown, up to $(ND_3)_9ND_4^+$. The voltage gradient across the first two TOF lenses is 200 V/cm. Taken with permission from ref. 90.

Metastable evaporation of the solvating molecules can take the stable cluster to below critical size, and subsequently lead to fission. Alternatively, Coulomb explosion may result. This latter case arises when the then multiply charged clusters are suddenly generated in clusters of sizes far below the corresponding critical size. These processes are characterized by dynamics proceeding on the femtosecond time scale, and in general it has been difficult to investigate the kinetic energy release for such processes. It should be noted, however, that this has been determined in a few of the afore-referenced cases, especially for multiply charged metal clusters.

Recent observations are beginning to shed new light on these phenomena, where femtosecond techniques have been particularly valuable. For example, a study of molecular clusters of ammonia have revealed the ejection of highly charged nitrogen atoms up to N^{+5} upon irradiation.[68,90] The ejection of highly charged species in the TOF lens assembly, together with the refocusing of the backward ejected peaks and the acceptance angle of the charged particle detector, leads to the appearance of split peaks in the time of arrival spectra. In addition to ejected nitrogen ions, intact ammonia clusters also displaying peak splittings indicative of Coulomb explosion have been well studied; see Figure 10a and b.

The observations show that the kinetic energy distributions arise from asymmetric charge distributions within the clusters, as evidenced by the differing kinetic energy release for clusters of light compared to heavy ammonia (Figure 11). Related work with HI clusters[91] has revealed the ejection of highly charged iodine atoms, with iodine up to +17 being observed, but only in the presence of clusters (see Figure 12). Similar observations have come from studies of acetone clusters[92] where highly charged carbon and oxygen atoms are found to be produced.

B. Evaporative Dissociation
Dynamics of Cluster Ions

Studies of clusters of ammonia have become an important test case for unraveling the factors governing the dynamics of dissociation and also elucidating the origin of magic numbers. During the course of a systematic study of the dissociation of ammonia clusters following their ionization by laser multiphoton ionization techniques, direct proof was obtained for the importance of the metastability of the cluster ions and of the extensive dissociation processes which influence the resulting cluster distributions.[33,35,42,70] In particular, a combination of collision-induced and unimolecular (evaporative) loss processes was observed for a range of cluster sizes up to more than 50 molecules. Direct evidence was obtained for the loss of as many as six monomer units from the protonated 9-mer following its production, at least two monomer units being lost directly by evaporation with the others evidently being lost due to collisional processes. Although all ammonia clusters undergo

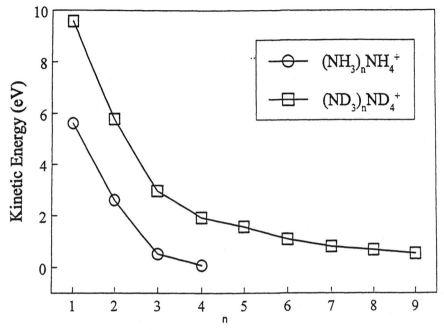

Figure 11. Average ion kinetic energies as a function of cluster size for ammonia clusters and deuterated ammonia clusters resulting from Coulomb explosion of clusters. The average kinetic energies are obtained from the splittings in the cluster ion time-of-flight peaks; see Figure 10. Taken with permission from ref. 90.

fragmentation following ionization, only the protonated pentamer displays an anomalously large abundance in the locality of its size range.

An example of the usefulness of the reflectron technique discussed earlier in this chapter is evident for the case of ammonia clusters in Figures 3b and c. Upon ionization, ammonia undergoes an internal ion–molecule reaction leading to protonated cluster ions, and concomitant evaporative unimolecular dissociation. This can be viewed in the context of equations 7–9 and the following:

$$[NH_4^+(NH_3)_{n-2}] \rightarrow NH_4^+(NH_3)_{n-2-m} + mNH_3 \tag{16}$$

Figure 3a shows a conventional time-of-flight spectrum obtained under total hard reflection conditions, while Figure 3b shows a spectrum obtained with the reflectron where a daughter ion (from the loss of one ammonia molecule during the flight in the field-free region prior to the reflectron) precedes each of the corresponding parents. By reducing the applied potential at the end of the reflectron (i.e. $U_k < U_t$ (see Figure 2b), only the lower kinetic energy products are reflected. The nondis-

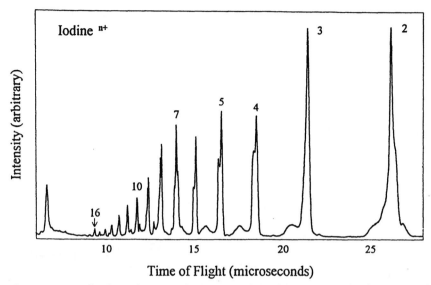

Figure 12. Coulomb explosion of hydrogen iodide clusters. Note the formation of I^{+16}; a small signal probably attributable to I^{+17} is seen in some experiments. Taken with permission from ref. 92.

sociating ions are eliminated from the spectrum as shown in the lower part of Figure 3c. Further reduction of the reflecting potential U_t (with $U_k < U_t$) improves the ability to discern small contributions arising from more extensive dissociation.

Recently, it has been shown by Castleman and coworkers[35,43] that by measuring the average kinetic energy release and unimolecular dissociation rate constant during ion decomposition, and by using the peak shape analysis method described in the experimental section, bond energies for cluster ions can be derived. This procedure is based on the fact that during the decomposition of a metastable parent ion in the field-free region, the internal energy of a parent ion is converted to the translation energy of the daughter ion. As a result, the translational energy distribution of the daughter ion is broader than that of the parent ion. The importance of using a reflectron time-of-flight mass spectrometer in studies of metastable unimolecular dissociation dynamics is that it enables simultaneous determination of the kinetic energy release and the dissociation rate. The method leads to values of high precision, which in combination with various theoretical approaches, enables a determination of cluster bond energies.

An important example of the application of this method is seen for the case of ammonia. Referring to Figure 13, the measured average kinetic energy release $<E_r>$ of metastable $(NH_3)_nH^+$ ($n = 4$–17) is seen to display a maximum value of

9.5 meV at $n = 5$ and a gradual decrease to a value of 5.0 meV for $n = 17$. The finding of a maximum in the kinetic energy release distribution for $(NH_3)_5H^+$ is consistent with recent KER measurements of $(NH_3)_nH^+$ ($n = 1–8$) obtained with a double focusing mass spectrometry method.[93] The largest amount of kinetic energy release is found to be associated with the parent cluster ion $(NH_3)_5H^+$ undergoing unimolecular decomposition. A small local maximum in the average kinetic energy release is also evident at $n = 12$. The use of these values in deducing thermochemical information is evident from equation 6 and is presented later in this chapter.

Related studies[94,95] were undertaken to investigate the reactions among various functional groups with a mixed cluster. Metastable unimolecular decomposition processes of the mixed ammonia–acetone cluster ions were also studied using a reflectron time-of-flight mass spectrometer in conjunction with multiphoton ionization. In a time window of a few tens of microseconds, the $\{(NH_3)_n \cdot (C_3H_6O)_m\}H^+$ ($n = 1$, $m = 2–5$) ions are found to lose one acetone moiety. On the other hand, the $\{(NH_3)_n \cdot (C_3H_6O)_m\}H^+$ ($n = 2–18$, $m = 1–5$) ions lose one ammonia moiety via unimolecular decomposition. The present results show that the cluster ions have a

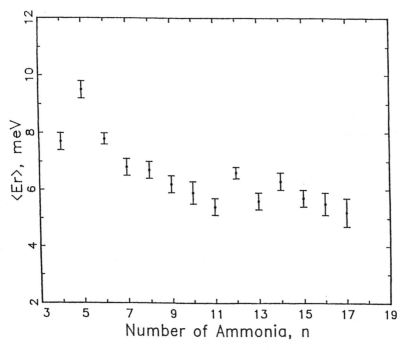

Figure 13. A plot of the measured average kinetic energy release $<E_r>$ during the metastable unimolecular decomposition of $(NH_3)_nH^+$, $n = 4–17$, as a function of cluster size. The technique involves use of the reflectron shown in Figure 2a. Taken with permission from ref. 2.

central core corresponding to NH_4^+, to which the other ligands making up the cluster are bound.

The operative unimolecular decomposition processes are independent of the formation method of the mixed neutral clusters, indicating that the rearrangement process following ion production and the ensuing ion–molecule reactions are rapid compared to the times of observation. The unimolecular decomposition processes are found to be independent of the formation method of the mixed neutral clusters in this study. It is generally inappropriate to relate the structures of the formed cluster ions to those of the neutral cluster precursors since there is extensive fragmentation and rearrangement upon ionization. If cluster ions prepared by ionizing neutral clusters formed by different methods were to have different structures, the structural effect does not play an important role in the metastable unimolecular decomposition processes.

C. Example of the Site Specific Blocking of a Reaction

Some direct evidence of how the presence of a solvent can influence a reaction is seen from data for the acetone system, where the solvent H_2O can dramatically affect the course of one of the (dehydration) reaction channels.[96,97] Observed major cluster ions resulting from prompt fragmentation following multiphoton ionization include $[(CH_3)_2CO]_m \cdot H^+$ ($m = 1$–15) $[(CH_3)_2CO]_m \cdot C_2H_3O^+$ ($m = 1$–17), and $[(CH_3)_2CO]_m CH_3^+$ ($m = 1$–10). In a time window of a few tens of microseconds, all of these three classes of cluster ions unimolecularly decompose, losing only one acetone monomer (designated T). The results of the study have shown that most of the internally excited $(T)_m^{+*}$ ions fragment to form $(T)_{m-1-x} \cdot H^+$, $(T)_{m-1-x} \cdot C_2H_3O^+$, $(T)_{m-1-x} \cdot CH_3^+$. There are three major prompt fragmentation processes which have been identified:

$$(T)_m^{+*} \rightarrow (T)_{m-1-x} \cdot H^+ + C_3H_5O + xT \tag{17}$$

$$(T)_m^{+*} \rightarrow (T)_{m-1-x} \cdot C_2H_3O^+ + CH_3 + xT \tag{18}$$

$$(T)_m^{+*} \rightarrow (T)_{m-1-x} \cdot CH_3^+ + C_2H_3O + xT \tag{19}$$

The evaporative loss of neutral fragments proceeds following ionization, in a manner analogous to the processes discussed above.

Interestingly, a reaction corresponding to the dehydration of $[(CH_3)_2CO]_m \cdot H^+$ and leading to the production of $[(CH_3)_2CO]_{m-2} \cdot C_6H_{11}O^+$ is observed for $m = 2$–6. The most striking finding is that the presence of water molecules in a cluster suppresses this dehydration reaction. Hence, this is an interesting example of a site-specific blocking of a reaction. This can be seen, for example, in the case of ion peaks at masses 157 and 215 (shown in Figure 14), which correspond to

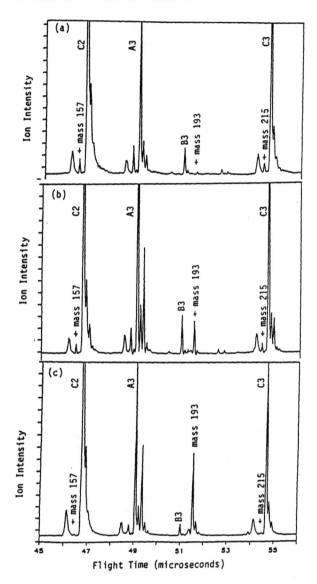

Figure 14. Reflectron-TOF spectra of acetone (T) cluster ions. $A_m \equiv (T)_m \cdot H^+$, $B_m \equiv (T)_m \cdot CH_3^+$, $C_m \equiv (T)_m \cdot C_2H_3O^+$. (a) 0.4% water in the acetone sample, $(T)_1 \cdot C_6H_{11}O^+$ (mass 157) and $(T)_2 \cdot C_6H_{11}O^+$ (mass 215) are seen, whereas ion signal corresponds to $(T)_3 \cdot H_3O^+$ (mass 193 is not found). (b) 0.7% water in the acetone sample, all peaks corresponding to masses 157, 215, and 193 are observed. (c) 1.0% water in the acetone sample, ion peaks corresponding to $(T)_1 \cdot C_6H_{11}O^+$ (mass 157) and $(T)_2 \cdot C_6H_{11}O^+$ (mass 215) are not seen; however, $(T)_3 \cdot H_3O^+$ (mass 193) is clearly identified. Neutral clusters are ionized at 355 nm using a pulsed Nd:YAG laser. Taken with permission from ref. 2.

$(T)_{m-2} \cdot C_6H_{11}O^+$ ($m = 3$ and 4) and are the dehydration products of the protonated species, $(T)_m \cdot H^+$. These $(T)_{m-2} \cdot C_6H_{11}O^+$ ions have been observed in previous studies of ion–molecule reactions[98] made on this system. However, Luczynski and Wincel[99] have demonstrated that protonated acetone can associate with $(CH_3)_2CO$ and H_2O molecules to form the proton-bound species $(T)_m \cdot H^+$ through sequential clustering reactions, and $(T)_{m-1} \cdot (H_2O) \cdot H^+$ (with m up to 6) through substitution reactions with water vapor processes. In addition, the $(T)_m \cdot H^+$, ($m \geq 2$) decomposes by eliminating a water molecule to form $(T)_{m-2} \cdot C_6H_{11}O^+$.

Figure 14a shows a portion of the TOF spectrum of the acetone cluster ions in the case where the water content in the acetone sample is 0.4% or less. There are no peaks corresponding to $(T)_3 \cdot (H_2O)^+$ (mass 192) or $\{(T)_3 \cdot (H_2O)\}H^+$ (mass 193), and one can readily identify the ion peaks at masses 157 and 215 corresponding to the $(T)_{m-2} \cdot C_6H_{11}O^+$ ($m \geq 2$) ions as clearly originating from the direct elimination of water molecules from the $(T)_m \cdot H^+$ via the reaction:

$$(T)_m \cdot H^+ \rightarrow (T)_{m-2} \cdot C_6H_{11}O^+ + H_2O \qquad (20)$$

Experiments conducted to study the influence of the presence of water in the cluster on the dehydration reactions were very revealing. Figure 14b shows the same portion of the TOF spectrum of the acetone cluster ions as that in Figure 14a, but with 0.7% water present in the acetone sample. Ion peaks corresponding to the $(T)_{m-2} \cdot C_6H_{11}O^+$ ($m = 3$ and 4; masses 157 and 215) and $\{(T)_3 \cdot (H_2O)\}H^+$ ion (mass 193) are evident. Additionally, Figure 14c also displays the same portion of the TOF spectrum of the acetone cluster ions where the water content in the acetone sample is 1.0%. Interestingly, the peak corresponding to the $\{(T)_3 \cdot (H_2O)\}H^+$ ion (mass 193) is evident, whereas those corresponding to the $(T)_{m-2} \cdot C_6H_{11}O^+$ ions are not seen (no peaks appear at either mass 157 or mass 215). The findings strongly suggest that the presence of water inhibits the dehydration mechanism of $(T)_m \cdot H^+$ cluster ions.

Although reaction 20 has been reported by several researchers[98–100] based on gas phase ion–molecule reactions, others[101] did not observe this reaction in cases where the neutral clusters of acetone were ionized by electron impact. A possible explanation put forth was the structures of the precursor ion $(T)_m \cdot H^+$ might be different in the two different experiments. However, the results of these newer studies show that the presence of water molecules in a cluster can significantly suppress the dehydration reaction, and although the earlier authors did not state the extent of water impurity in their system, if any, the new findings provide a plausible explanation for the discrepancy between the two studies mentioned above. The presence of water is believed to block a reaction site that otherwise enables the formation of the protonated mesityl oxide, $C_6H_{11}O^+$, to occur.[96] This finding not only suggests the probable reason for the discrepancy between several earlier studies, but, most importantly, provides evidence of another example of the influence of solvation on ion reactions in clusters.

D. Ligand Reactions

Protonated Systems

Most systems comprised of (at least partially) hydrogen-bonded constituents undergo similar processes, although some display additional rather interesting and revealing solvation driven competitive reaction channels. Consider, for example, the case of clusters comprised of methanol. Following multiphoton ionization (MPI), neutral methanol clusters are also found[34,102,103] to undergo a well-known ion–molecule reaction which leads to the production of protonated clusters and the evolution of CH_3O. In accord with observations for most other systems, there is a general trend for the evaporation rates to decrease with time after the initial ionization event and, for a given observational time window, to display an increase in rate with cluster size.

The cluster ions, $H^+(CH_3OH)_n$, are also found to undergo several other intracluster reaction pathways which show a dependence on the degree of aggregation. Two particularly interesting sequences of cluster ions are the protonated methanol clusters of the form $H^+(CH_3OH)_n$ and others of the form $H^+(H_2O)(CH_3OH)_n$, which become clearly evident after the peak corresponding to $n = 7$.

The origin of the sequence corresponding to protonated methanol peaks is a rapid intracluster proton transfer reaction following ionization of the neutral clusters. This reaction has a well-known bimolecular counterpart that proceeds at near collision rate:[104]

$$(CH_3OH)_n^+ \rightarrow H^+(CH_3OH)_{n-1} + CH_3O \qquad (21)$$

The formation of the sequence, $H^+(H_2O)(CH_3OH)_n$, is believed to proceed via:

$$[H^+(CH_3OH)_n]^* \rightarrow H^+(H_2O)(CH_3OH)_{n-3} + (CH_3)_2O + CH_3OH \qquad (22)$$

The operative mechanisms have been deduced by considering various peaks identified in the daughter spectrum employing the reflectron technique. One corresponds to a mass loss of 32 amu from a parent ion cluster that enters the drift region as $H^+(CH_3OH)_n$ via:

$$H^+(CH_3OH)_n \rightarrow H^+(CH_3OH)_{n-1} + CH_3OH \qquad (23)$$

Mass losses of more than one monomer unit appear as unresolved shoulders. The loss of up to five methanol monomers from the protonated octamer is observed. Another peak was found to correspond to loss of water from the protonated methanol dimer ion via the dehydration reaction,

$$H^+(CH_3OH)_2 \rightarrow (CH_3)_2OH^+ + H_2O \qquad (24)$$

a process requiring an induction time of at least several tenths of a microsecond[103] (also see ref. 94).

Of particular interest were two other sequences, one showing loss of 78 amu from $H^+(CH_3OH)_n$, indicating loss in the drift region of both a methanol monomer and C_2H_6O (dimethyl ether) for parent ions ($n = 4–9$; via equation 22). One final sequence was identified as representing the loss of one methanol moiety from the mixed cluster species $H^+(H_2O)(CH_3OH)_n$. If these clusters are also formed by reaction 22, they must initially contain 9 or more methanol molecules. Interestingly, reaction 24 which corresponds to the protonated methanol dimer ion eliminating H_2O while retaining dimethyl ether, recently has also been found in thermal reaction experiments and is attributable to solvation effects. Analogous water elimination reactions are not observed for the parent cluster ions larger than $n = 2$.

In the case of these large clusters, H_3O^+ is solvated more strongly by methanol than is protonated dimethyl ether, $(CH_3)_2OH^+$.[25] Hence for these clusters, water retention and dimethyl ether elimination leads to production of mixed clusters of form $H^+(H_2O)(CH_3OH)_n$. A "slower" elimination of dimethyl ether in the drift region is observed (along with CH_3OH) from cluster sizes $H^+(CH_3OH)_n$ ($n = 4–9$). Thus, a significant difference in reactivity with cluster size is observed; the smaller clusters lose dimethyl ether on a longer (field-free region) timescale, while the larger clusters undergo comparatively rapid loss of C_2H_6O in the ion lens.

Metal Ion–Solvent Reactions

Another interesting example, where solvation is believed to influence the energy barrier to reaction, is that of methanol bound to alkali metal ions.[105] Early results conducted in beam experiments suggested the likely occurrence of a dehydration reaction, though thermalization of the reactants could not be assured under the conditions of the experiment. More recent studies conducted[106] under well-thermalized flowtube conditions revealed a similar reaction, with a critical size cluster needed to initiate the reaction. A small inverse trend with cluster size is consistent with speculation that solvation of an ion leads to a small lowering of the barrier allowing the energetically favorable reaction to proceed at a measurable rate. The products are water and dimethyl ether, similar to the protonated system discussed in the foregoing subsection; (see Figure 15).

Polymerization Induced by Metal Ion Centers

Other work related to metal ion reactions with clustering ligands revealed the extent to which titanium and vanadium cations can lead to significant dehydrogenation[107–110] of a number of simple organic molecules such as acetylene, ethylene, and propylene even at room temperature. This work led to the identification of chemical reactions in which even simple alkanes and alkenes can become converted to organic polymers. Most interesting were observations of the dehydrogenation reactivity of Ti^+ and V^+, as well as their coordinated complexes toward ethylene at thermal energies, where the ligands are acetylene or ethylene for Ti^+ and ethylene for V^+. We have observed that coordinated Ti^+ and V^+ complexes show quite different reactivity from that of the bare metal ions, and that the effects of ligands

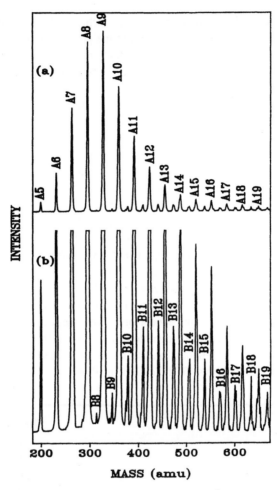

Figure 15. (a) Mass spectrum of alcohol clusters about Li⁺. Data taken at $T = 150$ K and (b) same as (a) but with an expanded intensity scale: An = Li⁺(CH₃OH)ₙ; Bn = Li⁺(CH₃OH)ₙ(H₂O). Taken with permission from ref. 106.

on the reactivity depend strongly on their number, as well as the type of the metal ion. The three-ligand coordinated species of both metal ions display higher dehydrogenation reactivity than the one- and two-ligand coordinated complexes.

The chemistry of the multistep reactions of Ti⁺ with propylene was also examined. Bare Ti⁺ exhibits an active reactivity toward breaking C–H bonds of the alkene molecules. But, the coordination of ligands on Ti⁺ was found to dramatically alter its dehydrogenation reactivity. For propylene, the multistep reactions were terminated at the fourth step, whereas the reactions with ethylene molecules were found to proceed far beyond the fourth step. Over 20 ethylene molecules have been observed to react with Ti^+, with their products attaching to the ion center. Based on various experimental results, we believe that the polymerization of ethylene induced by Ti^+ in the gas phase follows essentially the same mechanism by which the polymerization of olefin proceeds in the condensed phase. According to the basic concepts of Cossee's theory, we have developed a simple model to interpret the findings. The model suggests that a titanacyclebutane is formed in a six-coordinated Ti(I) complex to initiate the ethylene polymerization. Then, additional ethylene molecules can coordinate onto the vacant coordinating site of Ti^+, followed by the interposition of the coordinated ethylene from Ti^+ to the alkyl group. The reaction patterns are indicative of a continuing stepwise mechanism leading to the formation of a larger and larger ethylene polymer.

In a similar way, electronically excited states of these same species were found to display a large reactivity toward the breaking of CH, OH, NH and CO bonds in water, ammonia and alcohol molecules. For each of the primary reactions, we measured the reaction product distribution and the bimolecular rate constants. All the reactions were observed to proceed by way of bimolecular reaction mechanisms. Ti^+ displays a very high reactivity toward breaking C–H, O–H, N–H, and C–O bonds in these reactants. A dehydrogenation reaction is found to be the only channel in all of these reactions except ones with methanol. The order of the dehydrogenation rate constants is found to be the same as the inverse order of the bonding strength of X–H bonds, and this can be explained on the basis of dehydrogenation mechanisms. In the case of methanol, Ti^+ is also able to break

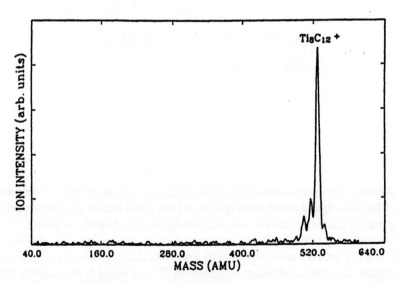

Figure 16. Mass distribution of $Ti_mC_n^+$ clusters generated from the reactions of titanium with CH_4. Note the "super magic" peak corresponding to $Ti_8C_{12}^+$. Taken with permission from ref. 92.

C–H and O–H bonds to form other products, although TiOH$^+$ is the dominant product formed from breaking the C–O bond.

The results indicate that excited-state ions are necessary before complete dehydrogenation of these molecules can occur. Indeed, related studies involving plasma

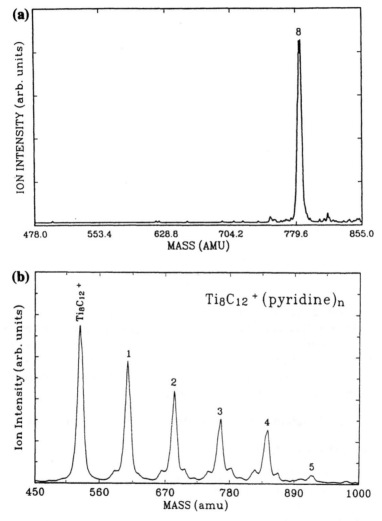

Figure 17. (a) Mass spectra of products arising from reactions of Ti$_8$C$_{12}^+$ with methanol. The number stands for the number of methanols associating onto Ti$_8$C$_{12}^+$, note that association reactions terminate at the eighth step. (b) Under similar conditions, the clustering of pyridine truncates at $n = 4$. (a) Taken with permission from ref. 115; (b) Taken with permission from *NATO ASI Series on Large Clusters of Atoms and Molecules*; Kluwer Academic: Dordrecht, 1996, pp 371–404.

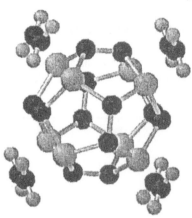

Figure 18. Model showing the possible bridging of four π-bonding molecules (C_2H_4) across two metal atoms of each pentagonal ring. Taken with permission from *NATO ASI Series on Large Clusters of Atoms and Molecules*; Kluwer Academic: Dordrecht, 1996, pp 371–404.

reactions leading to the full dehydration of small hydrocarbons led to the discovery of metallocarbohedrenes[111–114] (or Met-Cars) in our laboratory (see Figure 16). This led to a new class of molecular clusters and their ions for investigation. Note that the cationic Met-Cars readily associate eight polar molecules[115] to each metal site (Figure 17a) and take up four π-bonding molecules[116] (Figure 17b) possibly through bonding analogous to the adsorption of similar molecules to metal surfaces (see Figure 18). This subfield is now also growing at a rapid pace as a separate area of study. In summary, based on our studies, the presence of transition metal ions can be expected to lead to significant polymerization of organic molecules which can lead to sites for new particle formation and growth. Numerous related findings are now becoming available from other laboratories.[117]

E. Solvation Effects on the Reactions of Other Ion Cores

Hydrated Hydroxyl Anions

Thermal reaction techniques enable a quantification of the influence of solvation on reactivities.[1,2,19] One particular reaction which is a good example of how solvation can affect the nature of a core ion reaction site comes from a study[118] of the interaction of OH⁻ with CO_2. The gas-phase reaction between the individual species is quite exothermic and can only take place by a three-body association mechanism. The reaction proceeds very slowly in the liquid phase and has been calculated[119] to have a barrier of about 13 kcal mol⁻¹. In biological systems, the reaction rate is enhanced by about 4 orders of magnitude through the enzyme carbonic anhydrase. Recent studies carried out in our laboratory provide detailed information on the influence of hydration on the reaction kinetics, and support the suggested role played by the enzyme in facilitating this reaction.

Typically, hydration can have a pronounced effect on reaction thermodynamics, but in the case of $OH^-(H_2O)_n$ the clustering of a number of water molecules is still insufficient to cause the reaction to become endothermic. In fact, owing to the formation of very stable products, namely $HCO_3^-(H_2O)_m$, the reaction is still very exothermic even for the replacement of as many as three water ligands by a CO_2 molecule. Thus an explanation for the difference shown in Figure 19a between the experimentally measured rate constants and theoretically predicted values based on a parameterized trajectory theory[120] must be attributable to the reaction kinetics. A general formulation of the reaction can be written[118] in terms of a Lindemann-type mechanism:

$$CO_2 + OH^-(H_2O)_n \rightleftarrows \{OH^-(H_2O)_n(CO_2)\}^* \qquad (25)$$

$$\{OH^-(H_2O)_n(CO_2)\}^* + He \rightarrow OH^-(H_2O)_n(CO)_2 + He \qquad (26)$$

$$\{OH^-(H_2O)_n(CO_2)\}^* \rightarrow HCO_3^-(H_2O)_m + (n - m)(H_2O) \qquad (27)$$

The intermediate reaction complexes (after formation with rate constant, k_1), can undergo unimolecular dissociation (k_{-1}) back to the original reactants, collisional stabilization (k_s) via a third body, and intermolecular reaction (k_r) to form stable products $HCO_3^-(H_2O)_m$ with the concomitant displacement of water molecules. The experimentally measured rate constant, k_{exp}, can be related to the rate constants of the elementary steps by the following equation, through the use of a steady-state approximation on $\{OH^-(H_2O)_nCO_2\}^*$:

$$k_{exp} = k_1\{k_s[He] + k_r\}/\{k_{-1} + k_s[He] + k_r\} \qquad (28)$$

Equation 28 can be simplified according to four possible limiting situations, but for the experimental conditions used, the following applies,

$$k_{-1} \geq k_r \text{ and } k_s[He], \text{ and } k_r \geq k_s[He]$$

then:

$$k_{exp} = k_1 k_r/k_{-1} \qquad (29)$$

The experimental findings bear on the hydrolysis dynamics of CO_2. In the basic solution, OH^- can react directly with CO_2 to form HCO_3^-:

$$CO_2(aq) + OH^-(aq) \rightarrow HCO_3^-(aq) \qquad (30)$$

The second-order rate constant k_{OH}^- for reaction 31 has been measured to be $8.5 \times 10^3 \text{ M}^{-1} \text{ s}^{-1}$,[121] while the rate constant for the enzymatic hydrolysis of CO_2 has been measured to be 7.5×10^7 M^{-1} s^{-1},[122] or 1.2×10^{-13} cm^3 s^{-1} in the units employed by gas-phase chemists, which is about 4 orders of magnitude larger than $k_{OH^-}^-$. By contrast, the gas-phase collision limit for CO_2 with a large anion is about 10^{-10} cm^3 s^{-1} and the experimentally observed value decreases with increased hydration. It has been found[123] that the key aspect of the enzyme catalysis is the presence of a zinc atom which breaks a water molecule into H^+ and OH^- which can then react rapidly with CO_2 to form HCO_3^-.

It is interesting to compare the reactions of CO_2 and water clusters of OH^- with the results of the analogous reactions involving SO_2.[124] Both reactions are highly exothermic. Trends in hydration are not much different from that for the CO_2 case, at least for small clusters. Nevertheless, as seen in Figure 19b, SO_2 displays a very large reactivity for all cluster sizes over the full range investigated. Indeed, the trends are in good accord with computations made employing results of trajectory studies made by Su and Chesnavich.[120]

Measurements made in our laboratory, as well as those by others, reveal that SO_2 interacts with bare OH^- via an association reaction. However, the reaction mechanism changes from one of association to switching upon hydration:

$$OH^-(H_2O)_n + SO_2 \rightarrow HSO_3^-(H_2O)_m + (n-m)(H_2O) \quad n \geq 1 \qquad (31)$$

With additional water ligands bound to the OH^-, the reaction enthalpy becomes progressively less negative owing to the net stabilization of OH^- by water molecules. For example, in the case where all the water ligands are replaced by an SO_2 molecule during the reaction:

$$OH^-(H_2O)_n + SO_2 \rightarrow HSO_3^- + nH_2O \qquad (32)$$

The reaction enthalpy switches from being exothermic to being endothermic between $n = 3$ and $n = 4$. In hydrated clusters, only reactions leading to partial replacement of the water molecules maintain thermodynamic exoergicity:

$$OH^-(H_2O)_4 + SO_2 \rightarrow HSO_3^-(H_2O) + 3(H_2O) \qquad (33)$$

Our observations are in accord with this fact.

In the present experiment, even when $n = 59$, the anionic clusters $OH^-(H_2O)_n$ are still found to react very fast toward SO_2, approaching the gas collision limit of about 10^{-9} m^3 s^{-1}. Since SO_2 is a strong reductant in basic solution,[125] no kinetic data are directly available for the reaction:

$$SO_2(aq) + OH^-(aq) \rightarrow HSO_3^-(aq) \qquad (34)$$

However, the rate constant for the following reaction has been measured:[126]

$$SO_2(aq) + H_2O(aq) \rightarrow HSO_3^-(aq) + H^+(aq) \qquad (35)$$

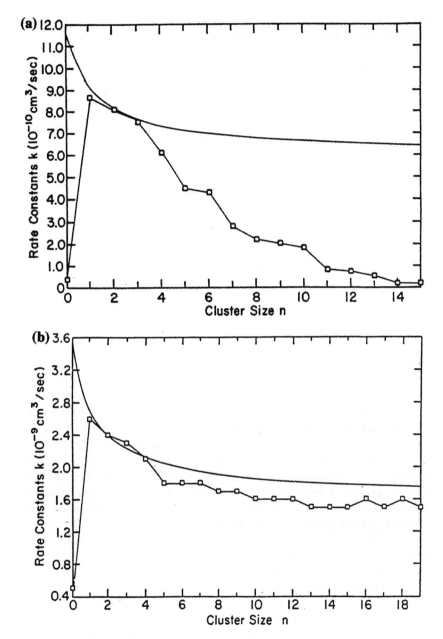

Figure 19. (a) Cluster size dependence of the rate constants for the reactions of CO_2 with the large hydrated hydroxyl anions at $T = 130$ K: $\Diamond$, experimental values for $OH^-\cdot(H_2O)_n$; —, calculated values for $OH^-\cdot(H_2O)_n$. (b) Dependence of rate constants on cluster size for the reactions of $OH^-(H_2O)_n$ with SO_2 at $T = 135$ K. Taken with permission from ref. 19.

It is $3.4 \times 10^6 \, M^{-1} \, s^{-1}$ ($5.6 \times 10^{-15} \, cm^3 \, s^{-1}$), which is about 10 orders of magnitude faster than the comparable reaction with CO_2.

The rate constant for reaction[35] is expected to be larger than $3.4 \times 10^6 \, M^{-1} \, s^{-1}$, not only because reactions involving ions usually have much smaller activation energies than those involving only neutrals, but also because the very stable products $HSO_3^-(H_2O)_n$ are formed. The observation that association becomes the dominant reaction channel at large cluster sizes in the present experiments indicates the possibility of formation of the products $OH^-(H_2O)_n(SO_2)$. It is interesting to note that in solution the $SO_2(OH^-)$ form exists and the $[SO_2(OH^-)]/[HSO_3^-]$ ratio is about five at 20 °C.[127]

Usually, hydration of the negative ions will make the reactions more endothermic owing to the stabilization of the reactant ions. However, since the products of the reaction between SO_2 and $X^-(H_2O)_n$ are so stable, the reactions accommodate their negative enthalpies by "boiling off" a certain number of water molecules from the products.[128]

Reactions of NO$_x$ with Hydrated Cluster Ions

Other studies have been conducted to ascertain the influence of solvation on the reactions of nitrogen oxides.[129] The results are of basic interest in the context of solvation phenomena, and also provide data of interest in the field of atmospheric science. We have devoted particular attention to unraveling the reactions of nitrogen oxides with water clusters over ranges of temperatures and varying degrees of hydration. Some of these may take place in the solution phase of an aerosol particle or cloud droplet, or in other cases they are believed to occur on the surfaces of condensed phases such as the aerosols and/or ice crystals. An extensive series of investigations have been focused on reactions of N_2O_5. Ones occurring with protonated and deuterated water clusters $X^+(X_2O)_{n=3-30}$, X = H or D, were studied at temperatures ranging from 128 to 300 K using a fast flow apparatus. For cluster ions with $n \geq 5$, the reaction $X^+(X_2O)_n + N_2O_5 \rightarrow X^+(X_2O)_{n-1} \, XNO_3 + XNO_3$ was observed to occur at temperatures below 150 K. At somewhat larger values of n, which were acquired at temperatures of about 130 K, the product ions $X^+(X_2O)_{n-2}(XNO_3)_2$ and $X^+(X_2O)_{n-3}(XNO_3)_3$ were also observed. The rate constants of the thermal energy reactions of N_2O_5 with $X^+(X_2O)_{n=5-21}$ were found to display both a size and pressure dependence. This work gives the first experimental evidence for the reaction between large protonated water clusters ($n \geq 5$) and N_2O_5 under laboratory conditions.

In order to elucidate the importance of reactions with anionic species, reactions of N_2O_5 with the anions $X^-(H_2O)_n$, where $0 \leq n \leq 5$ and X = O, OH, O_2, HO_2, and O_3, at several temperatures within the range 164–298 K were also studied. The ions $O^-(H_2O)_{n=0-2}$, $OH^-(H_2O)_{n=0-2}$, $O_2^-(H_2O)_{n=0,1}$, $HO_2^-(H_2O)_{n=0,1}$, and O_3^- react with N_2O_5 at the collision rate to form mainly bare NO_3^-; the larger hydrates of the reactant anions lead to production of the hydrated species $NO_3^-(H_2O)_m$, which can react further with N_2O_5 to form $NO_3^- \cdot HNO_3$.

Another series of reactions of potential importance involves the reaction of N_2O_5 with the $NO_2^-(H_2O)_{n=0-2}$, $NO_3^-(H_2O)_{n=1-2}$, and $NO_{n=2,3}^-HNO_2$ ions.[130,131] Studies were made at various temperatures within the range 167–298 K. All these ions were observed to react quite efficiently with N_2O_5, giving rise to the main product ion $NO_3^-\cdot HNO_3$. This finding is of atmospheric interest since these reactions may occur in the atmosphere and are potentially relevant to the chemistry of reactive nitrogen species involved in ozone destruction.

Observation[132] that switching reactions can be driven by solvation represented an important finding in the early field of cluster ion reactions. An especially significant study was that of the hydration of NO^+:

$$NO^+ + H_2O + M \rightarrow NO^+\cdot H_2O + M$$

$$NO^+\cdot H_2O + H_2O + M \rightarrow NO^+\cdot(H_2O)_2 + M$$

$$\vdots \qquad \vdots$$

$$NO^+(H_2O)_3 + H_2O \rightarrow H_3O^+(H_2O)_2 + HNO_2 \qquad (36)$$

Recently, similar findings[133] have been made in studies of the photoionization of NO containing water clusters[134] which established that the cluster reaction analogous to the last step in reaction 36 occurs for $n \geq 4$.

Metal Ion Solvation: Magnesium–Water System

There are other examples where solvation influences ion core transformations. For example, studies of the hydration of alkaline earth ions such as Mg^+, reveal that at progressively higher degrees of hydration, namely at $n = 5$, chemical transformation to $MgOH^+$ occurs[135–137] (see Figures 20a and b). Photoinduced chemical reactions have also been found to yield $MgOH^+$ in these clusters as well.[138] Theoretical work has provided evidence for a substantial shift in the charge distribution upon the progressive solvation of these metals with hydrogen bonded species.[135,138] The findings suggest that the energy required to lose a hydrogen atom decreases with an increase in the number of water molecules attached to the magnesium cation, and is attributable to a change in the oxidation state. This results in strong interactions with the water ligands as the clusters begin to more closely resemble a doubly charged magnesium cation solvated with water and OH^-.

F. Insights into Competition Between Ion Core Transformations, Ligand Switching Reactions, and Association Reactions

Cluster Formation and Switching Reactions of Ligated Metallic Atomic Ions

We have conducted studies to investigate general factors influencing the kinetics of cluster formation.[139,140] Extensive experiments were carried out with Na^+, partly

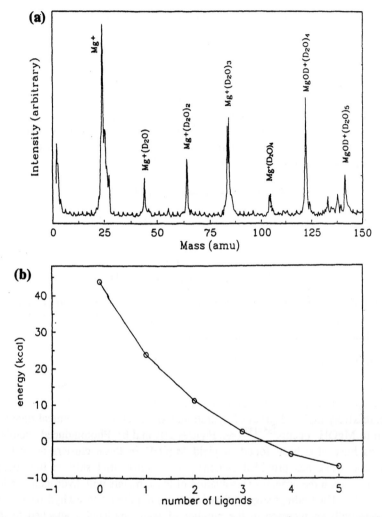

Figure 20. (a) Mass spectrum showing the reaction of Mg$^+$ with D$_2$O. (b) Energy required to form MgOH$^+$ as a function of degree of solvation. Taken with permission from ref. 135.

because of its simple electronic character and ease of comparison with theoretical calculations. Thermal rate coefficients have been measured for a wide range of association reactions with species including NH$_3$, ND$_3$, CH$_3$OH, SO$_2$, CO$_2$, and HCl. There remains considerable interest in the role of the third body in the collisional stabilization of clusters during their formation, as well as in the nature of the temperature dependence of the reaction. Studies were carried out in our laboratory which displayed a significant dependence on the mass of the third body.

An investigation of the association of the same molecules with Na^+ over a range of temperatures showed that the rate coefficient is proportional to T^{-n} as expected. The present findings of the investigations have provided a test of the available theoretical models and the usual procedures for determining n were not found to be valid. The data were found to be generally in accord with our suggestion that, for weakly bound systems, n is closely related to the number of new vibrational modes in the association complex.

The reactions of the bare sodium ion with all neutrals were determined to proceed via a three-body association mechanism and the rate constants measured cover a large range from a slow association reaction with NH_3 to a near-collision rate with $CH_3OC_2H_4OCH_3$ (DMOE). The lifetimes of the intermediate complexes obtained using parameterized trajectory results and the experimental rates compare fairly well with predictions based on RRKM theory. The calculations also accounted for the large isotope effect observed for the more rapid clustering of ND_3 than NH_3 to Na^+.

Data on switching reactions[141] are valuable for understanding a wide range of phenomena involving the substitution of ligands near a core ion such as occurs in cluster ions in the atmosphere, and as a result of the dynamics of ligand exchange in electrolyte solutions. Investigations of association reactions provide data on the rates of cluster formation, and can be used to test theoretical models. The kinetics and mechanisms of the reactions of $Na^+ \cdot (X)_n$ ($n = 0–3$; X = water, ammonia, and methanol reacting with CH_3CN, CH_3COCH_3, CH_3CHO, CH_3COOH, CH_3COOCH_3, NH_3, CH_3OH, and DMOE) were studied at ambient temperature under different flow-tube pressures. All the switching (substitution) reactions proceed at near-collision rate and showed little dependence on the flow-tube pressure, the type and size of the ligand, or on the type of core ions involved. Due to the steric effect caused by its large size, the switching reaction of DMOE is somewhat slower than other neutrals, which is contrary to the situation in the association reaction with bare sodium ions.

Kinetics of Proton Transfer in Molecule–Cluster Ion Interactions

The production of large water clusters[142–144] under thermal reaction conditions enables investigation of reactions at sites of selected degrees of hydration. This provides insight into reactions in the condensed state. During the course of these studies, large water clusters bound to H^+ or OH^-, with sizes ranging up to 50 molecules, were produced and various[42,142,145] reactions were carried out with them at well-defined temperatures using the fast-flow reactor technique. One test case for which extensive studies were made concerned their reactions with CH_3CN. While for protonated water clusters, CH_3CN undergoes a proton transfer reaction at small sizes and a replacement reaction at intermediate sizes, for large water clusters an association mechanism was found to dominate.

With regard to the hydrogen-containing anions, all of them were found to undergo a proton transfer reaction. The small hydrates were found to react with CH_3CN at near collision rate via proton-transfer and ligand-switching mechanisms; further hydration greatly reduces the reactivity due to the thermodynamic instability of the products compared to the reactants. In addition to proton transfer, our work shows that the hydrated anions of the oxygen atom also undergo a reaction to form $CHCN^-$ with further reactions to form CH_2CN^-.

During the course of the work, we also conducted further studies of the reactions of proton hydrates with CH_3COCH_3 and CH_3COOCH_3. The reaction mechanisms were found to change from proton transfer to ligand switching and ultimately to an association process, which would be equivalent to adsorption in the case of bulk systems.

Ligand Exchange: Reactions of Protonated Water Clusters with Nitric Acid

Interest in the interaction of water and nitric acid has arisen from several considerations involving such widely diverse problems as determining nitric acid uptake by water droplets and ice particles, to questions concerning the co-condensation of water and nitric acid to form polar stratospheric clouds[146] and related ones about nitric acid incorporation in protonated water clusters existing in the upper atmosphere. Crutzen and Arnold suggested[147] that,

$$H^+(H_2O)_{n-1}HNO_3 + H_2O \rightarrow HNO_3 + H^+(H_2O)_n \tag{37}$$

only become important, and exothermic, for degrees of hydration in excess of 6.

To the best of our knowledge, prior to our recent study, no data have been available for the interaction of hydrated clusters for $n > 4$ with HNO_3 under thermal conditions.[148] The successful production of large water cluster ions in a flow tube in our laboratory provided the capability to directly probe reactions and the stability of large hydrated nitric acid cluster ions under thermal conditions. The findings reveal that the reactions only take place beyond a specific cluster size at low temperature, yielding the products $D^+(D_2O)_n(DNO_3)$ ($n \geq 5$). (It should be noted that most work on these systems is conducted with deuterated molecules to avoid problems in species identification due to mass degeneracies for certain combinations of water and nitric acid.)

In the case of large water cluster ions $D^+(D_2O)_n$ (n up to 30), the addition of nitric acid led to efficient reactions taking place beyond a specific cluster size. A portion of the spectrum is presented in Figure 21a at different degrees of solvation. At even larger cluster sizes, mixed cluster ions of the form $D^+(D_2O)_n(DNO_3)_m$ (n is approximately ≥ 8 and 13 for $m = 2$ and 3, respectively) were also detected (see Figure 21b). Incorporation of a second and third nitric acid molecule is efficient for $n = 8$ and 13, respectively, even with the addition of low concentrations of DNO_3. Extensive studies over a range of pressures and temperatures revealed that switching rather than association reactions dominate. The results of the present study provide direct evidence for the stability of nitric acid water hydrates under conditions similar to those existing in polar stratospheric clouds. In our study, the first

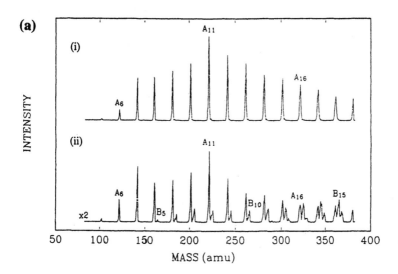

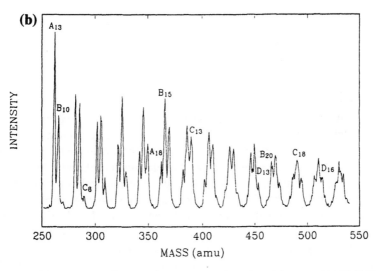

Figure 21. (a) Reactions of water cluster ions with nitric acid at $T = 150$ K: (*i*) Water cluster ions without reactant gas added; (*ii*) Cluster ion distribution with additions of reactant gas. $A_n = D^+(D_2O)_n$; $B_n = D^+(D_2O)_n(DNO_3)$. (b) Reactions of water clusters with nitric acid at larger cluster sizes, $T = 150$ K. $A_n = D^+(D_2O)_n$; $B_n = D^+(D_2O)_n(DNO_3)$; $C_n = D^+(D_2O)_n(DNO_3)_2$; $D_n = D^+(D_2O)_n(DNO_3)_3$. Taken with permission from ref. 148.

appearance of ionic nitric acid hydrate was at $D^+(D_2O)_5(DNO_3)$, where there is a dramatic onset of nitric acid accommodation in water clusters. This leads some credence to the suggestions of Crutzen and Arnold regarding the possible role of ion clusters in stratospheric nucleation processes under certain conditions.

G. Metal Cluster and Metal Oxide Anion Reactions

Cluster Reactivities: Reactivities and the Electronic Shell Model

Currently there is extensive interest in the reactivities of metal clusters. Studies in this area have been motivated, in part, by recognition that the results for certain systems give insight into the nature of reactions on metal surfaces. Also, such findings serve to further elucidate the evolution of the geometric and electronic states of matter of finite size with increasing size toward the infinite solid. The Jellium model, which considers the interaction of free electrons confined in a uniform potential of smeared-out positive charge, leads to a shell structure for the electron filling. This model has been found to be useful in accounting for the ionization potentials and electron affinity trends of clusters among certain metals, as well as the often observed magic numbers in these systems. It occurred to us that shell closings might also become manifested in affecting trends of reactivity with cluster size.

In this context, a series of studies in our laboratory focused on unraveling the chemistry of aluminum anion reactions[149–151] with oxygen. Aluminum anions (Al_n^-, $n = 5 \rightarrow 40$) were generated using laser vaporization and equilibrated to room temperature in a flow-tube apparatus. The behavior of these species, in general, shows a propensity to lose the neutral oxides. In sharp contrast, several anions— Al_{13}^-, Al_{23}^-, and Al_{37}^- —show special inertness to these etching reactions and, indeed, the etching reactions even contribute to their further production.

Aluminum was considered an important system for testing the models because in the solid state it is well represented by a free electron model, and with only three valence electrons per atom, it is readily amenable to theoretical treatment. In terms of the Jellium model, shell closings are expected corresponding to $n = 13, 23$, and 37 (representing 40, 70, and 112 electrons)[152] and $n = 11, 13, 19, 23, 35$, and 37 (representing 34, 40, 58, 70, 106, and 112 electrons) for two different potential wells, respectively. We undertook a study of the anion reactivities under thermal conditions to test these predictions. An odd/even alternation, where even-atom clusters react more quickly than successive odd ones, was apparent in the kinetic data. This is attributed to electron pairing within the cluster. Experimental observations of the unreactivity and even the production of Al_{13}^- and Al_{23}^- are consistent with the theoretical predictions of shell closings (see Figure 22).

Data taken in another mass range also showed the inertness of Al_{37}^-. While it was not produced by reaction of larger clusters, of which there are few, it was, however, found to be unreactive over a wide oxygen concentration range. In general, our results support the electron droplet Jellium model, although some anomalies (for

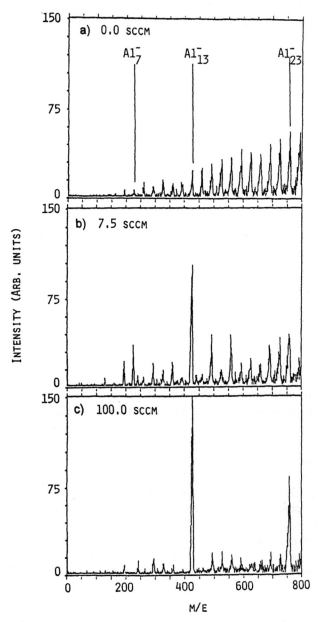

Figure 22. Series of mass spectra showing progression of the etching reaction of aluminum anions with oxygen. Note that magic number clusters corresponding to electron shell closings for 40 and 70 electrons (Al_{13}^- and Al_{23}^-) appear as the terminal product species of reactions with oxygen at flow rates of (a) 0.0, (b) 7.5, and (c) 100.0 standard cubic centimeters per minute (sccm). Taken with permission from ref. 92.

example, the comparatively slow reactivities of $n = 7$ and 22) suggest that other effects, such as geometry, also contribute to the local behavior of some of the clusters studied. It is interesting that the "magic" anion clusters fit best with the Jellium model.

The disappearance of the aluminum clusters is evidently by etching reactions:

$$Al_n^- + O_2 \rightarrow Al_{n-2}^- + Al_2O_2 \qquad (38)$$

$$Al_n^- + O_2 \rightarrow Al_{n-1}^- + AlO_2 \qquad (39)$$

We expect that these are the reactions which occur, both because they are exothermic and because Al_2O_2 and AlO_2 are species found in the gas phase from vaporization of alumina.

In order to further elucidate the factors governing the unusual stability of anion clusters, we undertook work on co-clustered species of aluminum with transition metals, in particular with niobium and vanadium.[151] A new laser vaporization source technique was employed which enabled mixed metal clusters to be selectively obtained and studied, each containing one transition metal atom embedded/attached to the other anions. Following their introduction into the flow tube, O_2 was added through the reactant gas inlet and the reaction products monitored. As in the case of the bare aluminum anion systems, all of the clusters rapidly reacted to form stable smaller clusters at the expense of the larger ones. In the case of niobium, other prominent species which even dwarfed the Al_{13}^- (found to be the most stable species in pure Al cluster oxidation work) were the formation of Al_4Nb^- and Al_6Nb^-. It is interesting to note that the Al_4Nb^- species is an 18-electron system, again being in accord with predictions of shell closings and nonreactivity based on the Jellium model. However, the Al_6Nb^- system deviates from these predictions. Further work with vanadium did not reveal the stable aluminum tetramer mixed cluster, but did show considerable stability of Al_6V^-. If every valence electron is counted as in the previous examples, VAl_6^- has 24 electrons which is not a predicted shell closing. (Note, however, if one of vanadium's electrons is promoted by hybridization, providing one free s-electron, the species then becomes a 20-electron system, and its lack of reactivity would be in general accord with the Jellium model.)

The Jellium model is evidently a valuable guide to the reactivity patterns and related electronic structure of metal alloy clusters, although it probably does not fully account for the behavior of even very free-electron metals. The interaction of the electronic orbitals of aluminum with those of the transition metals may be sufficiently strong to enable all of the electrons to contribute to the reactivity behavior. This is evident by there not being any observable difference in behavior for systems containing $4s^2$ and $5s^1$ electrons, and the general observation that odd–even electron number accounts for the reactivity of these alloy systems. Finally, it is interesting to speculate whether there might not be a structural form of VAl_6^- which involves electron hybridization and hence would behave as an

18-electron system; this may account for the lack of reactivity of this species. These various findings are leading us to new avenues of research to further elucidate the nature of these reactions.

Some related work has been undertaken with aluminum cations[150] for purposes of comparison to the anion work. Bimolecular rate constants were measured for the disappearance of Al^+ through Al_{33}^+. We did not observe a slow rate for the reaction of Al_3^+ which would be expected to have a closed electronic shell and hence would be expected to be comparatively unreactive. However, in the case of the cations, oxidation often leads to the retention of an oxide unit on the cation in contrast to the anion work.

Acid–Base Chemistry of Metal Oxide Anions

Other studies pertaining to reactions of metal oxide anions[153,154] have provided insights into analogies between gas-phase chemistry and acid–base reactions. In this work, anions of the monomer, dimer, trimer, tetramer, and pentamer of WO_3 were formed along with higher oxides of each species including WO_x^-, $(x = 3, 4, 5)$, $W_2O_x^-$, $(x = 6, 7, 8, 9)$, $W_3O_x^-$, $(x = 9, 10, 11)$, $W_4O_x^-$, $(x = 12, 13)$, and $W_5O_x^-$, $(x = 15)$. Interestingly, WO_3^- was found to react with H_2O to form $H_2WO_4^-$ while WO_4^- and WO_5^- were relatively unreactive. Oxide anions of molybdenum, tantalum, and niobium were produced and studied in a similar fashion. During the course of the work investigations of reactions of acid molecules with a few of these oxide anions[153] were conducted. The chemistry of reactions with HCl proved to be complex, but the findings yielded considerable insight into the nature of the species. For example, reactions of TaO_3^-, $TaO_2(OH)_2^-$, and TaO_5^- with HCl were made for comparison with those for the corresponding niobium species, which were easier to analyze because of the single isotope of Nb. Analogous to the niobium case, all three tantalum oxide anions lead to the same analogous products. The results show that whenever there is an OH unit bound to the metal oxide reactant, Cl from HCl replaces the OH unit which is then given off as H_2O. In the absence of OH units on the metal center, HCl adds in to form an OH group and a Cl bound to the metal atom. For $MO_2(OH)_2^-$ (here M represents Ta or Nb) the sequence of the product peaks was found to be: $MO_2(OH)Cl^-$, $MO_2Cl_2^-$, and $MO(OH)Cl_3^-$, followed by $MOCl_4^-$. It is of interest to note that at low acid concentrations, for niobium the second and fourth products ($NbO_2Cl_2^-$ and $NbOCl_4^-$) are the major ones, while for tantalum the third and fourth products dominate ($TaO(OH)Cl_3^-$ and $TaOCl_4^-$). This suggests a difference between Ta and Nb for the conversion of $MO_2Cl_2^-$ to $MO(OH)Cl_3^-$ and that the tantalum species reacts faster at this step. The requirement here is that the metal center change from four coordinate ($MO_2Cl_2^-$) to five coordinate ($ML(OH)Cl_3^-$). Tantalum and niobium are both group VB elements, and this reactivity difference is almost certainly due to the additional electronic shell for tantalum. Confirming experiments were conducted with DCl, demonstrating an interesting reaction mechanism in the gas phase which has an analogy with acid–base reactions known to occur in the condensed phase.

Molecular Adsorption on Metal Cluster Ions

Understanding the physical and chemical properties of metal and metal compound aggregates has important consequences in surface chemistry, for example, with particular reference to unraveling the basics of catalytic behavior. The intrinsic behavior of these species with regard to the evolution of solid or condensed states from those in the gas phase is also of interest. It was with some of these motivations in mind that we initiated a study of the kinetics of reaction of small molecules with metal cluster ions at thermal energies. Studies of the reactivity of neutral metal clusters at thermal energies using fast-flow reactors are abundant, but investigations of analogous thermal ion reactions are scarce. The study of ions is attractive because there are none of the fragmentation problems generally associated with ionizing neutrals. In addition, direct mass selection enables the details of the reaction mechanisms to be unraveled.

The interaction of carbon monoxide with metal surfaces has been studied extensively. With regard to bonding, adsorption sites, and catalytic processes, the influence of this species on surface chemistry is far reaching. The investigation[155] focused on the association of CO to small copper cluster cations ($Cu^+ - Cu_{14}^+$). Laser vaporization was used to create an ensemble of copper cluster ions, which were then thermalized at high pressures and subsequently allowed to react with carbon monoxide in a flow-tube apparatus. The CO molecule sticks to the copper cluster surface and shows a rapid increase in the association rate as a function of cluster size, clearly displaying transition from atomic to bulk behavior (see Figure 23). Indeed, the kinetics changed from exhibiting low-pressure termolecular behavior for the small clusters ($n < 5$ or 6) to a surface or bulk-like regime for the larger clusters. This transition is characterized by a change in the pressure dependence of the termolecular rate constant, where a collision limited rate is approached for $n \geq 7$.

The overall trend is explainable, at least qualitatively, with simple unimolecular decay theory. When a correction term is incorporated into the theoretical collision limit to account for the intrinsic surface site reactivity in the bulk, excellent agreement is obtained with the overall reactivity. In considering the overall trends, certain clusters, namely Cu^+, Cu_3^+, and Cu_9^+ in particular, displayed somewhat anomalous reactivities which might be associated with structural and/or electronic effects.

With respect to the thermodynamic stability of metal clusters, there is a plethora of results which support the spherical Jellium model for the alkalis as well as for other metals, like copper. This appears to be the case for cluster reactivity, at least for etching reactions, where electronic structure dominates reactivity and minor anomalies are attributable to geometric influence. These cases, however, illustrate a situation where significant addition or diminution of valence electron density occurs via loss or gain of metal atoms. A small molecule, like carbon monoxide,

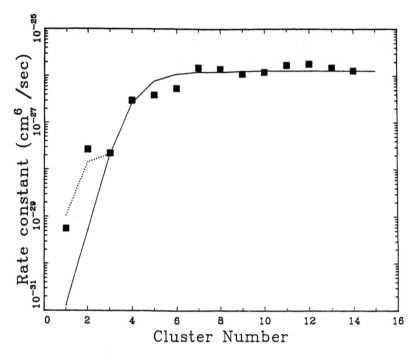

Figure 23. Plot of experimental (■) and theoretical three-body rate constants as a function of cluster size for the clustering of one CO molecule to copper clusters, Cu_n^+. Note the dramatic increase in reactivity (almost four orders of magnitude) within the first seven atom additions to the clusters. The overall trend represents a transition from termolecular to effective bimolecular behavior. The solid line (theory) was obtained assuming a loose transition state while the dotted line shows the results for a tight transition state for monomer and dimer only (upper limit). Taken with permission from ref. 155.

sticking to a cluster surface represents a much smaller net change of electron density and, therefore, geometric factors are expected to play a more important role. Our results show Jellium-like behavior is manifested in the bare cluster abundance spectrum while, in the reactivity pattern, only minor effects are exhibited and then only at small cluster sizes. Therefore, at least for the smaller species, cluster reactivity appears to involve a complex interplay between electronic and geometric factors. These investigations provide further insight useful in the development of simple theories to predict how cluster geometry and electronic structure govern reactivity.

During the course of related work, we identified the extent to which such metal clusters can take up CO through coordination reactions. The work not only shed light on the details of the coordination reactions, but also provided insight into the electronic and reactive characteristics of the metal clusters themselves. The usefulness of the approach was established through studies of the chemistry and kinetics of size-selected cobalt cluster $(Co)_n^+$ ($n = 2$ to 8) reactions with CO.[156] All reactions were found to be association reactions and their absolute rate constants, which were determined quantitatively, were found to have a strong dependence on cluster size. Similar to the cases of reactions with many other reactants such as H_2 and CH_4, Co_4^+ and Co_5^+ displayed a higher reactivity toward the CO molecule than clusters of neighboring size. The multiple-collision conditions employed in the present work have enabled a determination of the maximum coordination number of CO molecules bound onto each Co_n^+ cluster. It is found that the tetramer tends to bond 12 CO molecules, the pentamer 14 CO, hexamer 16 CO, and so on (see Figure 24, for example). The results are interpreted in terms of Lauher's calculations and the polyhedral skeletal electron pair theory.[157] All the measured maximum coordination numbers correlate extremely well with the predictions of these theories, except for the trimer where the measured number is one CO less than the predicted value. The good agreement between experiment and theory enables one to gain some insight into the geometric structures of the clusters. Based on the present findings, the cobalt tetramer cation is interpreted to have a tetrahedral structure, the pentamer a trigonal bipyramid, and the hexamer an octahedral structure (see Table 1).

The size-dependent chemistry and kinetics of gas-phase reactions of Co_n^+ ($n = 2$ to 9) with O_2 were also examined.[158] Co_n^+ are observed to display a very high

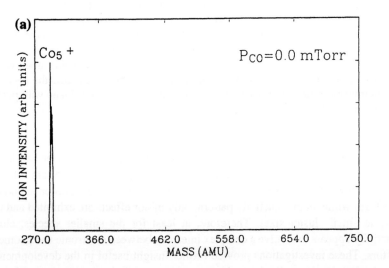

Figure 24. (a) Mass spectrum of Co_5^+ injected into the reaction cell for study. (b) Mass spectrum resulting from the reactions of Co_5^+ with CO at a CO partial pressure of 0.12 mtorr. The peaks labeled by n correspond to $Co_5^+CO_n$. (c) Mass spectrum resulting from the reactions of Co_5^+, with CO at a CO partial pressure of 2.0 mtorr. The labeled peaks correspond to $Co_5^+(CO)_n$. Note the truncation at $n = 14$. Taken with permission from ref. 156.

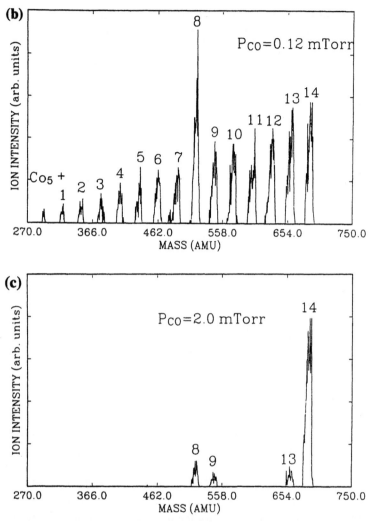

Figure 24. (Continued)

reactivity toward O_2, and the clusters tend to undergo successive oxidation reactions. The bimolecular reaction rate constants measured for the primary reactions display a strong correlation between size of the clusters and their reactivity. As in the case of reactions with other reactant molecules such as CO, clusters containing four and five cobalt atoms exhibit a higher reactivity toward oxygen than $n = 2, 3,$ 6 corresponding to neighboring clusters. The primary reactions result mainly in a replacement of a Co atom by one O_2. Except for a few cases, most of the oxide clusters react with oxygen via either switching or attachment pathways. The

Table 1. The Measured and Calculated Maximum Coordination
Number of CO, and the Predicted Structure for each of Co_n^+ [a]

	Co_n^+ Measured Number	Calculated Number	Predicted Structure
2	8	8	Linear
3	10	11	Triangle
4	12	12	Tetrahedral
5	14	14	Trigonal bipyramid
6	16	16	Octahedral
7	19	19	Pentagonal bipyramid
8	20	20	Capped pentagonal bipyramid

Note: [a]Taken by permission from ref. 156.

successive oxidation reactions of Co_n^+ terminate when the oxide clusters which are formed have stoichiometric structures corresponding to $(CoO)_4(CoO_2)_n^+$ ($n = 0$ to 3), or $Co_2O_{4,5}^+$, or $CO_3O_{4,5}^+$.

H. Intracluster Reactions Initiated through Photon Absorption

Resonant Enhanced Ionization

Resonance enhanced ionization is a valuable tool in the field of clusters, both for spectroscopic studies of solvation shifts, and as a way to initiate ion reactions of a specific nature. Applying this to the case of ammonia bound to the chromophore phenyl acetylene (PA), protonated ammonia clusters of specific size have been observed[159] to form following the resonant enhanced ionization of the clusters $PA \cdot (NH_3)_n$ in the O_0^0 region of the $S_1 - S_0$ transition of the (unclustered) PA. A detailed systematic investigation of the spectroscopic shifts upon the resonant enhanced ionization of $PA-NH_3$ clusters using a laser-based TOF apparatus has revealed rather different trends depending on whether the clusters were produced by coexpansion or by attaching PA to the preformed ammonia clusters.[159,160] Of particular interest was the finding that ionization also leads to the formation of protonated ammonia clusters of specific sizes, namely $H^+(NH_3)_n$, with $n \geq 2$.

The observed $H^+(NH_3)_n$ and $H^+(NH_3)_n(PA)$ clusters are thought to be formed in a two-step reaction sequence taking place after ionization of the $PA(NH_3)_n$ cluster. The first step is a charge transfer (CT) reaction between the resonantly ionized PA^+ and the NH_3 molecules in the cluster. The second step is an intracluster ion–molecule reaction (ICIMR) of the charged ammonia cluster leading to the formation of an $(n - 1)$ protonated cluster ion; this has been previously established for NH_3 clusters[33] and is sufficiently exothermic for fragmentation of the cluster.

The proposed reaction sequence is as follows:

$$\text{(PA)(NH}_3)_n \xrightarrow{h\nu} (\text{PA}^+)(\text{NH}_3)_n + e^-, \tag{40}$$

$$(\text{PA}^+)(\text{NH}_3)_n \xrightarrow{\text{CT}} (\text{PA})(\text{NH}_3)_n^+, \tag{41}$$

$$(\text{PA})(\text{NH}_3)_n^+ \xrightarrow{\text{ICIMR}} \text{PA} + \text{NH}_2 + \text{H}^+(\text{NH}_3)_{n-1-x} + x\text{NH}_3, \tag{42}$$

$$(\text{PA})(\text{NH}_3)_n^+ \xrightarrow{\text{ICIMR}} \text{NH}_2 + \text{H}^+(\text{NH}_3)_{n-1-x}\text{PA} + x\text{NH}_3. \tag{43}$$

In this reaction scheme, the magnitude of x is related to the excess energy of the ionization process.

The observed overlap between the spectral features which appear in the one-color TOF mass channels corresponding to $\text{PA(NH}_3)_n$ clusters, where $n \geq 5$, with those which appear in masses corresponding to the ammonium cluster ions obtained in the same cluster attachment experiments, enabled a determination of the cluster size where the reaction[42] first becomes exoergic. This assignment is made possible by the different spectral shifts in the case of ammonia–PA clusters produced by coexpansion versus attachment in a pick-up (collision cell) source , and the shift which this induces in the corresponding spectrum of the resulting protonated ammonia cluster ion products. The formation of these protonated cluster ions starting with the charge transfer from PA^+ to the attached ammonia clusters, sets an upper limit for the appearance potential of $(\text{NH}_3)_5^+$ equal to the IP of PA which is 8.8 eV. This is consistent with the ionization potentials estimated from a thermo-dynamic data using the Born–Haber cycle.[97] Based on the arguments given above, the adiabatic ionization potential of $(\text{NH}_3)_5$ is assigned to be (only slightly) below 8.8 eV and that of $(\text{NH}_3)_4$ to be above this value. These findings demonstrate another aspect of intercluster reactions, and also show how thermochemical information can be acquired under favorable conditions.

Ionization Through Intracluster Penning Processes

Another interesting situation which demonstrates the role of solvation on cluster reactions, with particular relevance to identifying mechanisms of possible impor-tance in the condensed phase, concerns the initiation of electron transfer reactions. Interesting examples include those initiated through Penning ionization.[161]

Clusters provide interesting systems for investigating the influence of solvation on ionization and concomitant electron transfer processes, including Penning ionization, in terms of their gas-phase counterparts. Analogous processes in paraxylene (PX) bound to $\text{N(CH}_3)_3$ were studied[86] following the absorption of photons through the perturbed S_1 state of paraxylene into high Rydberg states. An interesting comparison is provided by results of studies involving adducts of paraxylene bound to NH_3 with those involving trimethylamine, since the ionization potential of paraxylene is less than that of ammonia but greater than that of trimethylamine.[162] In the case of $PX \cdot NH_3$, ionization by adsorption of a second photon into the perturbed S_1 state of paraxylene was found to begin near the ionization threshold of paraxylene, and to lead to the expected cluster ion $PX \cdot NH_3^+$; NH_3^+ is also observed at 1.8 eV above this threshold and NH_4^+ is seen in two-color experiments made at high fluence of the ionizing laser.

By contrast, the absorption of photons into high Rydberg states of paraxylene below its ionization potential in $PX \cdot N(CH_3)_3$ was found to produce primarily $N(CH_3)_3^+$, with $H^+N(CH_3)_3$ as a minor product. No $PX \cdot N(CH_3)_3^+$ ion was detectable. One conclusion is that photoexcitation of paraxylene leads to an intercluster ionization process bearing analogy to Penning ionization, where the perturbed high Rydberg states of paraxylene interact with the partner molecule $N(CH_3)_3$. Another more interesting observation, was the finding of a slow ionization process as evident in the time-of-flight peak shapes shown in Figure 25. Since the laser interacts with the molecules in the first of a two-field acceleration region, a long tail is only possible when the ionization process is slow. (It should be noted that fragmentation leads to a knee in the peak shape, and not a long tail as observed in the figure.) Interestingly, the process is substantially slower when the energy of the ionizing

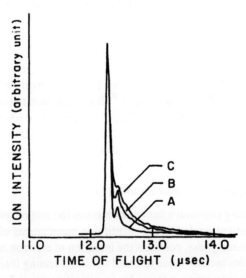

Figure 25. Ion mass peaks at different two-photon energies. Broadenings of trimethylamine (TMA$^+$) ion peaks as a function of the ionization energy. **A:** $h\nu_2 \geq 3.875$ eV; **B:** $h\nu_2 = 3.688$ eV; **C:** $h\nu_2 = 3.607$ eV. Excitation energy of paraxylene (PX) in the S_1 state = 3.90 eV. The broadenings in **B** and **C** correspond to time constants of 160 ± 20 and 200 ± 20 ns, respectively. Peaks corresponding to TMA·H$^+$ are also observable. Taken with permission from *Int. J. Mass Spectrom. Ion Proc.* **1994**, *131*, 233–264.

photon is decreased. Questions arise whether the slow step is associated with the proton transfer channel (i.e. the $(CH_3)_3NH^+$ product) or an electron transfer process (i.e. the $(CH_3)_3N^+$ product), but careful measurements with deuterated species revealed that the tail is largely associated with the electron transfer and not the proton transfer process. A plausible explanation is that a large geometry change is involved in the formation of the trimethylamine ion during the Penning-like ionization process. It is worth noting that Hatano[163] has found orientational effects in the liquid phase, where the motion is restricted. This can lead to a significant reduction in the rate of a process which is also believed to proceed via Penning ionization.

IV. GAINING INSIGHTS INTO CLUSTER THERMOCHEMISTRY AND STRUCTURES

Studying the formation of cluster ions and their ensuing ionization dynamics provides methods of ascertaining their thermodynamical properties, and in favorable cases, evidence of structures and the origins of magic numbers which appear in cluster distributions.

A. The Origin of Magic Numbers

Elucidating the origin of magic numbers has been a problem of long-standing interest, made accessible through the use of the laser-based reflectron TOF technique and evaporative ensemble theory. Three test cases are considered, first protonated ammonia clusters where $(NH_3)_4 \cdot NH_4^+$ has been found to be especially prominent, and then two other cases are considered, one involving water cluster ions and another rare gas clusters.

Protonated Clusters—Ammonia

As discussed above, clusters of ammonia have become an important system for unraveling the origin of magic numbers. During the course of a systematic study of the dissociation of ammonia clusters following their ionization by laser multiphoton ionization techniques, direct proof was obtained for the importance of the metastability of the cluster ions and for the extensive dissociation processes which influence the resulting cluster distributions.[33] Although essentially all ammonia clusters undergo fragmentation following ionization, among proximate species only the protonated pentamer displays an anomalously large abundance in its size range.[35,43,70]

Employing a modified QET/RRK statistical analysis,[164,165] binding energies for large cluster ions can be readily determined.[43] In this method, the deduced binding energy values must be scaled to some determined value (the best available[33] for $(NH_3)_5H^+$ is used here), whereupon the binding energies of all other clusters are found to be in good agreement with the reported literature values for cluster ions

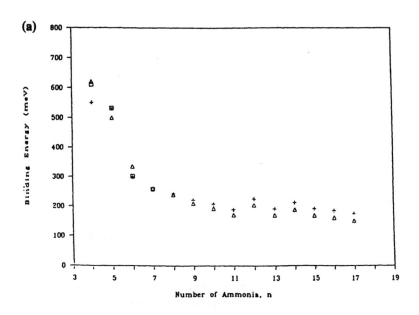

(b) Relative BE of Ammonia Cluster Ions

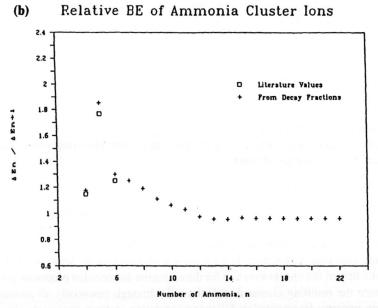

Figure 26. (a) A plot of calculated binding energies of $(NH_3)_nH^+$, $n = 4–17$, as a function of cluster size. Δ are values deduced using Klots' evaporative ensemble model; + are values deduced using Engelking's modified QET/RRK model and the kinetic energy measurements shown in Figure 13; $\square$ are literature values based on high-pressure mass spectrometry. (b) Relative binding energies for $(NH_3)_nH^+$ deduced from a study of the metastable decay fractions and employing the evaporative ensemble approach; equations 5 and 6. (a) Taken with permission from *J. Chem. Phys.* **1990**, *93*, 2506–2512. (b) Taken with permission from ref. 22.

$(NH_3)_nH^+$ ($n = 4$, 6 and 7). However, from equation 6 and the measured kinetic energy release values, the absolute bond energies also can be directly determined. Figure 26a demonstrates the applicability of this method in deducing bond energies for the ammonia system. The trend of the determined binding energies of metastable cluster ions $(NH_3)_nH^+$ is seen to undergo a progressive decrease from $n = 4$–5, display a precipitous drop from 5 to 6, and then slowly decreases thereafter. There are hints of stable structures at $n = 12$ and 14, which are slightly more stable than neighboring ones. The agreement with literature values determined by more conventional methods is excellent. Related techniques have been employed for a variety of cluster ions, establishing, for example, that magic numbers generally arise due to the stability of the cluster ions rather than the parent neutral clusters (see Figure 26b as an example for the ammonia system).

Rare-Gas Cluster Ions

Early studies of rare-gas clusters also attributed special magic numbers in these systems species to especially stable geometric structures existing in the neutral distribution,[166,167] but this interpretation was questioned[168] when recognition was given to the possibilities of fragmentation and metastable dissociation following the ionization processes employed in the investigations.

The metastable processes which typically can occur following cluster ionization have provided the basis for the design of experimental studies that have finally resolved the controversies. Considerations of unimolecular dissociation and the evaporative canonical ensemble[169] paved the way for determining bond energies of cluster systems through determinations of dissociation fractions versus cluster size in various time windows. A one-to-one correspondence between the relative bond energies of proximate species in the cluster distribution and the observed magic numbers in the xenon system has definitively established that magic numbers arise in these systems due to inherent thermochemical stability of the cluster ions per se; see Figures 27a and b.

Structure and Bonding of Water Clusters

The study of water clusters has been a subject of long-standing interest,[142,170–182] particularly because of the prominent magic number $(H_2O)_{20}\cdot H_3O^+$ which is seen in cluster distributions obtained by a wide range of experimental techniques. While this and related magic numbers have been reported for protonated water clusters under sputtering and beam expansion conditions, it has been a rather controversial subject as to whether or not these represented special structures or merely constituted the freezing-in of kinetic processes due to the nature of non-equilibrium production techniques. To partially resolve this issue we studied the quasi-equilibrium distribution of water clusters in a flow-tube reactor.[142] Their self-assembly and stability at thermal conditions proves that magic numbers of protonated hydrates, corresponding to $n = 21$, 24, 26, 28 and 38, represent species with special thermodynamic stability (see Figure 28a).

TOF Spectrum of Xenon Clusters

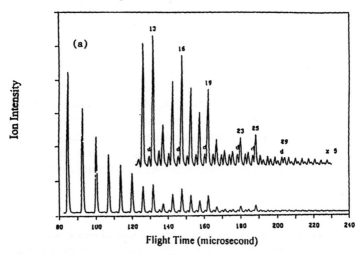

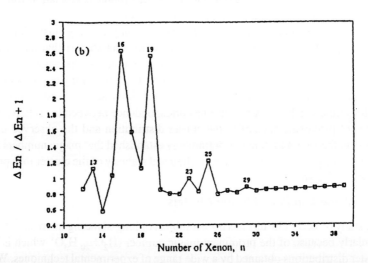

Figure 27. (a) A reflectron time-of-flight mass spectrum of xenon cluster ions Xe$_n^+$, n = 6–39. The peaks on the left side labeled as "d" are the daughter ions which are produced in the field-free region. The peaks on the right side labeled according to their sizes are the parent ions which do not dissociation in the field-free region. (b) Relative binding energies of $\Delta E_n/\Delta E_{n+1}$ deduced by fitting the calculated values to the measured ones (equations 5 and 6) are plotted as a function of cluster sizes, n assuming $\gamma = 25$ and $C_n = (2.77n - 6)$ k_b, where k_b is the Boltzmann constant. It is evident that the local maxima match exactly the magic numbers observed in the mass spectrum shown in (a). (a) Taken with permission from ref. 92. (b) Taken with permission from *J. Chem. Soc., Faraday Trans.* **1990**, *86*, 2389, 2452–2455, 2495–2522.

Thereafter, we succeeded in producing negative cluster ions of water of the form $OH^-(H_2O)_n$ with n ranging from 0 to 46.[144] To the best of our knowledge, these represented the first successful attempts to produce negative cluster ions containing OH^- for sizes greater than $n = 4$. In contrast to the results of the large protonated water clusters, the magic numbers observed under thermal conditions are not as prominent as those observed in the analogous cations, probably due to less strong bonding in the anion systems. Nevertheless, the main magic number is the same, namely $OH^-(H_2O)_{20}$; see Figure 28b. This suggested that the charge might reside at the center of a clathrate-like structure.

An important development from our laboratory has been the development of a "titration" technique[181] for use in elucidating the structures of clusters from which hydrogen atoms or protons extend. This work provided the first definitive evidence concerning the structure of the magic number for the protonated water cluster $(H_2O)_{21}H^+$. Many speculations existed in the literature concerning the stability of this species and the reasons for its observation in studies ranging from those of expansions of neutral water molecules and subsequent electron impact ionization, to expansion of protons in molecular beams, and even following the sputtering of ions from ice-like surfaces. Our work using the titration method now has established that the 21mer is a clathrate-like cage with an encaged ion (see Figure 29). As an extension of our investigations of the structures of clusters, we undertook studies of their interaction with various metal ions, also deducing structures that are analogous to clathrate hydrates known to exist in the condensed phase.

In order to contribute to further understanding of the origin of magic numbers, especially through providing thermochemical information on water clusters, we undertook investigations[183] of the thermochemical properties of water cluster ions. One of these was based on a TOF reflectron technique whereby the decay fractions of dissociating clusters could be measured and the binding energies derived using the evaporative ensemble model.[58–61] Then, through an investigation of the steady-state distribution of clusters in a flow tube reactor under well-defined thermal conditions, we ascertained intensity ratios and concentrations which enabled us to deduce values of their free energy of formation. By combining these measurements,[183] we were able to ascertain bond energies and entropies as a function of cluster size for species ranging from 6 to 28 water molecules; see Figure 30.

Considering the various thermochemical data suggests that the "magic" number is likely to be due to an entropic effect. To test this hypothesis, a series of simple calculations based on standard statistical mechanical equations, and ones employing the liquid drop model[184,185] were made to estimate entropy and enthalpy trends with cluster size. The calculations were found to be consistent with the experimental findings. Related studies on mixed alcohol–water systems have provided confirming evidence of the nature of the magic numbers in water clusters.[186]

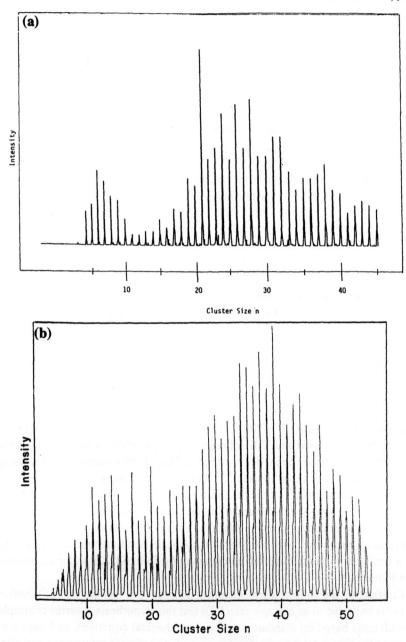

Figure 28. (a) Mass spectrum of protonated water clusters $H^+(H_2O)_n$ ($n = 4$–45) at 119 K and 0.3 torr He in a flow tube reactor. Note the prominence of $H_3O^+(H_2O)_{20}$ even under quasi-equilibrium conditions. (b) Mass-spectrometric abundance of OH^- $(H_2O)_n$ produced under thermal conditions. Note a magic number at $n = 20$, though not as prominent as for the case of H_3O^+ hydrates. Taken with permission from ref. 92.

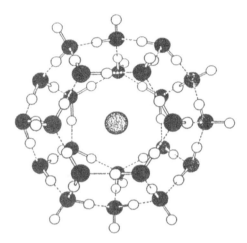

Figure 29. Water clusters with H_3O^+ or metal ions (depicted by a shaded circle) encaged inside the clathrate $(H_2O)_{20}$. Taken with permission from *Int. J. Mass Spectrom. Ion Proc.* **1994,** *131,* 233–264.

B. Influence of Solvation on the Composition of Protonated Clusters

Mixed Cluster Systems

Studies of ion solvation, structure, and stability of mixed clusters are also possible using the reflectron TOF mass spectrometer technique. In small hydrogen-bonded cluster ions, it is known that stability is due to the nature and location of the ligand molecules which bond to the central proton or protonated species, for example, $(CH_3COCH_3)_2H^+$,[187] $(CH_3OCH_3)_3H_3O^+$,[188] and $(NH_3)_4NH_4^+$.[70,189] It is interesting to consider the location of the proton in the general case for systems of widely varying proton affinity. For example, it is instructive to compare a situation where the proton affinity of molecule X in a mixed cluster ion such as $(NH_3)_n(X)_mH^+$ is less than that of another such as ammonia, with the alternative case where the proton affinity of molecule X is larger.

In all three mixed cluster ions systems, the intensity distributions show[94] that there is a maximum at $n + m = 5$. Results of metastable decomposition studies[95] of the mixed cluster ions were determined to be:

$$(NH_3)_n(X)_mH^+ \rightarrow (NH_3)_n(X)_{m-1}H^+ + X \qquad \text{for } n = 1 \qquad (44)$$

$$(NH_3)_n(X)_mH^+ \rightarrow (NH_3)_{n-1}(X)_mH^+ + NH_3 \qquad \text{for } n \geq 2 \qquad (45)$$

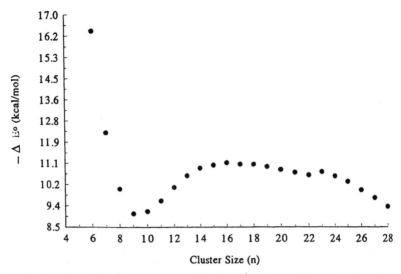

Figure 30. Relative binding energies of $(H_2O)_nH^+$. Taken with permission from *J. Chem. Phys.* **1993**, *99*, 8009–8015.

The proton affinity of ammonia is 204.0 cal mol^{-1} and those of CH_3COCH_3, CH_3CN, and CH_3CHO are 196.7 cal mol^{-1}, 188.4 cal mol^{-1}, and 186.6 cal mol^{-1}, respectively.[190] The results clearly indicate that NH_4^+ is the core ion and this is expected considering their relative proton affinities. The four available hydrogen-bonding sites become occupied in order for the first solvation shell to become completed. The four ligands can include any combination of ammonia and molecule X. The loss of an ammonia molecule resulting from the metastable decomposition for $n \geq 2$ indicates that X is more strongly bonded to the NH_4^+ than is another NH_3. This is not surprising since all molecules considered here have higher dipole moments and polarizabilities than ammonia, which leads to them having a greater ion–dipole and ion-induced dipole interaction.

Two other molecules considered in the same context, but with proton affinities larger than ammonia, were pyridine (C_5H_5N) and trimethylamine (TMA) $(CH_3)_3N$ whose proton affinities are 220.8 cal mol^{-1} and 225.1 cal mol^{-1}, respectively.[190] Metastable decomposition studies of $NH_3(C_5H_5N)_mH^+$ ($m = 1–5$) yield the following results:

$$NH_3(C_5H_5N)_mH^+ \rightarrow (C_5H_5N)_mH^+ + NH_3 \qquad \text{for } m < 4 \qquad (46)$$

$$NH_3(C_5H_5N)_mH^+ \rightarrow NH_3(C_5H_5N)_{m-1}H^+ + C_2H_5N \qquad \text{for } m \geq 4 \qquad (47)$$

For $m < 4$, the higher proton affinity of pyridine leads to its stronger bonding to the proton, whereby the ammonia molecule is lost upon metastable dissociation. The loss of pyridine in the case of $m \geq 4$ can be accounted for by the fact that a central NH_4^+ core ion is formed which provides four sites for hydrogen bonding to the ligands. Hence the structure is dictated by the net energetics of bonding, a finding consistent with the previous proposed stable structure for the cluster ion $(NH_3)_n(X)_m H^+$.

Other studies conducted on mixed protonated clusters of ammonia bound with TMA showed that the ion intensity distributions of $(NH_3)_n(TMA)_m H^+$ [191] display local maxima at $(n,m) = (1,4)$, $(2,3)$, $(2,6)$, $(3,2)$, and $(3,8)$. Observation that the maximum ion intensity occurs at $(n,m) = (1,4)$, $(2,3)$, and $(3,2)$ indicates that a solvation shell is formed around the NH_4^+ ion with four ligands of any combination of ammonia and TMA molecules. In the situation where the maximum of the ion intensity occurs at $(n,m) = (2,6)$ and $(3,8)$, the experimental results suggest that another solvation shell forms which contains the core ions $[H_3N–H–NH_3]^+$ (with six available hydrogen-bonding sites) and $[H_3N–H(NH_2)H–NH_3]^+$ (with eight available hydrogen-bonding sites). The observed metastable unimolecular decomposition processes support the above solvation model.

Examples of cluster distributions, and structures related to these are shown in Figures 31a and b. In considering the implications of the findings, it must be realized that the systems are very floppy and the resulting clusters are not expected to display a rigid structure. Other studies were conducted on ternary systems comprising the foregoing species bound with water. The observations revealed the general validity of the above concepts.

Experiments made at higher degrees of aggregation have provided strong evidence[192] for ring-like structures for mixed neutral clusters. For example, under a wide variety of experimental conditions, mixed cluster ions display a maximum intensity at $m = 2(n + 1)$ when $n \leq 5$ for $(NH_3)_n \cdot (M)_m H^+$, and $m = n + 2$ when $n \leq 4$ for $(H_2O)_n \cdot (M)_m H^+$; M is a proton acceptor such as acetone, pyridine, and trimethylamine. These findings reveal that the cluster ions with these compositions have stable solvation shell structures as discussed above.

A breakdown of the pattern is seen to occur at $n > 5$ for the ammonia system and $n > 4$ for the water system, with the most intense peaks occurring for species with one molecule less than the expected pattern, i.e. $m = 2(n + 1) - 1$ when $n = 6$ for $(NH_3)_n \cdot (M)_m H^+$ and $m = (n + 2) - 1$ when $n = 5$ for $(H_2O)_n \cdot (M)_m H^+$. These results are consistent with suggestions that hydrogen-bonded ring structures form. As the clusters grow to larger and larger sizes, the structures evidently are more stable when they undergo rearrangement or ring closure (see Figure 32).

In other related studies,[186] we determined the structures of mixed water–methanol clusters; the intensity distributions of $(H_2O)_n(CH_3OH)_m H^+$ showed magic numbers at $n + m = 21$, $0 \leq m \leq 8$ due to the enhanced stabilities of the dodecahedral cage structures in the mixed clusters. Studies of the metastable dissociation of $(H_2O)_n(CH_3OH)_m H^+$ $(n + m < 40)$ provided evidence that the dissociation channels

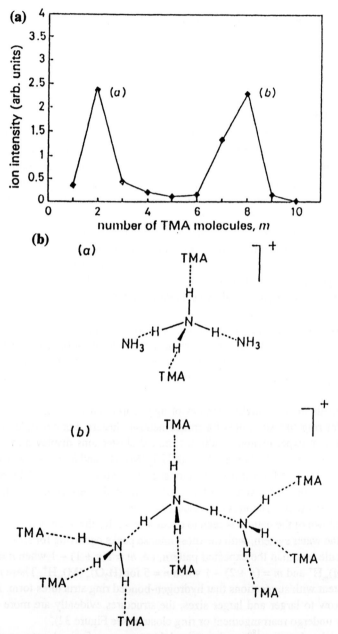

Figure 31. (a) Ion intensity distribution of $(TMA)_m(NH_3)_3H^+$. (b) Proposed structure for (a) $(TMA)_2(NH_3)_2H^+$ and (b) $(TMA)_8(NH_3)_3H^+$. Taken with permission from *J. Chem. Soc., Faraday Trans.* **1990**, *86*, 2389, 2452–2455, 2495–2522.

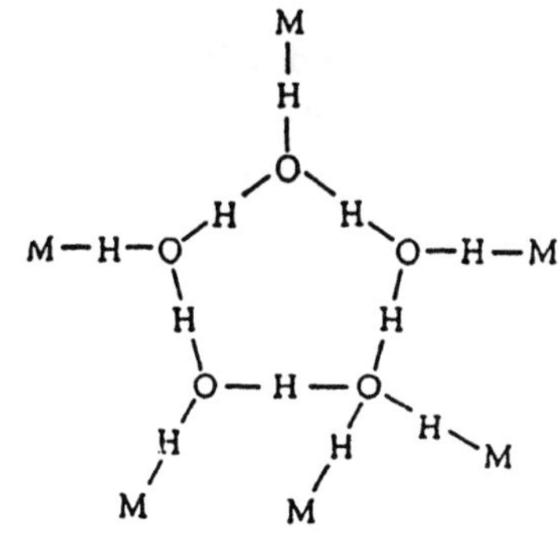

```
                    M
                    |
                    H
                    |
                  .O.
               H.´    `.H
        M—H—O´            `O—H—M
            \              /
            H              H
             \            /
              O — H — O
             /         \  `.H
            H           H    `M
           /           /
          M           M
```

Figure 32. Structure compatible with the observed magic number corresponding to $(H_2O)_5(M)_6H^+$. Taken with permission from *Int. J. Mass Spectrom. Ion Proc.* **1994,** *131,* 233–264.

are governed by the competition of effects dominated by dipole–dipole and dipole polarizability rather than those arising due to the hydrogen-bonded nature of the cluster, and the cluster structure.

An important aspect of studying metastable dissociation of these clusters is that the measurements enable a determination of the surface composition of mixed systems. This is important in designing experiments to study the heterogeneous chemistry of aqueous systems. For example, the loss channel of H_2O is found to be open to all $(H_2O)_n(CH_3OH)_mH^+$ except $(H_2O)(CH_3OH)_mH^+$ for which the water loss is relatively small. For the water-rich composition mixed clusters, the results show that water molecules have a tendency to build a cage structure in the cluster size region $m + n = 21$, with $0 \leq m \leq 8$.

V. SUMMARY AND OUTLOOK FOR THE FUTURE

The field of cluster science has grown explosively in recent years and has become one of the major subjects in chemical physics. This attention has been driven in part by the availability of new experimental techniques to study clusters, but more importantly due to the fact that they serve to bridge the gas and the condensed phases. This subject has been the main theme of the current review. While most of the attention has been given only to results from our own laboratory, even then one can see how extensive the field has become. The availability of experimental tools, such as the fast-flow reactor, that enables rate measurements and equilibrium determinations under well thermalized conditions, provide a very useful means of deducing rate constants. These give insight into the nature of solvation-driven reactions and at the same time new information on ones of practical value in wide ranging fields from atmospheric science to biochemistry. Studies in molecular beams, especially in conjunction with the reflectron time-of-flight technique coupled with pulsed lasers, have enabled the detailed dynamics of cluster reactions and dissociation to be interrogated at the molecular level. The findings provide insights into the detailed mechanisms involved, especially when studied with ultrafast pump-probe techniques. This is undoubtedly an area which will continue to expand and yield new knowledge on unexpected effects of solvation on ion–molecule reactions that have, so far, only been touched upon.

ACKNOWLEDGMENTS

Financial support by the Atmospheric Sciences Division of the National Science Foundation (Grant No. ATM-9321660); the Division of Chemical Sciences, Office of Basic Energy Sciences, Office of Energy Research of the Department of Energy (Grant No. DE-FGO2-92ER14258); and the Air Force Office of Scientific Research (Grant No. F49620-94-1-0162) is gratefully acknowledged.

REFERENCES

1. Castleman, Jr., A. W.; Keesee, R. G. *Chem. Rev.* **1986**, *86*, 589–618.
2. Castleman, Jr., A. W. In *Clusters of Atoms and Molecules II*; Haberland, H., Ed.; Springer-Verlag: Berlin, Heidelberg, 1994, pp 77–133.
3. Castleman, Jr., A. W.; Keesee, R. G. *Ann. Rev. Phys. Chem.* **1986**, *37*, 525–50.
4. Castleman, Jr., A. W.; Keesee, R. G. *Accts. Chem. Res.* **1986**, *19*, 413–419.
5. Castleman, Jr., A. W.; Keesee, R. G. *Science* **1988**, *241*, 36–42.
6. Jortner, J.; Pullman, A.; Pullman, B. (Eds). *Large Finite Systems*; 1987, Vol. 20.
7. Jena, P.; Rao, B. K.; Khanna, S. N. *Physics and Chemistry of Small Clusters*; NATO ASI Series, 1987, Vol. 158.
8. Benedek, G.; Martin, T. P.; Pacchioni, G. (Eds.). *Elemental and Molecular Clusters*; Springer-Verlag: Berlin, 1988; *Surface Sci.* **1985**, *156*, Part 1 and Part 2.
9. Schuster, P.; Wolschann, P.; Tortschanoff, K. In *Chemical Relaxation in Molecular Biology*; Pecht, I;. Rigler, R., Eds.; Springer-Verlag: 1977, pp 107–190.
10. Castleman, Jr., A. W.; Märk, T. D. *Adv. Atomic Molecular Phys.* **1984**, *20*, 65–172.
11. Castleman, Jr., A. W.; Märk, T. D. In *Gaseous Ion Chemistry/Mass Spectrometry*; John Wiley and Sons, New York, 1986, Chapter 12, pp 259–303.
12. Castleman, Jr., A. W.; Keesee, R. G. In *Physics and Chemistry of Small Clusters*; Jena, P.; Rao, B. K.; Khanna, S. N., Eds. NATO ASI Series B: Physics, 1987, Vol. 158, pp 565–582.
13. Castleman, Jr., A. W.; Keesee, R. G. In *Elemental and Molecular Clusters in Materials Science*; Benedik, G.; Martin, T. P.; Pacchioni, G., Eds.; Springer Verlag, 1987, pp 307–328.
14. Castleman, Jr., A. W.; Keesee, R. G. In *The Structure of Small Molecules and Ions*; Naaman, R.; Vager, Z., Eds.; Plenum: New York, 1988, pp 133–145.
15. Keesee, R. G.; Castleman, Jr., A. W. In *Ion and Cluster Spectroscopy and Structure*; Maier, J., Ed.; Elsevier Science: 1989, pp 275–327.
16. Keesee, R. G.; Castleman, Jr., A. W. In *Atomic and Molecular Clusters*; Bernstein, E. R., Ed.; Elsevier Science: 1990, pp 507–550.
17. Castleman, Jr., A. W. *Int. J. Quantum Chemistry* **1991**, *25*, 527–544.
18. Castleman, Jr., A. W. In *Clustering Phenomena in Atoms and Nuclei*; Brenner, M.; Lönnroth, T.; Malik, F. B., Eds.; Springer-Verlag: Berlin, 1992, pp 99–109.
19. Castleman, Jr., A. W. In *Advances in Mass Spectrometry*; Elsevier Science: 1992, Vol. 12, pp 167–189; see also *Int. J. Mass Spectrom. Ion Proc.* **1992**, *118/119*, 167–189.

20. Stanley, R. J.; Castleman, Jr., A. W. In *Molecular Dynamics and Spectroscopy by Stimulated Emission Pumping*; Dai, H. L.; Field, R. W., Eds.; World Scientific: 1995, pp 689–730.
21. Castleman, Jr., A. W.; Wei, S. *Annu. Rev. Phys. Chem.* **1994**, *45*, 685–719.
22. Wei, S; Castleman, Jr., A. W. In *Chemical Reactions in Clusters*; Bernstein, E. R., Ed.; Oxford University. In press.
23. Keesee, R. G.; Castleman, Jr., A. W. *J. Phys. Chem. Ref. Data* **1986**, *15*, 1011–1071.
24. Kebarle, P. *Ann. Rev. Phys. Chem.* **1977**, *28*, 445.
25. Hunton, D. E.; Hofmann, M.; Lindeman, T. G.; Castleman, Jr., A. W. *Chem. Phys. Lett.* **1983**, *96*, 328–332.
26. Hunton, D. E.; Hofmann, M.; Lindeman, T. G.; Castleman, Jr., A. W. *J. Chem. Phys.* **1985**, *82*, 134–150.
27. Hunton, D. E.; Hofmann, M.; Lindeman, T. G.; Albertoni, C.; Castleman, Jr., A. W. *J. Chem. Phys.* **1985**, *82*, 2884–2895.
28. Ferguson, E.; Albertoni, C. R.; Kuhn, R.; Chen, Z. Y.; Keesee, R. G.; Castleman, Jr., A. W. *J. Chem. Phys.* **1988**, *88*, 6335–6340.
29. Albertoni, C. R.; Castleman, Jr., A. W.; Ferguson, E. E. *Chem. Phys. Lett.* **1989**, *157*, 159–163.
30. Chen, Z. Y.; Albertoni, C. R.; Hasegawa, M.; Kuhn, R.; Castleman, Jr., A. W. *J. Chem. Phys.* **1989**, *91*, 4019–4025.
31. Chen, Z. Y.; Cogley, C. D.; Hendricks, J. H.; May, B. D.; Castleman, Jr., A. W. *J. Chem. Phys.* **1990**, *93*, 3215–23.
32. Levinger, N. E.; Ray, M. L.; Alexander, M. L.; Lineberger, W. C. *J. Chem. Phys.* **1988**, *89*, 5654; Levinger, N. E.; Ray, D.; Murray, Mullin, A. S.; Schulz, C. P.; Lineberger, W. C. *J. Chem. Phys.* **1988**, *89*, 71.
33. Echt, O.; Dao, P. D.; Morgan, S.; Castleman, Jr., A. W. *J. Chem. Phys.* **1985**, *82*, 4076–4085.
34. Morgan, S.; Castleman, Jr., A. W. *J. Phys. Chem.* **1989**, *93*, 4544–4550.
35. Wei, S.; Tzeng, W. B.; Castleman, Jr., A. W. *J. Chem. Phys.* **1990**, *93*, 2506–2512.
36. Wei, S.; Tzeng, W. B.; Castleman, Jr., A. W. *J. Am. Chem. Soc.* **1991**, *113*, 1960–1969.
37. Walder, G.; Foltin, M.; Stefanson, T.; Castleman, Jr., A. W.; Märk, T. D. *Int. J. Mass Spectrom. Ion Proc.* **1991**, *107*, 127–134.
38. Foltin, M.; Walder, G.; Castleman, Jr., A. W.; Märk, T. D. *J. Chem. Phys.* **1991**, *94*, 810–11.
39. Foltin, M.; Walder, G.; Mohr, S.; Scheier, P.; Castleman, Jr., A. W.; Märk, T. D. *Z. Phys. D.* **1991**, *20*, 157–159.
40. Wei, S.; Tzeng, W. B.; Castleman, Jr., A. W. *J. Phys. Chem.* **1991**, *95*, 5080–5085.
41. Wei, S.; Kilgore, K.; Tzeng, W. B.; Castleman, Jr., A. W. *J. Phys. Chem.* **1991**, *95*, 8306–8309.
42. Yang, X.; Castleman, Jr., A. W. *J. Chem. Phys.* **1991**, *95*, 130–134.
43. Wei, S.; Tzeng, W. B.; Castleman, Jr., A. W. *J. Chem. Phys.* **1990**, *92*, 332–339.
44. Chen, Z. Y.; May, B. D.; Castleman, Jr., A. W. *Z. Phys. D.* **1993**, *23*, 239–246.
45. Ferguson, E. E.; Fehsenfeld, F. C.; Schmeltekopf, A. L. *Adv. At. Mol. Phys.* **1969**, *5*, 1.
46. Smith, D.; Adams, N. G. In *Gas Phase Ion Chemistry*, Bowers, M. T., Ed.; Academic: 1979, Vol. 1, p 1.
47. Albritton, D. L. *At. Data Nucl. Data Tables* **1978**, *22*; Ikezoe, Y.; Matsuoka, S.; Takebe, M.; Viggiano, A. *Gas Phase Ion-Molecule Reaction Rate Constants Through 1986*; Maruzen: Japan, 1987.
48. Upschulte, B. L.; Shul, R. J.; Passarella, R.; Keesee, R. G.; Castleman, Jr., A. W. *Int. J. Mass Spectrom. Ion Proc.* **1987**, *75*, 27–45.
49. Upschulte, B. L.; Shul, R. J.; Passarella, R.; Keesee, R. G.; Castleman, Jr., A. W. *Int. J. Mass Spectrom. Ion Proc.* **1988**, *85*, 277–285.
50. Castleman, Jr., A. W.; Sigsworth, S.; Leuchtner, R. E.; Weil, K. G.; Keesee, R. G. *J. Chem. Phys.* **1987**, *86*, 3829–3835.
51. Shul, R. J.; Passarella, R.; Upschulte, B. L.; Keesee, R. G.; Castleman, Jr., A. W. *J. Chem. Phys.* **1987**, *86*, 4446–4451.

52. Shul, R. J.; Passarella, R.; Yang, X. L.; Keesee, R. G.; Castleman, Jr., A. W. *J. Chem. Phys.* **1987**, *87*, 1630–1636.
53. Wiley, W. C.; McLaren, I. H. *Rev. Sci. Instrum.* **1955**, *26*, 1150.
54. Durant, J. L.; Rider, D. M.; Anderson, S. L.; Proch, F. D.; Zare, R. N. *J. Chem. Phys.* **1984**, *80*, 1817.
55. Kuhlewind, H.; Boesl, U.; Weinkauf, R.; Neusser, H. J.; Schlag, E. W. *Laser Chem.* **1983**, *3*, 3.
56. Karataev, V. I.; Mamyrin, B. A.; Shmikk, D. V. *Sov. Phys. Tech. Phys.* **1972**, *16*, 1177; Mamyrin, V. A.; Karataev, V. I.; Shmikk, D. V.; Zauglin, V. A. *Sov. Phys. JETP* **1973**, *37*, 45.
57. Anderson, J. B.; Andres, R. P.; Fenn, J. B. *Adv. Chem. Phys.* **1966**, *10*, 275; Hagena, O. F. *Surf. Sci.* **1981**, *106*, 101.
58. Klots, C. E. *Z. Phys. D* **1991**, *21*, 335; *ibid* **1991**, *20*, 105.
59. Klots, C. E. *J. Chem. Phys.* **1985**, *83*, 5854.
60. Klots, C. E. *Nature* **1987**, *327*, 222.
61. (a) Klots, C. E.; *Z. Phys. D* **1987**, *5*, 83; (b) Klots, C. E. Kinetic methods for quantifying magic. *East Coast Symposium on the Chemistry and Physics of Clusters and Cluster Ions*; Baltimore, MD, 14–16 January 1991.
62. Zewail, A. H. *Femtochemistry: Ultrafast Dynamics of the Chemical Bond*; World Scientific: Singapore, 1994, Volumes I and II.
63. Wei, S.; Purnell, J.; Buzza, S. A.; Stanley, R. J.; Castleman, Jr., A. W. *J. Chem. Phys.* **1992**, *97*, 9480–9482.
64. Wei, S.; Purnell, J.; Buzza, S. A.; Castleman, Jr., A. W. *J. Chem. Phys.* **1993**, *99*, 755–757.
65. Castleman, Jr., A. W.; Wei, S.; Purnell, J.; Buzza, S. A. In *Reaction Dynamics in Clusters and Condensed Phases*; Kluwer Academic, Dordrecht: Holland, 1994, pp 1–12.
66. Purnell, J.; Wei, S.; Buzza, S. A.; Castleman, Jr., A. W. *J. Phys. Chem.* **1993**, *97*, 12530–12534.
67. Buzza, S. A.; Wei, S.; Purnell, J.; Castleman, Jr., A. W. *J. Chem. Phys.* **1995**, *102*, 4832–4841.
68. Wei, S.; Purnell, J.; Buzza, S. A.; Snyder, E. M.; Castleman, Jr., A. W. In *Femtosecond Chemistry*; Manz, J.; Wöste, L., Eds.; Springer-Verlag: Germany, 1994, pp 449–474.
69. Snyder, E. M.; Purnell, J.; Wei, S.; Buzza, S. A. ; Castleman, Jr., A. W. *Chem. Phys.* **1996**, *207*, 355–366.
70. Echt, O.; Stanley, R. J.; Dao, P. D.; Castleman, Jr., A. W. *Deutsche Bunsen-Gesellschaft fur Physikalische Chemie* **1984**, *88*, 217–219.
71. Shinohara, H.; Nishi, N.; *Chem. Phys. Lett.* **1987**, *141*, 292.
72. Shinohara, H.; Nishi, N.; Washida, N. *J. Chem. Phys.* **1985**, *83*, 1939.
73. Kaiser, E.; de Vries, J.; Steger, H.; Menzel, C.; Kamke, W.; Hertel, I. V. *Z. Phys. D* **1991**, *20*, 193.
74. Ceyer, S. T.; Tiedemann, P. W.; Mahan, B. H.; Lee, Y. T. *J. Chem. Phys.* **1979**, *70*, 14.
75. Stephan, K.; Futrell, J. H.; Peterson, K. I.; Castleman, Jr., A. W.; Wagner, H. E. *Int. J. Mass Spectrom. Ion Phys.* **1982**, *44*, 167.
76. Misaizu, F.; Houston, P. L.; Nishi, N.; Shinohara, H.; Kondow, T.; Kinoshita, M. *J. Phys. Chem.* **1989**, *93*, 7041.
77. Misaizu, F.; Houston, P. L.; Nishi, N.; Shinohara, H.; Kondow, T.; Kinoshita, M. *J. Chem. Phys.* **1993**, *98*, 336.
78. Cao, H.; Evleth, E. M.; Kassab, E. *J. Chem. Phys.* **1984**, *88*, 6680.
79. Tomoda, S. *Chem. Phys.* **1986**, *110*, 431.
80. Gellene, G. I.; Porter, R. F. *J. Phys. Chem.* **1984**, *88*, 6680.
81. Hertel; I. V. Personal communication.
82. Ziegler, L. D. *J. Chem. Phys.* **1985**, *82*, 665.
83. Breen, J. J.; Peng, L. W.; Willberg, D. M.; Heikal, A.; Cong, A.; Zewail, A. H. *J. Chem. Phys.* **1990**, *92*, 805.
84. Syage, J. A. In *Femtosecond Chemistry*; Manz, J.; Wöste, L., Eds.; VCH: Weinheim, Germany and New York, 1995, pp 475–496.

85. Kreisle, D.; Echt, O.; Knapp, M.; Recknagel, E.; Leiter, K.; Maerk, T. D.; Saenz, J. J.; Soler, J. M. *Phys. Rev. Lett.* **1986**, *56*, 1551.
86. Brechignac, C.; Cahuzac, Ph.; Carlier, F.; de Frutos, M.; Leygnier, J.; Roux, J. Ph. *J. Chem. Phys.* **1995**, *102*, 763; Brechignac, C.; Cahuzac, Ph.; Carlier, F.; de Frutos, M. *Phys. Rev. Lett.* **1994**, *72*, 1636.
87. Shukla, A. K.; Moore, C.; Stace, A. J. *Chem. Phys. Lett.* **1984**, *109*, 324; Stace, A. J. *Phys. Rev. Lett.* **1988**, *61*, 306.
88. Scheier, P.; Maerk, T. D. *J. Chem. Phys.* **1987**, *86*, 3056; Lezius, M.; Maerk, T. D. *Chem. Phys. Lett.* **1989**, *155*, 496; Scheier, P.; Maerk, T. D. *Chem. Phys. Lett.* **1988**, *148*, 393; Scheier, P.; Stamatovic, A.; Maerk, T. D. *J. Chem. Phys.* **1988**, *88*, 4289; Scheier, P.; Dunser, B.; Maerk, T. D. *Phys. Rev. Lett.* **1995**, *74*, 3368.
89. Kreisle, D.; Leiter, K.; Echt, O.; Maerk, T. D. *Z. Phys. D* **1986**, *3*, 319.
90. Snyder, E. M.; Wei, S.; Purnell, J.; Buzza, S. A.; Castleman, Jr., A. W. *Chem. Phys. Lett.* **1996**, *248*, 1–7.
91. Purnell, J.; Snyder, E. M.; Wei, S.; Castleman, Jr., A. W. *Chem. Phys. Lett.* **1994**, *229*, 333–339.
92. Buzza, S. A.; Snyder, E. M.; Card, D. A.; Folmer, D. E.; Castleman, Jr., A. W. *J. Chem. Phys.* **1996**, *105*, 7425–7431.
93. Lifshitz, C.; Louage, F. *J. Phys. Chem.* **1989**, *93*, 5633.
94. Tzeng, W. B.; Wei, S.; Neyer, D. W.; Keesee, R. G.; Castleman, Jr., A. W. *J. Am. Chem. Soc.* **1990**, *112*, 4097–4104.
95. Tzeng, W. B.; Wei, S.; Castleman, Jr., A. W. *Chem. Phys. Lett.* **1990**, *166*, 343–52.
96. Tzeng, W. B.; Wei, S.; Castleman, Jr., A. W. *J. Am. Chem. Soc.* **1989**, *111*, 6035–40; "Erratum," *J. Am. Chem. Soc.* **1989**, *111*, 8326.
97. Castleman, Jr., A. W.; Tzeng, W. B.; Wei, S.; Morgan, S. *J. Chem. Soc., Faraday Trans.* **1990**, *86*, 2417–2426.
98. Sieck, W.; Ausloos, P. *Radiat. Res.* **1972**, *52*, 47.
99. Luczynski, Z.; Wincel, H. *Int. J. Mass Spectrom. Ion Phys.* **1977**, *23*, 37.
100. MacNeil, K. A. G.; Futrell, J. H. *J. Phys. Chem.* **1972**, *76*, 409.
101. Stace, A. J.; Shukla, A. K. *J. Phys. Chem.* **1982**, *86*, 865.
102. Morgan, S.; Castleman, Jr., A. W. *J. Am. Chem. Soc.* **1987**, *109*, 2867–70.
103. Morgan, S.; Keesee, R. G.; Castleman, Jr., A. W. *J. Am. Chem. Soc.* **1989**, *111*, 3841–45.
104. Bowers, M. T.; Su, T.; Anicich, V. G. *J. Chem. Phys.* **1973**, *58*, 5175; Bass, L. M.; Cates, R. D.; Jarrold, M. F.; Kirchner, N. J.; Bowers, M. T. *J. Am. Chem. Soc.* **1983**, *105*, 7024.
105. Draves, J. A.; Lisy, M. J. *J. Am. Chem. Soc.* **1990**, *112*, 9006.
106. Zhang, X.; Castleman, Jr., A. W. *J. Am. Chem. Soc.* **1992**, *114*, 8607–8610.
107. Guo, B. C.; Castleman, Jr., A. W. *J. Am. Soc. for Mass Spectrom.* **1992**, *3*, 464–466.
108. Guo, B. C.; Castleman, Jr., A. W. *Int. J. Mass Spectrom. Ion Proc.* **1992**, *113*, R1–R5.
109. Guo, B. C.; Kerns, K. P.; Castleman, Jr., A. W. *J. Phys. Chem.* **1992**, *96*, 4879–4833.
110. Guo, B. C.; Castleman, Jr., A. W. *J. Am. Chem. Soc.* **1992**, *114*, 6152–6158.
111. Guo, B. C.; Kerns, K. P.; Castleman, Jr., A. W. *Science* **1992**, *255*, 1411–1413.
112. Guo, B. C.; Wei, S.; Purnell, J.; Buzza, S.; Castleman, Jr., A. W. *Science* **1992**, *256*, 515–516.
113. Wei, S.; Guo, B. C.; Purnell, J.; Buzza, S.; Castleman, Jr., A. W. *J. Phys. Chem.* **1992**, *96*, 4166–4168.
114. Wei, S.; Guo, B. C.; Purnell, J.; Buzza, S.; Castleman, Jr., A. W. *Science* **1992**, *256*, 818–820.
115. Guo, B. C.; Kerns, K. P.; Castleman, Jr., A. W. *J. Am. Chem. Soc.* **1993**, *115*, 7415–7418.
116. Deng, H. T.; Kerns, K. P.; Castleman, Jr., A. W. *J. Am. Chem. Soc.* **1996**, *118*, 446–450.
117. Castleman, Jr., A. W. In *Applications of Organometallic Chemistry in the Preparation and Processing of Advanced Materials*; Harrod, J. F.; Laine, R. M., Eds.; NATO ASI Series E: Appl. Sci.; Kluwer Academic: The Netherlands, 1995, Vol. 297, pp. 269–281.
118. Yang, X.; Castleman, Jr., A. W. *J. Am. Chem. Soc.* **1991**, *113*, 6766–6771.
119. Pinsent, B. R. W.; Pearson, L.; Roughton, F. J. W. *J. Chem. Sec., Faraday Trans.* **1959**, *52*, 11512.

120. Su, T.; Chesnavich, W. J. *J. Chem. Phys.* **1982**, *76*, 5183.
121. Martin, R. B. *J. Inorg. Nucl. Chem.* **1976**, *38*, 511.
122. Khalifah, R. G. *Proc. Natl. Acad. Sci. USA* **1973**, *70*, 1986.
123. Woolley, P. *Nature* **1975**, *258*, 677.
124. Yang, X.; Castleman, Jr., A. W. *J. Phys. Chem.* **1991**, *95*, 6182–6186.
125. Cotton, F. A.; Wilkinson, G. *Advanced Inorganic Chemistry*, 5th edn.; Wiley, New York, 1975.
126. Eigen, M.; Kustin, K.; Maass, G. *Z. Phys. Chem.* **1961**, *30*, 130.
127. Horner, D. A.; Connick, R. E. *Inorg. Chem.* **1986**, *25*, 2414.
128. Pochan, J. M.; Stone, R. G.; Flygare, W. H. *J. Chem. Phys.* **1969**, *51*, 4278.
129. Wincel, H.; Mereand, E.; Castleman, Jr., A. W. *J. Phys. Chem.* **1994**, *98*, 8606–8610.
130. Wincel, H.; Mereand, E.; Castleman, Jr., A. W. *J. Phys. Chem.* **1995**, *99*, 1792–1798.
131. Wincel, H.; Mereand, E.; Castleman, Jr., A. W. *J. Phys. Chem.* **1995**, *102*, 9228–9234.
132. Fehsenfeld, F. C.; Mosesman, M.; Ferguson, E. E. *J. Chem. Phys.* **1971**, *55*, 2120; Lineberger, W. C.; Puckett, L. J. *Phys. Rev.* **1969**, *187*, 286; Puckett, L. J.; Teague, M. W. *J. Chem. Phys.* **1971**, *54*, 2564; Howard, C. J.; Rundle, H. W.; Kaufman, F. *J. Chem. Phys.* **1971**, *55*, 4772; French, M. A.; Hills, L. P.; Kebarle, P.; *Can. J. Chem.* **1972**, *51*, 456.
133. Stace, A. J.; Winkel, J. F.; Lopez Martens, R. B.; Upham, J. E. *J. Phys. Chem.* **1994**, *98*, 2012.
134. Poth, L.; Shi, Z.; Zhong, Q.; Castleman, Jr., A. W. *Int. J. Mass Spectrom. Ion Proc.* **1996**, *154*, 35–42.
135. Harms, A. C.; Khanna, S. N.; Chen, B.; Castleman, Jr., A. W. *J. Chem. Phys.* **1994**, *100*, 3540–3544.
136. Misaizu, F.; Sanekata, M.; Tsukamoto, K.; Fuke, K.; Iwata, S *J. Phys. Chem.* **1992**, *96*, 8259.
137. Yeh, C. S.; Willey, K. F.; Robbins, D. L.; Pilgrim, J. S.; Duncan, M. A. *Chem. Phys. Lett.* **1992**, *196*, 233; Willey, K. F.; Yeh, C. S.; Robbins, D. L.; Duncan, M. *J. Chem. Phys.* **1992**, *96*, 9106.
138. Sanekata, M.; Misaizu, F.; Fuke, K.; Iwata, S.; Hashimoto, K. *J. Am. Chem. Soc.* **1995**, *117*, 747; Watanabe, H.; Iwata, S.; Hashimoto, K.; Misaizu, F.; Fuke, K. *J. Am. Chem. Soc.* **1995**, *117*, 755.
139. Passarella, R.; Castleman, Jr., A. W. *J. Phys. Chem.* **1989**, *93*, 5840–45.
140. Guo, B. C.; Conklin, B. J.; Castleman, Jr., A. W. *J. Am. Chem. Soc.* **1989**, *111*, 6506–6510.
141. Yang, Xiaolin; Castleman, Jr., A. W. *J. Chem. Phys.* **1990**, *93*, 2405–2412.
142. Yang, X.; Zhang, X.; Castleman, Jr., A. W. *Int. J. Mass Spectrom. Ion Proc.* **1991**, *109*, 339–354.
143. Yang, X.; Castleman, Jr., A. W. *J. Am. Chem. Soc.* **1989**, *111*, 6845–46.
144. Yang, X.; Castleman, Jr., A. W. *J. Phys. Chem.* **1990**, *94*, 8500–8502; "Erratum," *J. Phys. Chem.* **1990**, *94*, 8974.
145. Yang, X.; Zhang, X.; Castleman, Jr., A. W. *J. Phys. Chem.* **1991**, *95*, 8520–8524.
146. Webster, C. R.; May, R. D.; Toohey, D. W.; Avallone, L. M.; Anderson, J. G.; Newman, P.; Lair, L.; Schoeberl, M. R.; Elkins, J. W.; Chan, K. R. *Science* **1993**, *261*, 1130; Toon, O.; Browell, E.; Gary, B.; Lait, L.; Livingston, J.; Newman, P.; Pueschel, R.; Russell, P.; Schoeberl, M.; Toon, G.; Traub, W.; Valero, F. P.; Selkirk, H.; Jordan, J. *Science* **1993**, *261*, 1136; Wilson, J. C.; Jonsson, H. H.; Brock, C. A.; Toohey, D. W.; Avallone, L. M.; Baumgardner, D.; Dye, J. E.; Poole, L. R.; Woods, D. C.; DeCoursey, R. J.; Osborn, M.; Pitts, M. C.; Kelly, K. K.; Chan, K. R.; Ferry, G. V.; Lowenstein, M.; Podolske, J. R.; Weaver, A. *Science* **1993**, *261*, 1140; P. S. Zurer, *C&E News*, May 34, **1993**, *8*; Molina, M. J.; Zhang, R.; Wooldridge, P. J.; McMahon, J. R.; Kim, J. E.; Chang, H. Y.; Beyer, K. D. *Science* **1993**, *261*, 1418.
147. Crutzen, P. J.; Arnold, F. *Nature* **1986**, *324*, 651.
148. Zhang, X.; Mereand, E. L.; Castleman, Jr., A. W. *J. Phys. Chem.* **1994**, *98*, 3554–3557.
149. Leuchtner, R. E.; Harms, A. C.; Castleman, Jr., A. W. *J. Chem. Phys.* **1989**, *91*, 2753–2754.
150. Leuchtner, R. E.; Harms, A. C.; Castleman, Jr., A. W. *J. Chem. Phys.* **1991**, *94*, 1093–1101.
151. Castleman, Jr., A. W.; Harms, A. C.; Leuchtner, R. E. *Z. Phys. D.* **1991**, *19*, 343–346.
152. de Heer, W. A.; Knight, W. D. In *Elemental and Molecular Clusters, Springer Series in Material Science 6*; Benedek, G.; Martin, T. P.; Pacchioni, G., Eds.; Springer-Verlag: Berlin, Heidelberg, pp 45–63, 1988.
153. Sigsworth, S. W.; Castleman, Jr., A. W. *J. Am. Chem. Soc.* **1992**, *114*, 10471.

154. Keesee, R. G.; Chen, B.; Harms, A. C.; Castleman, Jr., A. W. *Int. J. Mass Spectrom. Ion Proc.* **1993**, *123*, 225–231.
155. Leuchtner, R. E.; Harms, A. C.; Castleman, Jr., A. W. *J. Chem. Phys.* **1990**, *92*, 6527–6537.
156. Guo, B. C.; Kerns, K. P.; Castleman, Jr., A. W. *J. Chem. Phys.* **1992**, *96*, 8177–8186.
157. See: Lauher, J. W.; *J. Am. Chem. Soc.* **1978**, *100*, 5305; McGlinchey, M. J.; Mlekuz, M.; Bougeard, P.; Sayer, B. G.; Marinetti, A.; Saillard, J. Y.; Jaouen, G. *Can. J. Chem.* **1983**, *61*, 1319; Mingos, D. M.; Slee, T.; Zhenyang, L. *Chem. Rev.* **1990**, *90*, 383; Mingos, D. M.; Wales, D. J. *J. Am. Chem. Soc.* **1990**, *112*, 930; Wade, K. *Adv. Inorg. Chem. Radiochem. P*, **1976**, *18*, 1; Mingos, D. M. *J. Chem. Soc., Dalton Trans. P*, **1974**, *133*; Mingos, D. M. *Acc. Chem. Res.* **1984**, *17*, 311; Mingos, D. M. *Chem. Soc. Rev.* **1986**, *15*, 31.
158. Guo, B. C.; Kerns, K. P.; Castleman, Jr., A. W. *J. Phys. Chem.* **1992**, *96*, 6931–6937.
159. Breen, J.; Tzeng, W.-B.; Keesee, R. G; Castleman, Jr., A. W. *J. Chem. Phys.* **1989**, *90*, 19–24.
160. Breen, J. J.; Kilgore, K.; Wei, S.; Tzeng, W.-B.; Keesee, R. G.; Castleman, Jr., A. W. *J. Chem. Phys.* **1989**, *90*, 11–18.
161. McDaniel, E. W. *Collision Phenomena in Ionized Gases*; John Wiley & Sons: 1964, p 260.
162. Dao, P. D.; Castleman, Jr., A. W. *J. Chem. Phys.* **1986**, *84*, 1434–1442.
163. Hatano, Y., Personal Communication; see also T. Wada, T.; Shinsaka, K.; Namba, H.; Hatano, Y. *Can. J. Chem.* **1977**, *55*, 2144.
164. Engelking, P. C. *J. Chem. Phys.* **1986**, *85*, 3103.
165. Engelking, P. C. *J. Chem. Phys.* **1987**, *87*, 936.
166. Echt, O.; Sattler, K.; Recknagel, E. *Phys. Rev. Lett.* **1981**, *47*, 1121.
167. Miehle, W.; Kandler, O.; Leisner, T. *J. Chem. Phys.* **1989**, *91*, 5940.
168. Haberland, H. *Surface Sci.* **1985**, *156*, 305.
169. Wei, S.; Shi, Z.; Castleman, Jr., A. W. *J. Chem. Phys.* **1991**, *94*, 8604–8607.
170. Hermann, V.; Kay, B. D.; Castleman, Jr., A. W. *Chem. Phys.* **1982**, *72*, 185–200.
171. Holland, P. M.; Castleman, Jr., A. W. *J. Chem. Phys.* **1980**, *72*, 5984–90.
172. Searcy, J. Q.; Fenn, J. B. *J. Chem. Phys.* **1974**, *61*, 5282.
173. Udseth, H.; Zmora, H.; Beuhler, R. K.; Friedman, L. *J. Phys. Chem.* **1982**, *86*, 612.
174. Beuhler, R. J.; Friedman, L. *J. Chem. Phys.* **1982**, *77*, 2549.
175. Lin, S. S. *Rev. Sci. Instrum.* **1973**, *44*, 516.
176. Dreyfuss, D.; Wachman, H. Y. *J. Chem. Phys.* **1982**, *76*, 2031.
177. Stace, A. J.; Moore, C. *Chem. Phys. Lett.* **1983**, *96*, 80.
178. Echt, O.; Kreisle, D.; Knapp, M.; Recknagel, E.; *Chem. Phys. Lett.* **1984**, *108*, 401.
179. Nagashima, U.; Shinohara, H.; Nishi, N.; Tanaka, H. *J. Chem. Phys.* **1986**, *84*, 209.
180. Kassner, Jr., J. L.; Hagen, D. E. *J. Chem. Phys.* **1976**, *64*, 1860.
181. Wei, S.; Shi, Z.; Castleman, Jr., A. W. *J. Chem. Phys.* **1991**, *94*, 3268–3270.
182. Shinohara, H.; Nagashima, U.; Tanaka, H.; Nishi, N. *J. Chem. Phys.*. **1985**, *83*, 4183.
183. Shi, Z.; Ford, J. V.; Wei, S.; Castleman, Jr., A. W. *J. Chem. Phys.* **1993**, *99*, 8009–8015.
184. Lee, N.; Keesee, R. G.; Castleman, Jr., A. W. *J. Colloid Interface Sci.* **1980**, *75*, 555.
185. Holland, P. M.; Castleman, Jr., A. W. *J. Phys. Chem.* **1982**, *86*, 4181–88.
186. Shi, Z.; Wei, S.; Ford, J. V.; Castleman, Jr., A. W. *Chem. Phys. Lett.* **1992**, *200*, 142–146.
187. Lau, Y. K.; Saluja, P. P. S.; Kebarle, P. *J. Am. Chem. Soc.* **1980**, *102*, 7429.
188. Hiraoka, K.; Grumsrud, E. P.; Kebarle, P. *J. Am. Chem. Soc.* **1974**, *96*, 3359.
189. Castleman, Jr., A. W.; Tang, I. N. *J. Chem. Phys.* **1975**, *62*, 4576.
190. Lias, S. G.; Liebman, J. F.; Levin, R. D. *J. Phys. Chem. Ref. Data* **1984**, *13*, 695.
191. Wei, S.; Tzeng, W. B.; Castleman, Jr., A. W. *J. Phys. Chem.* **1991**, *95*, 585–91.
192. Wei, S.; Tzeng, W. B.; Castleman, Jr., A. W. *Chem. Phys. Lett.* **1991**, *178*, 411–418.

THERMOCHEMISTRY OF SINGLY AND MULTIPLY CHARGED IONS PRODUCED BY ELECTROSPRAY

John S. Klassen, Yeunghaw Ho,

Arthur T. Blades, and Paul Kebarle

Advances in Gas-Phase Ion Chemistry
Volume 3, pages 255–318
Copyright © 1998 by JAI Press Inc.
All rights of reproduction in any form reserved.
ISBN: 0-7623-0204-6

ABSTRACT

Electrospray (ES) is a method with which ions present in a solution can be transferred to the gas phase. It has caused an explosive development in bioanalytical mass spectrometry and is also of great interest to the gas-phase ion chemist. It allows the production of multiply charged ion–molecule complexes involving most metal ions as well as the production of bio-ions such as multiply protonated peptides and deprotonated nucleic acids. A description of the ES method and mechanism is provided. Mass spectrometric apparatus is described with which ion–molecule equilibria involving ES produced ions can be determined. From these, one obtains the free energy $\Delta G°$, enthalpy $\Delta H°$ and entropy $\Delta S°$, changes for ion–molecule reactions such as ion–molecule association, proton transfer, electron transfer, and ligand transfer reactions. Thermochemical data obtained for the hydration reactions of a variety of ions are presented. Multiply charged ions such as $M^{2+}(H_2O)_n$ on desolvation become unstable and undergo charge reduction where the departing solvent molecule takes away one positive charge. Conditions for which charge reduction occurs are described for a variety of positive and negative ions. Also, apparatus is described with which collision induced dissociation (CID) product ion cross sections can be determined in the threshold region. From these, the threshold energy E_0 for the product ion formation is obtained. Results for the Na^+ and K^+ affinities to acetone, acetamide, N-methylacetamide, glycine, glycineamide, glycylglycine, and succinamide are presented and discussed. Thresholds for fragment ions obtained from CID of protonated peptides provide the activation energies for the decomposition reactions. These, combined with theoretical evaluations of the transition state energies, can lead to confirmation of proposed mechanisms for the decomposition pathways. Results for protonated glycine and oligoglycines are discussed.

I. INTRODUCTION: WHY IONS PRODUCED BY ELECTROSPRAY?

Gas-phase ion chemistry has applications in many areas; however there are two fields to which the connection may be viewed as most important: analytical mass spectrometry and ion chemistry in solutions. The beginnings of gas-phase ion chemistry can be seen in the work on the origin of electron ionization mass spectra which resulted in the quasi equilibrium theory of mass spectra,[1] and dealt with the kinetics of unimolecular dissociation of excited ions. Measurements of the energies involved, based on electron ionization and photoionization and appearance potentials,[2] represented the energetics complement to the theory of mass spectra. However, it was soon realized that the thermochemical data obtained from these measurements are also of significance in organic chemistry where the same ions may undergo similar reactions but in the presence of a solvent environment. An example of such early gas-phase ion work, aimed at correlation with solution, is the determination of the ionization energies of substituted benzyl free radicals which led to the findings that the ionization energies are linearly correlated to activation energies for solvolysis reactions in solution.[3]

The first systematic measurements of the reactions of ions with molecules in the gas phase were initiated largely by workers associated with analytical mass spectrometry.[4-6] It was the rapidly expanding area of ion–molecule reactions which led to the origin of Gas-Phase Ion Chemistry as a distinct field.[7] The discovery that ion–molecule equilibria in the gas phase can be determined by mass spectrometric techniques[8] led to an explosion of thermochemical measurements based on determination of equilibria by a variety of techniques.[9] Significantly, for the first time, information could be obtained on the thermochemistry of reactions which had solution counterparts of paramount importance such as acidities and basicities. These were obtained from: proton transfer equilibria such as,

$$CH_3CO_2^- + CH_2ClCO_2H = CH_3CO_2H + CH_2ClCO_2^- \tag{1a}$$

$$NH_4^+ + CH_3NH_2 = NH_3 + CH_3NH_3^+ \tag{1b}$$

electron transfer equilibria, corresponding to oxidation reduction reactions in solution,

$$C_6H_5NO_2^{\cdot-} + NO_2C_6H_4NO_2 = C_6H_5NO_2 + NO_2C_6H_4NO_2^{\cdot-} \tag{2}$$

and ion transfer to different ligands L, equilibria such as:

$$M^+OH_2 + CH_3OH = M^+CH_3OH + OH_2 \tag{3}$$

The relative gas-phase acidities or basicities obtained via equilibria (equation 1) were converted to absolute gas phase acidities,

$$AH = A^- + H^+ \qquad \Delta G^\circ_{Acid}(AH) \tag{4}$$

$$BH^+ = B + H^+ \qquad \Delta G^\circ_{Base}(B) \tag{5}$$

by calibration of the relative gas-phase acidity scales to one absolute value for one given acid AH which had been determined by some other technique. The same approach was used also for obtaining absolute gas-phase basicities from equilibria of the type in equation 1b, electron affinities from equilibria as in equation 2, and M^+ affinities from equilibria as in equation 3. Literally thousands of thermochemical data have been obtained by this approach.[10-15]

When the gas-phase reactions, such as the relative acidities or basicities were compared with their counterparts in solution (in a solvent such as water) it was generally found[16,17] that the energetics in the solvent were strongly affected by solvation effects and particularly the solvation of the ionic reactants. Relationships between the gas-phase and solution-phase reactions and the solvation energies of the reactants are generally obtained through thermodynamic cycles. From the cycle,

$$
\begin{array}{ccc}
AH(g) + H_2O\,(g) & \xrightarrow{\;\;\Delta G_2^{\circ}\;\;} & A^-\,(g) + H_3O^+\,(g) \\[2mm]
\Big\uparrow \Delta G_3^{\circ} & & \Big\downarrow \Delta G_4^{\circ} \qquad\qquad (6)\\[2mm]
AH\,(aq) + H_2O\,(aq) & \xrightarrow{\;\;\Delta G_1^{\circ}\;\;} & A^-\,(aq) + H_3O^+\,(aq)
\end{array}
$$

one can obtain an expression for the acid dissociation constant K_a (AH) in aqueous solution:

$$\Delta G_1^{\circ} = 2.3\ RT\ pKa(AH) = \Delta G_2^{\circ} + \Delta G_3^{\circ} + \Delta G_4^{\circ}$$

$$= \Delta G_{acid}^{\circ}\,(AH) - \Delta G_{base}^{\circ}\,(H_2O) + \Delta G_{solv}^{\circ}(A^-)$$

$$+ \Delta G_{sol}^{\circ}\,(H_3O^+) - \Delta G_{solv}^{\circ}(AH) + \Delta G_{evap}^{\circ}\,(H_2O) \qquad (7)$$

ΔG_{acid}° (AH) and ΔG_{base}° (H$_2$O) are the free energy changes for the gas-phase reactions (equations 4 and 5).

The solvation energies, ΔG_{sol}°, which in the present example are for solvation in aqueous solution, correspond to the transfer process for species X from the gas phase to solution:

$$X(g) \rightarrow X(aq) \qquad\qquad \Delta G_{hyd}^{\circ}(X) \qquad (8)$$

Because the acid dissociation constants in solution are generally known, and also, often the solvation energies of the neutral AH have been determined by neutral gas-phase solution equilibria, equation 8 can be used to obtain the sum of the solvation energies $\Delta G_{sol}^{\circ}(A^-) + \Delta G_{sol}^{\circ}$ (H$_3$O$^+$). When this approach is applied to a series of acids AH, values for the solvation energies ΔG_{sol}° (A$^-$), relative to the constant ΔG_{sol}° (H$_3$O$^+$), can be obtained. Such thermodynamic determinations of the solvation energies can be very revealing of the factors which govern differences between gas-phase and solution reactivities of ions.[16,17a]

Even when solvation energies for the ions are available problems of interpretation remain. The factors which cause given observed solvation energy differences may not be obvious. Thermochemical data obtained from the determination of sequential gas-phase ion–solvent molecule equilibria can be extremely useful for this purpose.

Gas-phase ion–solvent molecule equilibria such as reaction 9,

$$BH^+ + H_2O = BH^+(H_2O) \qquad (0,1)$$

$$BH^+(H_2O) + H_2O = BH^+(H_2O)_2 \qquad (1,2) \qquad\qquad (9)$$

$$BH^+(H_2O)_{n-1} + H_2O = BH^+(H_2O)_n \qquad (n-1,n)$$

involving both positive and negative ions and a variety of solvent molecules in addition to H_2O have been determined by ion–molecule equilibria techniques.[8,15] It has turned out that thermochemical data, such as $\Delta G^\circ_{n-1,n}$ and $\Delta H^\circ_{n-1,n}$ can in general be obtained from $n = 1$ to $n < 10$. Comparison of these sequential ion–solvent molecule data (and their sums which lead to $\Delta G^\circ_{0,n}$) with the total solvation energies in the given solvent, have shown[17b] that the ion–molecule solvation energies, $\Delta G^\circ_{0,n}$, reproduce the major trends observed in the total solvation energies, ΔG°_{sol}, even already at low n values, i.e. $n < 10$. The advantage of observing solvation trends when the interactions involve one or a few solvent molecules, compared with interactions with the liquid solvent molecules are obvious. The problem is shifted from the obscure area of the liquid state to the area of ion–molecule, i.e. molecule–molecule, interactions which are familiar to the chemist.

A major part of modern theoretical investigations[18–20] of ionic reactions in solution has followed the same approach as the experimental investigations described above. Quantum mechanical calculations are made for each of the reacting species at the molecular level.[18] From energies of the reactants, one can then evaluate the energy change for the reactions such as shown in equations 1–3. The effect of the solvent can then be included by calculations of the interactions of a given molecular or ionic species with one and more than one solvent molecules.[19–21] These are performed at the ab initio level and the results are used for development of pair potential functions with which the interactions of each ionic or neutral reactant with a large number of solvent molecules can be modeled by Monte Carlo simulations or molecular dynamics (MD) techniques.[20,21] The availability of the gas-phase ion–molecule thermochemical data described above has played a significant role in the development of the theoretical methods. The experimentally determined thermochemical values[9–15] have provided a guide to the level at which the ab initio calculations had to be performed to achieve agreement with the experimental results.[17–19] The quoted theoretical work[18–21] is only a very small but representative sample of a very large number of publications in this area.

As already mentioned, the experimental studies were restricted to singly charged ions. While singly charged ions are involved in the majority of the important condensed-phase ion chemistry, there are many important processes involving multiply charged ions. Thus, doubly charged ions such as Mg^{2+}, Ca^{2+}, Fe^{2+}, Co^{2+}, Ni^{2+}, Cu^{2+}, etc. are of paramount importance in condensed-phase chemistry and biochemistry.

Multiply charged "naked" metal ions are not difficult to produce in the dilute gas phase. This is illustrated by the availability of experimental higher ionization potentials for the atoms of the elements. The difficulties arise when other chemical matter is near the multiply charged ion. For example, the general method used for generating singly charged ion clusters like $M^+(H_2O)_n$ is to generate M^+ in a gas phase where ligand molecule (H_2O) vapor is present. The formation of $M^+(H_2O)_n$ then proceeds spontaneously by third body dependent association reactions (see equation 9). This method may not work for M^{2+} species when the second ionization energy of M, $IE(M^+)$ is larger than the ionization energy of the ligand L, as in the case for Mg where $IE(Mg^+) = 15.0$ eV while $IE(H_2O) = 12.6$ eV. Little information is available on the outcome of gas-phase reactions where $IE(M^+) > IE(L)$. Charge transfer to the ligand leading to M^+ and L^+ or occurrence of a reaction such as formation of $MOH^+ + H_3O^+$, when the ligand and the third body are H_2O, are reaction channels that could compete with, or completely replace, the formation of the stabilized, $M^{2+}(L)_n$ ion.

Some results for reactions $M^{2+} + L$ have been reported.[22-24] These involved[22] the relatively low $IE(M^+)$ species, Ca^{2+} and Ba^{2+}. Reactions of Ti^{2+} with the high $IE(L)$ alkanes have also been examined,[23] and similar work has also been reported[24] for Nb^{2+} and La^{2+}. The results are of considerable interest. Many of the reactions where $IE(M^+)$ was relatively high led to charge separation, i.e., the products were two singly charged ions. It appears that only a limited range of M^{2+} ions can be produced with the methods used and that many of the ligands will lead to reactions corresponding to charge separation rather than ion–ligand association.

Since doubly and triply charged ions exist in solution, preparation of gas-phase ions such as $M^{2+}L_n$ might be possible by "transfer" of these ions from the liquid solution to the gas phase. Until recently, this alternative would have been considered impossible. However, recent mass spectrometric techniques developed for analytical purposes such as fast atom bombardment (FAB),[25-27] thermospray,[28,29] and electrospray,[30] have demonstrated that inorganic ions M^{2+} attached to ligands can be produced in the gas phase when the corresponding ions are present as electrolytes in a solution which is subjected to the given process. Due to a long-standing interest in initiating multiply charged gas-phase ion chemistry and thermochemical measurements, we became involved in research involving these techniques.[27,31] Only very low intensities of $M^{2+}L_n$ ions are observed with FAB[26] and then only when large multidentate complexing agents are used, presumably because these stabilize the charge by extensive charge delocalization. More promising is thermospray, and Röllgen and coworkers[29] have reported the observation of $Ca^{2+}(H_2O)_n$ and

$Sr^{2+}(H_2O)_n$ (for $n = 3$–12). However, as mentioned above, these ion hydrates are relatively easy to produce since $IE(H_2O) > IE(M^+)$.

A vast variety of multiply charged ions can be produced by electrospray. This technique comes closest to a "true transfer" method of ions from solution to the gas phase. The "transfer" of ions from solution to the gas phase is a highly endothermic and endoergic process. The minimum energy required to completely free the ion from the solvent corresponds to the reverse of the solvation energy of the ion (see equation 8). For an ion like sodium, $-\Delta G^{\circ}_{hyd}(Na^+) \approx 100$ kcal/mol.[32] A major distinction between electrospray and the other methods such as FAB and the recently developed matrix-assisted laser desorption ionization (MALDI)[33] lies in the rate at which the energy required to desolvate the ions is supplied. In FAB and MALDI the energy supply is highly localized and delivered over a very short time. Thus the process is associated with high internal excitation of the ion and nearest neighbor solvent molecules. On the other hand, in electrospray, the energy for desolvation is supplied over a long period of time (milliseconds) and essentially only thermal energy at relatively low temperatures is used. The same is also true for thermospray, but much lower ion yields are obtained with that method.

Electrospray affords transfer to the gas phase of a wide variety of ions dissolved in a wide variety of solvents. The ions include singly and multiply charged inorganic ions such as the alkali ions—Li^+, Na^+, K^+, Cs^+, Rb^+—the alkaline earths—Mg^{2+}, Ca^{2+}, Sr^{2+}, Ba^{2+}—singly and doubly charged transition metal ions complexed to solvent molecules or to other mono and polydentate ligands; anions of inorganic and organic acids such as NO_3^-, Cl^-, $H_2PO_4^-$, HSO_4^-, SO_4^{2-}; as well as singly and multiply protonated organic bases such as amines, alkaloids, peptides and proteins and singly and multiply deprotonated organic acids or organophosphates such as the nucleic acids DNA and RNA. The solvents that can be used include practically all polar solvents, be they protic solvents like water, methanol, and ethanol or aprotic solvents like acetone acetonitrile and dimethylsulfoxide.

It is not an exaggeration to say that electrospray has introduced a new era, not only for the analytical mass spectroscopist, but also for the more physically oriented researcher interested in physical measurements involving the above ions, which are of such great importance in condensed-phase ion chemistry. In particular, gas-phase ions produced by electrospray allow, for the first time, thermochemical measurements involving ions of biochemical significance such as protonated peptides, deprotonated nucleotides, and metal ion complexes with peptides and proteins. It is to be expected that such data will be of importance in the development of theoretical modeling of the state of these systems in the condensed phase.[34,35]

We have discussed above some of the applications of gas-phase ion thermochemical data to ionic reactions in solution. However the new analytical "ion-transfer from solution to the gas-phase" techniques have also created an application for these data in the new analytical mass spectrometry. In fact, much of the background knowledge required for this new analytical mass spectrometry, and particularly MALDI and electrospray, is the gas-phase ion chemistry developed for applications to ions in solution! The very attempts to understand the mechanisms by which the gas-phase ions are produced in MALDI and electrospray are closely tied to the gas phase–solution ion chemistry background.[31b,36] Also many of the analytical applications are better understood on the basis of such knowledge. For example, many analytes of interest such as carbohydrates and other oligomers and polymers are not ionic in solution, yet they can be observed mass spectrometrically by MALDI and electrospray as alkali ion adducts and particularly Na^+ adducts. These adducts can be observed because the ions interact strongly with certain functional groups in the oligomer and information as to which groups will lead to strong interactions has been answered by the gas-phase ion chemistry studies of metal cation affinities[15] (see also equation 3).

II. ELECTROSPRAY: METHOD AND MECHANISM

A. Description of Overall Process

Electrospray (ES) existed long before its application to mass spectrometry (MS). It is a method of considerable importance for the electrostatic dispersion of liquids and creation of aerosols. The interesting history and notable research advances in that field are very well described in Bailey's book: "Electrostatic Spraying of Liquids."[37] Much of the theory concerning the mechanism of the charged droplet formation was developed by researchers in this area. The latest works can be found in a special issue[38] of the *Journal of Aerosol Science* devoted to ES.

The present section on ES method and mechanism and MS detection is based on two previous reviews of the subject[31b,36] and the reader interested in details and additional references in the literature might find these reviews useful. In the present account, we include information aimed at the investigator who is interested in performing physical measurements on ions produced by ES. We became convinced that a minimum of mechanistic information is necessary, on the basis of a personal experience. Some years ago, we persuaded a scientific colleague working with molecular beams to try to study ions produced by ES. He was enthusiastic about the possibilities and took off with modifications of his apparatus. Some months later he reported his disappointment. He needed the ions to be in helium gas and tried ES in helium. He saw no ions and gave up. We hope that the present account will show him why he could not observe ions in He and how to go about to obtain ions in helium.

There are three major steps in the production by ES of gas-phase ions from electrolyte ions in solution. These are: (1) production of charged droplets at the ES capillary tip; (2) shrinkage of the charged droplets by solvent evaporation and repeated droplet disintegrations leading ultimately to very small highly charged droplets capable of producing gas-phase ions; and (3) the actual mechanism by which gas-phase ions are produced from the very small and highly charged droplets.

As shown in the schematic representation of the charged droplet formation (Figure 1), a voltage, V_c, of 2–3 kV is applied to the metal capillary which is typically 0.2 mm o.d. and 0.1 mm i.d. and located 1–3 cm from the counter electrode. The counter electrode may be a plate with an orifice leading to the mass spectrometric sampling system or a sampling capillary mounted on a plate where the sampling capillary leads to the vacuum chamber and MS. A sampling system with such a sampling capillary is shown in Section III (Figure 4).

Because the ES capillary tip is very thin, the electric field, E_c, in the gas at the capillary tip is very high ($E_c \approx 10^6$ V/m). This field strength at the capillary tip, when the counter electrode is large and planar, can be evaluated with the approximate relationship,[39]

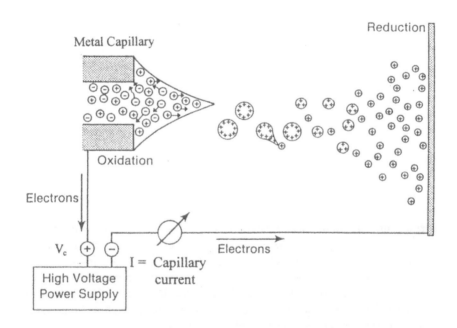

Figure 1. Schematic representation of processes leading to gas-phase ion formation from electrolyte ions present in solution. The high electric field imposed on the tip of the metal capillary (shown highly enlarged) which contains a solution causes enrichment of positive electrolyte ions at the meniscus of the solution. This net charge is pulled downfield expanding the meniscus into a filament which breaks up into very small droplets which are positively charged due to an excess of positive electrolyte ions. Solvent evaporation from droplets reduces volume of droplets and increases the electric field at droplet surface. Droplet fission leads to smaller charged droplets and ultimately to gas-phase positive ions. Charge balance is maintained in the electrospray device by electrochemical oxidation at positive electrode and reduction at negative electrode. From Kebarle, P.; Tang, L. *Anal. Chem.* **1993**, *654*, 973A, with permission.

$$E_c = 2V_c/[r_c\ln(4d/r_c)] \qquad (10)$$

where V_c is the applied potential, r_c is the capillary outer radius, d is the distance from capillary tip to the counter electrode, and ln stands for the natural logarithm.

The typical solution present in the capillary consists of a polar solvent in which electrolytes are soluble. As an example, we can use methanol as solvent and a simple salt like NaCl or BHCl, where B is an organic base, as the solute. Low electrolyte concentrations, 10^{-5}–10^{-3} mol/L (M), are typically used in electrospray mass spectrometry (ESMS). For simplicity we will consider only the positive ion mode in the subsequent discussion.

The field E_c, when turned on, will penetrate the solution at the capillary tip and the positive and negative electrolyte ions in the solution will move under the influence of the field until a charge distribution results which counteracts the imposed field and leads to essentially field-free conditions inside the solution. When the capillary is the positive electrode, positive ions will have drifted down-field in the solution, i.e., towards the meniscus of the liquid and negative ions will have drifted away from the surface. The repulsion between the positive ions at the surface overcomes the surface tension of the liquid and the surface begins to expand allowing the positive charges and liquid to move downfield. A cone forms, the Taylor cone[40], and if the applied field is sufficiently high, a fine jet emerges from the cone tip which breaks up into small, charged droplets (see Figure 1).

The droplets are positively charged due to the excess of positive electrolyte ions at the surface of the cone and the cone jet. Thus if the electrolyte present in the solution was NaCl, the excess positive ions at the surface would be Na^+ ions and paired Na^+ and Cl^- ions will be present inside the droplet. This mode of charging, which depends on the positive and negative ions drifting in opposite directions under the influence of the electric field, has been called the electrophoretic mechanism.[39,41–43] The charged droplets, produced by the cone-jet drift through the air downfield, towards the counter electrode. Solvent evaporation at constant charge leads to droplet shrinkage and an increase of the electric field normal to the surface of the droplets. At a given radius R and charge q, the force due to the repulsion of the surface charges becomes equal to the surface tension force of the liquid. This condition is expressed by Rayleigh's stability limit equation,

$$q_{Ry} = 8\pi(\varepsilon_0\gamma R_{Ry}^3)^{1/2} \qquad (11)$$

where ε_0 is the permitivity of vacuum and γ is the surface tension of the solvent. Evaporation leading to a decrease of radius below the Rayleigh value leads to droplet fission. Droplet fission followed by repeated evaporation and repeated droplet fission leads ultimately to gas-phase ions by processes that will be described in more detail later in this text.

The cone-jet mode described is only one of the possible ES modes. For a qualitative description of other modes and the cone-jet mode see Cloupeau.[41] The cone-jet mode is most often used in ESMS. It is also the best charact ized mode in the ES literature.[38] For recent treatments see Hendricks,[39] Smith,[42] Hayati,[43] and Fernandez de la Mora.[44]

Assuming that the charge separation is electrophoretic, at a steady operation of the electrospray, positive droplet emission will continuously carry off positive ions. The requirement for charge balance in such a continuous electric current device and the fact that only electrons can flow through the metal wire supplying the electric potential to the electrodes (Figure 1), lead to the supposition that the ES process must include an electrochemical conversion. In other words, the ES device can be viewed as a special type of electrolytic cell.[45,46] It is special because the ion transport does not occur through uninterrupted solution, as is normally the case in electrolysis. Part of the ion transport occurs through the gas phase where unipolar charged droplets and later gas-phase ions are the charge carriers. A conventional electrochemical oxidation reaction should be occurring at the positive electrode, i.e., at the liquid–metal interface of the capillary (Figure 1). This reaction should be supplying positive ions to the solution by converting atoms from the metal to positive metal ions which enter the solution and leave behind electrons (see equation 12a). The other alternative is the removal of negative ions present in the solution by an oxidation reaction as illustrated below (equation 12b) for aqueous solution.

$$M(s) \rightarrow M^{2+}(aq) + 2e^- \text{ (in metal)} \tag{12a}$$

$$4OH^-(aq) \rightarrow O_2(g) + 2H_2O + 4e^- \text{ (in metal)} \tag{12b}$$

Experimental evidence for the occurrence of such electrolytic reactions at the ES capillary electrode has been provided.[45,46]

B. Electrical Potential Required for Electrospray Onset and Interference from Electrical Gas Discharges

D.P.H. Smith[42] was able to derive an equation for the required electric field, E_{on}, at the capillary tip, which leads to the onset of electrospray:

$$E_{on} \approx \left(\frac{2\gamma \cos \theta}{\varepsilon_o r_c} \right)^{1/2} \tag{13}$$

This equation for the onset field, when combined with equation 10, leads to the potential, V_{on}, required for the onset of electrospray,

$$V_{on} \approx \left(\frac{r_c \gamma \cos \theta}{2\varepsilon_o} \right)^{1/2} \ln(4d/r_c) \tag{14}$$

where γ is the surface tension of the solvent, ε_o is the permitivity of vacuum, r_c is the radius of the capillary, and $\theta = 49°$, the half angle of the Taylor cone. Shown in

Table 1. Electrospray Onset Voltages V_{on} for Solvents with
Different Surface Tensions, γ

Solvent	CH_3OH	CH_3CN	$(CH_3)_2SO$	H_2O
γ (N/m^2)	0.0226	0.030	0.043	0.073
V_{on} (kV)	2.2	2.5	3.0	4.0

Table 1 is the surface tension for four solvents and the calculated electrospray onset potentials for $r_c = 0.1$ mm and $d = 40$ mm. The surface of the solvent with the highest surface tension (H_2O) is the hardest to stretch into a cone and a jet and this leads to the highest value for the onset potential V_{on}.

Experimental verification of equations 13 and 14 has been provided by Smith[42] and work from our laboratory.[47,48] For stable ES operation, one needs to go a few hundred volts higher than V_{on}. Using water as the solvent can lead to the initiation of an electric discharge from the capillary tip particularly when the capillary is negative, i.e. in the negative ion mode. The electrospray V_{on} is the same for both the positive and negative ion mode; however the electric discharge onset is lower when the capillary electrode is negative.[48]

The occurrence of an electric discharge leads to an increase of the capillary current I. Currents above 10^{-6} A are generally due to the presence of an electric discharge. A much more specific test is provided by the appearance of discharge characteristic ions in the mass spectrum. Thus, in the positive ion mode the appearance of protonated solvent clusters such as $H_3O^+(H_2O)_n$ from water or $CH_3OH_2^+(CH_3OH)_n$ from methanol indicates the presence of a discharge.[47] The protonated solvent ions are abundantly produced by ES in the absence of a discharge only when the solvent has been acidified, i.e. when H_3O^+ and $CH_3OH_2^+$ are present in the solution.

The presence of an electric discharge degrades the performance of ESMS, particularly so at high discharge currents. The electrospray ions are observed to be at much lower intensities than was the case prior to the discharge and the discharge generated ions appear with very high intensities.[47,48] It is likely that the discharge reduces the electric field near the capillary tip and this interferes with the charged droplet formation.

The high potentials required for electrospray show that air at atmospheric pressure is not only a convenient, but also a very suitable ambient gas for ES, particularly when solvents with high surface tension, like water, are to be subjected to electrospray. The oxygen in the air, which has a positive electron affinity, captures free electrons and acts as a discharge suppressor.

Electrospray in gases like helium or the other noble gases is not possible even with solvents like methanol which have a low surface tension. Electric discharges occur very readily in these gases since they do not capture electrons and are unable to reduce the electrons' kinetic energy significantly by inelastic collisions. ES could be possible[47] in the presence of an admixture with electron capture agents such as SF_6. A better solution is to produce the ions by ES in atmospheric air. The ions can then be transferred into a "reaction chamber" containing the noble gas. Such a device is described in Section III.B (see Figure 9). In the examples given there, the gas is not a noble gas, but nitrogen; however the principle is the same.

C. Dependence of Charged Droplet Current *I*, Droplet Radius *R*, and Charge Droplet *q* on Experimental Parameters

Theoretical[39] and semiempirical[44] equations have been derived which predict approximately, the current *I*, the radius *R*, and charge *q* of average initial droplets produced in the cone-jet mode by ES. Very useful are equations 15–17 which relate the above to important experimental variables,[44]

$$I \approx f\left(\frac{\varepsilon}{\varepsilon_0}\right)\left(\gamma K V_f \frac{\varepsilon}{\varepsilon_0}\right)^{1/2} \tag{15}$$

$$R \approx (V_f \varepsilon / K)^{1/3} \tag{16}$$

$$q \approx 0.7[8\pi(\varepsilon_0 \gamma R^3)^{1/2}] \tag{17}$$

where,

$$
\begin{align}
K &= \text{electrical conductivity of solution} \\
V_f &= \text{flow rate volume/time of solution through ES capillary} \\
\gamma &= \text{surface tension of solvent} \\
\varepsilon &= \text{permitivity of solvent} \\
\varepsilon_0 &= \text{permitivity of vacuum} \\
\varepsilon/\varepsilon_0 &= \text{dielectric constant of solvent}
\end{align}
$$

and $f(\varepsilon/\varepsilon_0)$ is a tabulated numerical function.[44] For example, the value of $f(\varepsilon/\varepsilon_0)$ is 18 for liquids whose dielectric constant is, $\varepsilon/\varepsilon_0 \geq 40$. The relationships were obtained with solutions having conductivities K larger than 10^{-4} S m^{-1}. For polar solvents like water and methanol and electrolytes that dissociate quite completely to ions, this corresponds to solutions with concentrations bigger than ~ 10^{-5} mol/L, i.e. a concentration range that is commonly used in ESMS. The flow rates used in experiments on which equations 15–17 are based[44] were below one µL/min and are also in the range of flow rates used in ESMS.

ES is normally operated at low total electrolyte concentrations, $C < 10^{-2}$ mol/L. For such concentrations a simple proportionality between the conductivity K and the concentration C holds:

$$K = \lambda_{0,m} C \tag{18}$$

The equivalent molar conductivity of the electrolyte, $\lambda_{o,m}$ introduces the dependence of the conductivity K on the specific nature of the electrolyte ions. $\lambda_{o,m}$ depends on the mobility of the ions in the solution and due to ion solvation, large and small ions have quite similar $\lambda_{o,m}$ values. Thus, $\lambda_{o,m}$ (LiCl) = 91, Ohm^{-1} cm^2 mol^{-1}, $\lambda_{o,m}$ (KCl) = 104, Ohm^{-1} cm^2, mol^{-1}. Combining equations 15 and 18 one obtains:

$$I \propto (\lambda_{o,m} C)^n \qquad n = 0.5 \qquad (19)$$

The exponent predicted by equation 15 is $n = 0.5$. However, also smaller coefficients down to $n = 0.3$ have been observed in experimental measurements.[51]

The current due to gas-phase ions is found[51] to be approximately proportional to the droplet current I, for electrolyte concentrations up to 10^{-3} mol/L. The droplet current I does not necessarily correlate with the gas-phase ion current when parameters other than the electrolyte concentration are changed. For example, an increase of the current I with the flow rate of the solution as predicted by equation 15 is observed experimentally. However, the gas-phase ion current may remain flat or even decrease with increasing flow rate.[45] This result is a consequence of the dependence of the radius of the initial droplets on the flow rate (see equation 16). The larger droplets produced at the higher flow rate lead to lower gas-phase ion yields.

D. Evolution of Initial Charged Droplets to Very Small Droplets which Produce Gas-Phase Ions

The detailed history of the evaporating and fissioning droplets will depend on the initial size and charge of the droplets produced by the electrospray. The important parameters determining the radius of the droplets (see above) were the flow rate V_f and the conductivity K of the solution (equation 16). Low flow rates and high conductivities lead to small droplets. Flow rates of a few µL/min and conductivities $K \approx 10^{-4}$ to 10^{-2} S m^{-1}, corresponding to electrolyte concentrations of 10^{-5} mol/L to 10^{-3} mol/L in polar solvents, lead to droplet radii of a few µm. It is also observed[44] that the charge of the initial droplets is not far from the Rayleigh limit. Thus, $q_o \approx 0.7 \, q_{Ry}$, is a typical value for the initial charge on the droplets (see equations 16 and 17).

The fission of the droplets at the Rayleigh limit has been studied with special methods.[49,50] The unstable droplet distorts into a teardrop shape and the resulting point of the drop emits a fine jet of very small droplets. These droplets have radii which are roughly 1/10 of the parent droplet radius. They carry away only ~2% of the parents mass but some 15% of the parents charge.[50] A droplet fission is shown in the insert of Figure 2. We shall call this type of fission, "jet fission." A schematic representation of the development of the droplet populations due to droplet evaporation and fission is shown in Figure 2. The time required for an initial droplet with radius R_o and charge q_o to reach the size R_1 that leads to the first fission can be estimated[51] with the use of expressions providing the rate of solvent evaporation from small droplets.[52] When relatively volatile solvents such as methanol, water, and acetonitrile are involved and the droplets are a few micrometers or less in diameter, the evaporation rate follows the surface evaporation limit law[52] which leads to a simple dependence[51] of the droplet radius on the time t,

$$\frac{dR}{dt} = -\frac{\alpha \bar{v}}{4\rho} \frac{p^\circ M}{R_g T} \tag{20a}$$

$$R = R_o - \frac{\alpha \bar{v} p^\circ M}{4\rho R_g T} t \tag{20b}$$

where,

$\bar{v}$	=	the average thermal velocity of the solvent vapor molecules
p°	=	the saturation vapor pressure of the solvent at the temperature of the droplet
M	=	molar mass of the solvent molecule
ρ	=	density of solvent
R_g	=	gas constant
T	=	temperature of droplet
α	=	condensation coefficient of solvent

and $\alpha \approx 0.04$ for water, ethanol and probably methanol.[51]

The time history of a parent droplet with initial radius $R_o = 1.0$ μm and charge $q_o = 10^{-14}$ C, where the solvent is methanol, the ambient temperature 35 °C and the droplet temperature 25 °C is depicted in Figure 2. The times Δt (μs) given in the figure were evaluated with equation 20b for methanol ($p^\circ = 100$ torr at 25 °C; $\alpha \approx 0.04$). It was assumed that jet fission occurs somewhat before the Rayleigh limit, when the charge is 0.8 q_{Ry}, and that the offspring droplets have radii equal to 0.1 the parent radius and remove 2% of the mass and 15% of the charge of the parent droplet. These assumptions are in line with experimental observations.[49,50] Also shown in Figure 2 (bottom right) is the fission of one of the offspring droplets creating a second generation of offspring.

Presently held theories assume that gas-phase ions are formed only from very small droplets. The theory attributed to Dole[53] assumes that gas-phase ions result from the evaporation of solvent from very small droplets, i.e. the second-generation offspring droplets of $R = 3$ nm and charges $N = 2$. Assuming that singly charged ions are present, evaporation of solvent from such a droplet until only several solvent molecules are left should lead to two singly charged gas-phase ions each stabilized by a few solvent molecules.

The rival theory of Iribarne and Thomson[54] assumes that gas-phase ions solvated by a few solvent molecules are emitted into the gas phase by charged droplets with radii $R < 10$ nm, which are somewhat below the Rayleigh limit. This "ion evapora-

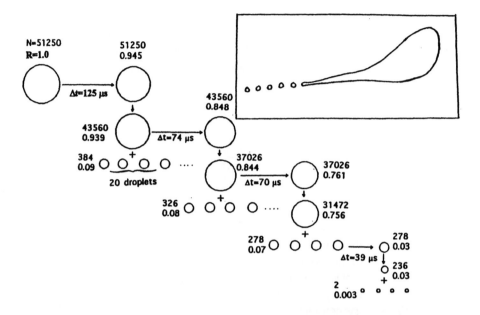

Figure 2. Schematic representation of time history of parent and offspring droplets. The droplet at top left is a typical droplet obtained from the spray at the capillary tip. N equals the number of elementary charges and R is the radius of the droplet in μm. Evaporation of solvent at constant charge leads to increase of the electric field due to the charge and to droplet fission at the Rayleigh limit. Typical fission shown in insert at top right. Offspring droplets which are much smaller than parent droplets are produced. These carry off ~2% of parent mass and ~15% of parent charge. Δt is the time between droplet fissions in μs. Only the first three successive fissions of a parent droplet are shown. The fission of an offspring droplet is shown at the bottom right of the figure. From Kebarle, P.; Tang, L. *Anal. Chem.* **1993**, *65*, 973A, with permission.

tion" suppresses further jet fission and thus also the abundant formation of the very small droplets required by the Dole theory.

Unfortunately, the conflict has not been resolved.[36] However from the standpoint of the experimentalist, many of the consequences of the two theories are similar. Both theories require very small droplets to generate gas-phase ions. The time requirement for the evolution to such droplets (see Figure 2) is in the hundreds of microseconds. The ambient gas is essential to provide the thermal energy for the evaporation. Solvents with low vapor pressure and condensation coefficients α may not be suitable or may require higher ambient temperatures.

To provide sufficient time for the droplet evolution, one can increase the distance between the ES capillary and the sampling orifice or sampling capillary. However, this reduces greatly the sampling efficiency since the charged droplet density and ion density decrease rapidly with distance from the ES capillary tip, due to the effect of space charge. A much better alternative is to use very low flow rates, obtaining smaller size initial droplets which require shorter times for evaporation. Flow rates of 1 μL/min require ES capillary tip to sampling capillary tip distances of ~1 cm. An extreme case of low flow rates is the "nanospray" method developed by Wilm and Mann[55] where the radius of the initial droplets is $R \approx 0.2$ μm and the ES capillary tip can be placed ~ 1 mm in front of the sampling orifice or right inside the sampling capillary, thus achieving a very high sampling efficiency.

E. Dependence of Gas-Phase Ion Intensities on the Concentrations of the Electrolyte Ions in Solution

The dependence of the observed gas-phase ion intensity of a given ion on the concentrations of the ions in solution is of interest to the investigator who intends to perform physical measurements involving the gas-phase ions. In general, one would be interested in maximizing this intensity.

The intensity of the protonated morphine ion observed in mass spectra when methanol solutions containing morphine hydrochloride at different concentrations were subjected to electrospray[51] is shown in Figure 3. We shall call this ion the analyte ion, A^+. The logarithmic plot used accommodates a wide range of ion intensities and concentrations. The analyte plot has a shape which is commonly observed.[40,53] A linear section with a slope ≈ 1 in the low concentration range up to ~ 10^{-6} M is followed by "saturation" and even a small decrease of intensity at the highest concentrations, 10^{-3} M. The key to understanding the complete analyte curve is the realization that the solution involved is not a single electrolyte system, because the solvent involved, unless special deionization procedures are applied, always contains impurity electrolyte. The impurities in the reagent grade methanol used[51] are mainly ammonium and sodium salts at a total concentration of ~ 10^{-5} mol/L. For analyte concentrations below 10^{-6} mol/L, ES is possible only because of the presence of impurity electrolyte,[45] which provides a minimum conductivity required for ES.

In the low concentration range of analyte A, the capillary current I, also shown in Figure 3, is carried by the dominant electrolyte, i.e. the impurities, Na^+ and NH_4^+, which are at a constant concentration. We shall call the impurity ions the B^+ ions. I is therefore constant in this range (see Figure 3A). The sum of the mass-analyzed total ion intensity is also constant in this range because it is dominated by the impurities B^+. Above an analyte A concentration of ~ 10^{-5} mol/L, the analyte begins to dominate and the total electrolyte concentration begins to increase. An increase of capillary current I occurs in this region as expected from equation 19. Because the molar conductivity λ_m° for different electrolytes M^+ generally changes by less than a factor of 2 over a wide range of electrolytes and the exponent n is very small, the current I is close to independent on the nature, i.e. the λ_m°, of the electrolyte. Changes of the capillary current I with concentration C must be

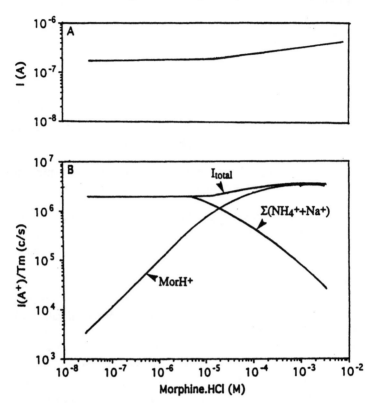

Figure 3. Dependence of analyte ion intensity on concentration. (A) Dependence of total current I on concentration M = mol/L of analyte ion, (Morphine)H^+, i.e., MorH$^+$, in solution. (B) Dependence of mass-analyzed MorH$^+$ ion current in counts/s on MorH$^+$ concentration. At low MorH$^+$ concentrations, [MorH$^+$] < 10^{-6} M, the dominant electrolyte in the solution are impurity ions Na$^+$ and NH$_4^+$. In this region MorH$^+$ intensity is proportional to [MorH$^+$] in solution. Mass-analyzed ion intensity was corrected for mass-dependent transmission Tm, of quadrupole. Concentration of morphine hydrochloride given in mol/L (M). From Kebarle, P.; Tang, L. *Anal. Chem.* **1993,** *65,* 973A, with permission.

considered because C can be changed over a much wider range. However, again due to small exponent n, these changes are still relatively small, i.e. a factor of $\approx$ 4–8 for a change of C by a factor of 100.

The shape of the mass-analyzed gas phase I_{tot} curve is seen to be very similar to that of the capillary current I (Figure 3). An approximate proportionality between the two currents when the concentration is increased is generally observed in this concentration range.[45,51] For an analyte concentration above 10^{-5} M, the intensity of the impurity ions B$^+$ is observed to decrease. This decrease is a consequence of the weak dependence of the capillary current I on the total electrolyte concentration.

Since the current is proportional to the total droplet charge, this means that the addition of A^+ to the solution leads only to a small increase in the droplet charge. However, in the charged droplets, the ions A^+ compete with the ions B^+ in the conversion process to gas-phase ions. A proportionality to concentrations of ions A^+ and B^+ in the droplets may be expected in this competition[51] so that an increase of the concentration of A^+ in the solution should lead to a decrease of production of gas-phase ions B^+, i.e. to a decrease of I_B. This is exactly what is observed in Figure 3.

It is evident from the data in Figure 3, that maximum gas-phase ion intensities for a given analyte ion are obtained with 10^{-4} to 10^{-3} mol/L analyte in the solution. The $\sim 3 \times 10^6$ ion counts/s observed at such concentrations are intensities after mass analysis at low resolution where the gas was admitted through a 100 μm orifice into a vacuum chamber which contains a triple quadrupole mass spectrometer.[51]

III. APPARATUS FOR THERMOCHEMICAL MEASUREMENTS INVOLVING ELECTROSPRAY-PRODUCED IONS

Apparatus, developed in this laboratory for two types of thermochemical measurements—(a) gas-phase ion molecule equilibria and (b) collision-induced dissociation (CID) threshold measurements—will be described. For both purposes, a triple quadrupole mass spectrometer is used. It is only the front end modifications that provide the conditions for (a) or (b).

A. Apparatus for Determination of Ion–Molecule Equilibria

The present description is based on previous publications from this laboratory[56–59] and the interested reader will find additional details and references in that work. Two different ion-source reaction chambers are used. One of these sources which operates at room temperature is shown in Figure 4. The second source, a variable temperature source will also be described. The electrospray generator and the ion-source reaction chamber are shown in Figure 4, while the mounting of the ion source and the front end of the mass spectrometer are shown in Figure 5.

The solution to be electrosprayed is passed through the electrospray capillary (ESC) by means of a motor driven syringe. Some of the spray containing the ions then enters the pressure reducing capillary (PRC) leading to the forechamber (FCH) of the ion source. The exit tip of the PRC directs the gas jet in a direction parallel to the bottom of the FCH, i.e. across the interface plate (IN). An orifice of 4 mm diameter in the interface plate connects the FCH to the reaction chamber (RCH). The ions in the jet exiting from the PRC are deflected out of the jet towards this orifice and into the RCH by means of an electric field applied across the FCH. A weak field is also applied across the RCH. At the bottom of the RCH a small orifice, 100 μm diameter, allowed some gas and ions to leak into the vacuum of the mass

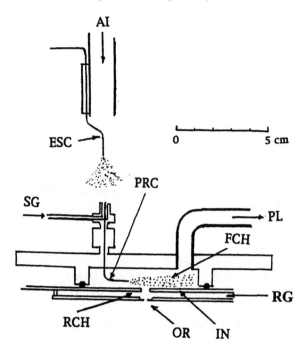

Figure 4. Ion source and reaction chamber for ion–molecule equilibria. Solution to be electrosprayed flows through elestrospray capillary ESC at ~1-2 µL/min. Spray and ions enter pressure reduction capillary PRC and emerge into forechamber FCH maintained at 10 torr by pump PL. Ions in gas jet, which exits PRC, drift towards interface plate IN under influence of drift field imposed between FCH and IN. Ions enter the reaction chamber RCH through an orifice in IN and can react with reagents in the reagent gas mixture RG. This flows into RCH and out of RCH to FCH where it is pumped away. Ions leaking out of RCH through orifice OR are detected with a mass spectrometer. To reduce the inflow of solvent vapors into the pressure reduction capillary PRC, a stream of dry air is directed through the pipe AI, at 60 L/min, and pure N_2 is directed at SG into the annular space at the entrance of the pressure reduction capillary, PRC. From Klassen, J. S.; Blades, A. T.; Kebarle, P. *J. Phys. Chem.* **1995**, *99*, 1509, with permission.

spectrometer. The pumping lead out of the FCH leads through a control valve to a forepump. The valve is adjusted to obtain a given constant pressure, generally 10 torr, in the FCH and the RCH.

A reagent gas mixture consisting of pure nitrogen, at 10 torr, as carrier and thermal bath gas, and the reagent L, at a given pressure between 1–80 m torr, was passed in slow flow through the reaction chamber. The measured flow rate when the FCH and RCH were at 10 torr, was typically 90 mL/min reagent gas mixture into the RCH. Some 2 mL/min of this flow escaped through the 100 µm orifice into

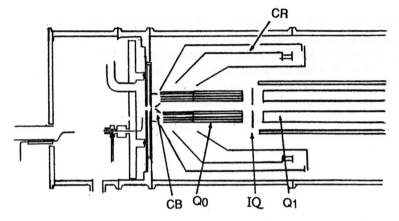

Figure 5. Front end of triple a quadrupole mass spectrometer with an ion source and reaction chamber (see Figure 4) mounted in front of an evacuated space containing skimmer cone CB, AC only quadrupole lens Q_0, and first quadrupole Q_1. The second and third quadrupoles, Q_2 and Q_3, are not shown. CR denotes cryopumping surfaces. IQ is an interquad lens.

the vacuum of the mass spectrometer housing while the rest, i.e. ~ 98 mL/min entered the FCH and was pumped out through the pumping lead (PL). The volume flow rates quoted are for gas at 700 torr. Thus the actual volume flow rate at 10 torr is 70 times higher.

The flow rate of gas entering into the PRC was 750 mL/min (at 700 torr). This gas would normally contain methanol vapor due to evaporation of the electrospray droplets. The presence of methanol vapor in the RCH is undesirable because this gas will interfere with the equilibria measured. Thus, the methanol will compete in ion–ligand equilibria by clustering to the ions. The methanol vapor entering the FCH could be reduced to negligible amounts by using two stages of vapor purging. The first stage is at the ESC where a fast stream of dry air, ~ 60 L/min (see Figure 4) was applied. The second stage is at the entrance tip of the PRC where a stream of reagent gas mixture, at ~ 2 L/min, was used in a counter current to the electrospray plume. The charged droplets and ions drift through the counter current to the PRC tip under the influence of the electric field between the ESC and PRC (Figure 4). The effectiveness of the purging stages could be easily demonstrated. When the reagent gas was L = H_2O, the ions observed in the absence of the purging gas flows were $M^+(CH_3OH)_x(H_2O)_y$ where x > y. In the presence of the purging gas this reversed to x << y, and generally x = 0 for more than ~80% of the M^+ ion clusters.

Cluster ions containing mostly methanol, i.e. x >> y are expected to be formed from the charged droplets in the plume of the ESC capillary. The exchange of methanol for water in the space between the ESC and PRC tips is facilitated not only by the removal of the methanol and supply of water vapor but also by the collisional activation of the ions due to the electric field which is present in the space between the ESC and PRC. The temperature of the gas and ions escaping through the orifice into the vacuum was assumed to be the same as the temperature of the bottom wall of the RCH near the orifice. This temperature was determined by a thermocouple attached to the wall near the orifice.

The PRC is of stainless steel, 0.7 mm o.d., 0.4 mm i.d., and 5 cm long. It was silver soldered into a 1.5 mm o.d. tubing for rigidity. The capillary exit is ~ 3 mm above the interface plate (IN) and 1 cm from the axis of the orifice leading to the RCH. The width of the FCH is 9 mm while that of the RCH is 4 mm. The solution flow rate through the ESC was 1.5–2 μL/min and the ESC tip was 2–3 cm away from the PRC tip (Figure 4). It was found most convenient to use methanol as the solvent. The desired ions, M^{z+} or BH^{+}, were obtained by adding 10^{-4} mol/L of the salts $M^{z+}(X^{-})_{y}$ or $BH^{+}X^{-}$ to the solvent. As was shown in the previous section (Figure 3) concentrations in this range lead to optimum ion intensities.

The mass spectrometer used (see Figure 5) is a Sciex TAGA 6000E triple quadrupole,[60] whose front end was modified as discussed above. The gas escaping from the orifice (OR) expands and is pumped by the cryosurfaces (CR; Figure 5) and a second set of cryosurfaces around Q_2, not shown in Figure 5. For the present measurements, where only a single mass filter is required, Q_1, and Q_2 were operated in the AC only mode and mass analysis was obtained with Q_3. The ions are detected with an ion counting system.

In normal analytical applications of the mass spectrometer, the offset potentials of the electrode CB and the Brubacker lens (AC only quadrupole Q_0, constructed of small, 1 mm diameter rods, so as to achieve high gas conduction) (Figure 5) are more negative than that of OR. This leads to maximum sensitivity, but because the gas pressure is relatively high in this region, the accelerated ions are activated by collisions and weakly bonded complexes such as ion hydrates can decompose in this region. To remove this problem, all measurements were performed with $V_{OR} = V_{CB} = V_{Q_0}$. The ions were accelerated only after Q_0, where the pressure is expected to be relatively low ($p \approx 10^{-5}$ torr). Quadrupoles are known to have mass dependent transmission. When used at a constant radial DC/AC ratio and constant ion accelerating voltage, ions of higher mass are transmitted with higher resolution and lower efficiency. In the present work, we corrected for this effect by determining the change of ion transmission (ion intensity) as a function of the selected DC/AC ratio for ions of a given m/z value.[56]

A potential drop of ~55 V between the PRC and IN (Figure 4) drives the ions into the RCH. This potential was adjusted so as to obtain a near maximum ion intensity. On the other hand, the potential between the IN and OR was deliberately selected to be small so as to allow the ions to reach near thermal energies. Generally, that potential drop was 10 V and the distance is 0.5 cm. At 10 torr this leads to an $E/p \approx 2$ V/(cm torr). Castleman and coworkers[61] have observed thermal behavior and measured equilibria at $E/p < 15$ V/(cm torr). Ion intensity ratios

$BH^+(H_2O)_n/BH^+(H_2O)_{n-1}$ measured in function of the IN-OR potential are shown in Figure 6A for B = octyl amine and $(n/n-1) = (1/0)$ and $(2/1)$. The ion ratios are found constant as would be expected for thermal ions at equilibrium and following Castleman's[61] results. The actual ion intensities, rather than the ratio, are shown in Figure 6B. These increase with increase of the IN-OR potential, an expected result since the ion transport to the orifice plate is improved by the higher drift field. A potential drop of IN-OR = 10 V was selected for the equilibria measurements because it leads to a fair intensity while keeping the ions well in the thermal range.

The ion-source reaction chamber with which equilibrium can be determined at different temperatures is shown in Figure 7. It operates very much on the same principles as the room temperature source. The forechamber FC and reaction chamber RC are now housed in a solid cylindrical copper block, CB. The copper block has four wells in which cartridge heaters are placed (not shown in Figure 7) and a narrow well for the thermocouple TC with which the temperature is determined. A copper thermal shield TS which is bolted to the water cooled flange acts as thermal shield protecting the cryosurfaces of the cryo-pumped triple quadrupole (see Figure 5). This source has an upper temperature limit of ~200 °C. Above this temperature, the cryopump efficiency becomes reduced. Better thermally shielded ion sources are in the testing stage. van't Hoff plots of the equilibrium constants, $K_{0,1}$, for the mono-hydration of the double-protonated diamines, $H_3N(CH_2)_pNH_3^{2+}$, obtained[62] at different temperatures with the ion source (Figure 7) are shown in Figure 8.

B. Apparatus for Determination of CID Thresholds

Two important conditions must be met if one is to obtain reliable determinations of CID thresholds which lead to good quality thermochemical data. The precursor ions, whose decomposition is to be studied must have a known internal energy such as a thermal energy distribution at a known temperature and the kinetic energy of the ions entering the collision cell must be well defined and have a narrow distribution. These two conditions are not easily met for ions produced by electrospray where the ions are obtained at atmospheric pressure. While the ions would be very quickly thermalized by collisions with the gas at atmospheric pressure, energy inhomogeneities can be easily induced in the atmospheric pressure to vacuum interface leading to the mass spectrometer particularly when electric fields are present at pressure conditions in the millitorr range.

In order to avoid such uncontrolled collisional activation, we chose to use apparatus which is closely related to the ion-source reaction chambers developed for thermal ion–molecule equilibria (see preceding Section A). In fact, sources like those shown in Figures 4 and 7 are well suited for providing thermalized ions; however they provide somewhat low ion intensities, typically some 50,000 counts/s of a given major ion in a mass spectrum after mass analysis. However, such intensities are completely sufficient for CID threshold measurements and the source

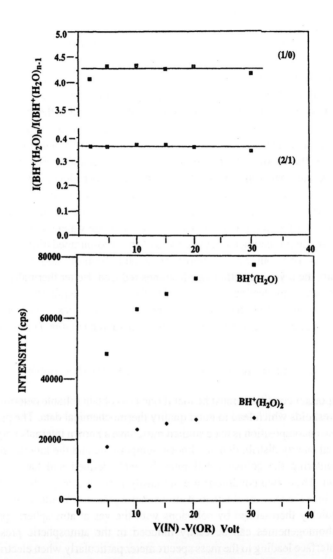

Figure 6. Dependence of intensities of ions $BH^+(H_2O)_n$ on drift potential in reaction chamber. The ion intensities decrease as drift potential decreases, however ion ratio shown in upper plot remains constant. B = *n*-octylamine. From Klassen, J. S.; Blades, A. T.; Kebarle, P. *J. Phys. Chem.* **1995**, *99*, 1509, with permission.

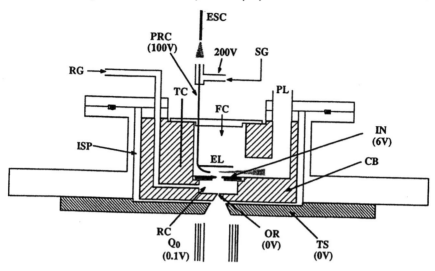

Figure 7. Ion source and reaction chamber for the study of ion-molecule reactions at different temperatures. Notation used is same as in Figure 4 except CB = copper block, EL = electrode attached to pressure reducing capillary, TC = thermocouple, TS = thermal shield, ISP = evacuated space reduces thermal conductivity from CB to flange. From Klassen, J. S.; Blades, A. T.; Kebarle, P. *J. Am. Chem. Soc.* **1996**, with permission.

in Figure 4 has been used for cross checking results obtained with the higher intensity source shown in Figure 9, which is used for the majority of the CID threshold measurements in our laboratory.[63-65] This source provides about 5 times higher intensities, i.e. ~200,000 count/s for the major ions in the mass spectrum.

The low-pressure source used in the apparatus (Figure 9) also has a forechamber, between CAP and IN, which now is called the low-pressure source (LPS) and a "reaction chamber" between IN and OR. However now the reaction chamber is supplied with pure N_2 and acts as an ion thermalization (IT) chamber. The pressure reduction capillary CAP is now coaxial with the orifice in IN leading to IT, and it is this coaxial arrangement that leads to higher ion intensities. Both LPS and IT are maintained at 10 torr. The drift field between IN and the orifice OR is kept low, E/p < 6 volt/cm torr, a value that is consistent with essentially thermal ions.

The ions entering the LPS through the CAP are generally clustered with a few solvent molecules, CH_3OH and H_2O, when methanol was used as solvent for the electrospray. For many of the threshold determinations it may be desirable to decluster the precursor ion. This is achieved by applying a large voltage (> 50 v) between CAP and IN. For an example of such declustering see Figure 2 in Ref. 63. Tests were made[63] to show that the presence of such a declustering field does not change the threshold curves of the product ions. The ions enter the vacuum of the

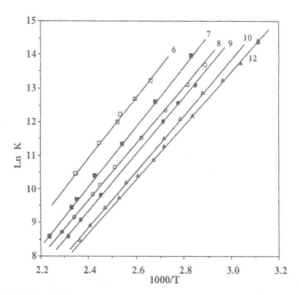

Figure 8. van't Hoff plots of equilibrium constant K, standard state 1 atm., for 0,1 equilibrium: $H_3N(CH_2)_pNH_3^{2+} + H_2O \cdot H_3N(CH_2)_0NH_3(H_2O)^{2+}$. Values of chain length p given beside each plot in the figure. Equilibrium constant K decreases as p is increased. From Klassen, J. A.; Blades, A. T.; Kebarle, P. *J. Am. Chem. Soc.* **1996**, with permission.

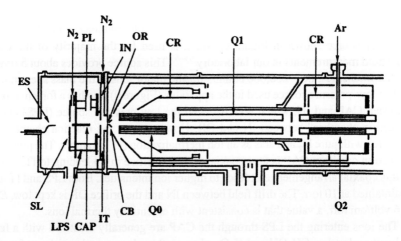

Figure 9. Apparatus for the determination of CID thresholds. Electrospray ions are produced as in Figures 4 and 7, however pressure reduction capillary CAP leading to 10 torr low pressure chamber LPS is coaxial with orifices leading to triple quadrupole. The IT chamber at 10 torr is used for ion thermalization. Collision chamber at Q_2 is usually used with collision gas Ar or Xe. The last quadrupole Q_3 is not shown on Figure.

triple quadrupole mass spectrometer through the orifice OR. The offset potentials of the electrodes CB and Q_0 are kept equal to OR in order not to induce collisional excitation of the ions in the space between OR and the first quadrupole Q_1.

The precursor ion is mass-selected with Q_1 and subjected to single-collision conditions, collision gases Ar or Xe, in the AC only quadrupole Q_2. The CID-produced ions are detected with Q_3 (not shown in Figure 9). For additional details see experimental parts of papers.[63–64]

The threshold curves obtained when the accelerating potential between OR and Q_2 was changed over a wide range from the onset of a given production to its maximum intensity were fitted with the theoretical cross section program, CRUNCH, developed by Armenrout and coworkers,[67–69] which leads to a determination of the threshold energies (see Section IV.C).

IV. DISCUSSION OF SELECTED RESULTS

A. Charge Reduction (Charge Separation) on Desolvation of Multiply Charged Ion–Solvent Molecule Clusters

Monoatomic Positive Ions

When a multiply charged ion solvated by several solvent molecules is subjected to heating or to vibrational excitation due to collisions with inert neutrals, ion desolvation occurs by successive loss of single solvent molecules, as illustrated below, for a double-charged ion and solvent molecules Sl:

$$M^{2+}(Sl)_n = M^{2+}(Sl)_{n-1} + Sl \tag{21}$$

For some ions the desolvation can proceed down to the naked ion M^{2+} while for others a reduction of the charge occurs at a given $n = r$ (r for reduction). The charge reduction may take different forms, the "simplest" of which is loss of Sl^+:

$$M^{2+}(Sl)_r = M^+(Sl)_{r-1} + Sl^+ \tag{22}$$

This type of charge reduction by charge transfer to the solvent molecule occurs in general when Sl are polar solvent molecules of aprotic character such as dimethylsulfoxide, dimethyl formamide, and acetonitrile. Protic solvents such as water lead to charge reduction which involves an intracluster proton transfer reaction:

$$M^{2+}(H_2O)_r = MOH^+(H_2O)_{r-k-2} + H_3O^+(H_2O)_k \tag{23}$$

Examples illustrating the reactions 21–23 are given in Figures 10–12. Shown in Figure 10 is the CID mass spectrum for the desolvation of $Ni^{2+}(H_2O)_{10}$. The sequence of product ions $Ni^{2+}(H_2O)_n$ where $n = 9$ to $n = 4$ illustrates the sequential solvent loss represented by equation 21. The CID spectra in Figure 11 demonstrate that for the $n = r = 4$, charge reduction via internal proton transfer (see equation 23)

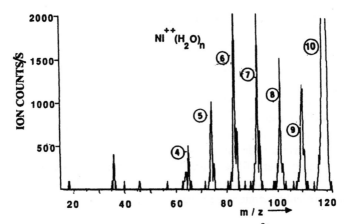

Figure 10. CID mass spectrum of precursor ion $Ni(H_2O)_{10}^{2+}$ illustrating formation of sequential H_2O loss down to $Ni(H_2O)_4^{2+}$. From Blades, A. T.; Jayaweera, P.; Ikonomou, M. G.; Kebarle, P. *Int. J. Mass Spectrom. Ion Proc.* **1990**, *101*, 325, with permission.

becomes competitive and dominant. The CID spectrum for the desolvation of $Cu^{2+}(DMSO)_3$ (Figure 12) illustrates charge reduction by charge transfer (see equation 22) when the dipolar aprotic DMSO is the solvent.

Table 2 gives a summary of results[70,71] which provide the solvent number r at which simple solvent loss and charge reduction become competitive. Charge reduction becomes dominant for solvent numbers lower than r. Ion clusters for

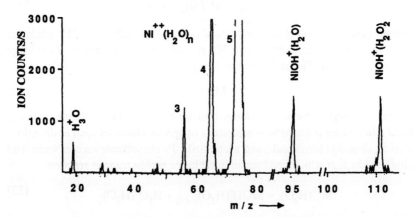

Figure 11. CID mass spectrum of precursor ion $Ni(H_2O)_5^{2+}$, which shows that $Ni(H_2O)_4^{2+}$ undergoes intramolecular proton transfer (see equation 23), which leads to $NiOH^+(H_2O)_2$ and H_3O^+. This reaction is competitive with simple H_2O loss leading to $Ni(H_2O)_3^{2+}$. From Blades, A. T.; Jayaweera, P.; Ikonomou, M. G.; Kebarle, P. *Int. J. Mass Spectrom. Ion Proc.* **1990**, *101*, 325 with permission.

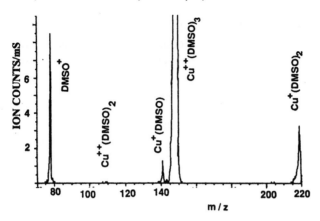

Figure 12. CID of precursor ion Cu(DMSO)$_3^{2+}$, which shows that this ion undergoes charge reduction by electron transfer (see equation 22 which leads to Cu(DMSO)$_2^+$ and DMSO$^+$. From Blades, A. T.; Jayaweera, P.; Ikonomou, M. G.; Kebarle, P. *J. Chem. Phys.* **1990**, *92*, 5900, with permission.

Table 2. Ligand Number at which Charge Reduction from M^{2+} to M^+ Occurs

Ion^e	L	$IE(L)^a$	r^b	$IE(M^+)$
		Doubly charged ions M^{2+}		
Be^{2+}	H_2O	12.6	$>15^c$	18.2
Mg^{2+}	H_2O		3	15.0
Ca^{2+}	H_2O		2	11.9
Sr^{2+}	H_2O		2	11.0
Ba^{2+}	H_2O		0	10.0
Mn^{2+}	H_2O		3	15.6
Fe^{2+}	H_2O		~4	16.2
Co^{2+}	H_2O		4	17.1
Ni^{2+}	H_2O		4	18.2
Cu^{2+}	H_2O		$>15^c$	20.3
Zn^{2+}	H_2O		~5	18.0
Mn^{2+}	DMF	9.1	$<1^d$	15.6
Cu^{2+}	DMF	9.1	3	20.3
Cu^{2+}	DMSO	9.0	3	20.3
		Triply charged ions M^{3+}		$IE(M^{2+})$
Y, La, Ce, Nd, Sm	H_2O	12.6	$>15^c$	19–23
Y, La, Ce, Nd, Sm	DMSO	9.0	3	19–23

Notes: [a]Ionization energy of ligand in eV, Lias.[14]

      [b]$M^{2+}(L)_r$ is the complex for which both simple loss of ligand and charge reduction (see reactions 21 to 23, Section IV.A), occur competitively on CID.

      [c]Based on observations that doubly charged ions M^{2+} could not be produced by electrospray when this ligand was present in the solution or in the vapor of electrospray or interface chamber.

      [d]Lowest ion observed on CID of Mn^{2+} (DMF)$_n$ was Mn^{2+} DMF.

      [e]Data from Blades et al.[70,71]

which r = 0 is indicated in Table 2, can be desolvated down to the naked ion without the occurrence of charge reduction. For the ions shown in Table 2, and L = H_2O, this was possible only for Ba^{2+}, although some very small yields of naked Sr^{2+} could be observed[70,71] even though charge reduction occurred at r = 2.

For L = H_2O, and doubly charged ions, charge reduction is observed (Table 2) at higher r when the ionization energy $IE(M^+)$ is high. Mg^{2+} and Cu^{2+}, which have highest $IE(M^+)$ of 18.2 and 20.3 eV, respectively, undergo charge reduction even when they are associated with a very large number of solvent molecules of r > 15. Because the charge reduction when L = H_2O does not involve direct electron transfer, but intracluster proton transfer, the strong dependence of r on $IE(M^+)$, observed in Table 2, needs to be explained.

The enthalpy change ΔH_{24} for the charge reduction reaction 24 can be obtained from a thermodynamic cycle,[71]

$$M^{2+}(H_2O)_r = MOH^+(H_2O)_{r-2} + H_3O^+$$

$$\Delta H_{24} = \Delta H_{r,0}(M^{2+}) - IE(M^+) - D(M^+-OH) - \Delta H_{r-2,0}(MOH^+)$$
$$+ D(H-OH) + IE(H) - PA(H_2O) \tag{24}$$

where $\Delta H_{r,0}(M^{2+})$ is the desolvation energy of the cluster $M^{2+}(H_2O)_r$ down to the naked ion and $\Delta H_{r-2,0}(MOH^+(H_2O)_{r-2})$ involves the same process but for $MOH^+(H_2O)_{r-2}$. D stands for the bond dissociation enthalpy, IE for ionization energy and PA for proton affinity. Equation 24 shows that the enthalpy change for the reaction decreases when $IE(M^+)$ is high and this may be considered as a partial explanation of the observed trend. However, it would be much more desirable to obtain information on the magnitudes of the activation enthalpy $\Delta H_4^{\ddagger}$. An analysis leading to approximate values for ΔH_{24} and $\Delta H_{24}^{\ddagger}$ is possible but only for the single case of Ca^{2+} because the required supporting thermochemical data are available only for this ion. Ca^{2+} was observed to undergo charge reduction for r = 2 (Table 2). Applying equation 24 to r = 2 and Ca^{2+}, one obtains,

$$\Delta H_{24}(Ca^{2+}, r = 2) = \Delta H_{2,0}(Ca^{2+}) - D(Ca^+-OH) + 11.5 \text{ eV} - 11.9 \tag{25}$$

where $D(H-OH) + IE(H) - PA(H_2O) = 11.5$ eV and IE $(Ca^+) = 11.9$ eV.

Theoretical calculations by Klobukowsky[74] with extended basis sets and electron correlation have provided the values,

$$Ca^{2+}(H_2O)_2 = Ca^{2+} + 2H_2O \qquad \Delta E_{2,0} = 4.26 \text{ eV} \tag{26a}$$

$$Ca^{2+}(H_2O)_1 = Ca^{2+} + H_2O \qquad \Delta E_{1,0} = 2.26 \text{ eV} \tag{26b}$$

while experimental measurements by Magnera et al.[72] lead to:

$$CaOH^+ = Ca^+ + OH \qquad \Delta E = 4.6 \text{ eV}. \tag{27}$$

Assuming that these values are close to the corresponding enthalpy changes and substituting into equation 25, one obtains the interesting result that reaction 24 is exothermic, i.e.,

$$\Delta H_{24} \; (Ca^{2+}, r = 2) = -0.74 \; eV \tag{28}$$

An estimate for the activation energy can be obtained by the consideration that for $r = 2$, the charge reduction reaction must proceed at a rate similar to that for the single solvent molecule dissociation, which means that the activation free energies for the two reaction are of similar magnitude. Neglecting the difference between $\Delta G^{\ddagger}$ and $\Delta H^{\ddagger}$ we obtain,

$$\Delta H_{24}^{\ddagger}(Ca^{2+}, r = 2) = \Delta H_{2,1} \; (Ca^{2+}) = 2.0 \; eV \tag{29}$$

where $\Delta H_{2,1}$ was obtained from the data given in equation 26 and use was made of the general observation that for ion declustering reactions, the activation enthalpy equals the reaction enthalpy, i.e. the reverse, clustering reactions proceed without activation energy.

The activation energy for the charge reduction reaction is due to two factors: the bond stretching and distortions of the originally near linear complex, so as to achieve the internal proton transfer and the increase of energy due to the Coulombic repulsion between the two charged products, a repulsion that leads to a release of kinetic energy on their separation.

For higher $Ca^{2+} \; (H_2O)_n$ clusters, the enthalpy change ΔH_{24} will increase due to increase of the $\Delta H_{n,0}^{\circ}$ energy which is not counterbalanced by the much smaller increase of $\Delta H_{n-2} \; (Ca^+OH)$; see equation 24. This effect will lead to positive values for ΔH_{24} and the charge reduction reaction will shut down.

One expects that an ion–solvate, which on desolvation undergoes a charge reduction reaction, cannot be produced by the process where one starts with the naked ion and solvent vapor and, at low temperatures, builds up to the ion solvate by sequential ion–solvent association reactions. This restriction may certainly be expected at relatively low total pressures where collisional cooling of the association complex is slow, i.e. in the kinetic third body-dependent range. Under these conditions, the energy released by the association reaction at a given $n = r$, will activate the charge reduction reaction and if the charge reduction reaction is faster, the simple association will be prevented. Exactly such a process must have occurred in the association reaction rate measurements by Spears and Fehsenfeld.[75] For Ca^{2+} these authors observed the formation of the first hydrate, $Ca^{2+} \; H_2O$, but could not obtain the second hydrate. Instead they observed the intracluster proton transfer reaction,

$$Ca^{2+}OH_2 + OH_2 \rightarrow Ca^{2+} \; (OH_2)_2{}^* \rightarrow CaOH^+ + H_3O^+ \tag{30}$$

leading to charge reduction. The results by these authors thus provide a conformation of the $r = 2$ for Ca^{2+} given in Table 2.

Examining Table 2, one comes to the conclusion that only Ba^{2+} $(H_2O)_n$ where $n > 1$ can be produced by the association reactions of M^{2+} with H_2O. For all the other ions only the monohydrate will be obtainable. For ions with high IE(M) values, even the monohydrate, $M^{2+}H_2O$ may not be obtained because of charge transfer reactions to H_2O (see equation 22). Other protic solvents will lead to charge reduction by proton transfer at different values of r. Only NH_3 has been examined.[71] It leads to much more facile charge reduction than H_2O. Many of the doubly charged ions that were observed as hydrates could not be observed as the equivalent clusters of NH_3.

For ammonia, the energy change equivalent to equation 29 for H_2O,

$$D(NH_2-H) + IE(H) - PA(NH_3) = 9.3 \text{ eV} \tag{31}$$

is almost 2 eV smaller than the equivalent expression for H_2O (see equation 25) and this difference is probably responsible for the observed effect. While some solvent molecules like NH_3 destablize the doubly charged ion, i.e. promote charge reduction, others such as the dipolar aprotic solvents lead to stabilization. Thus, the Cu^{2+} solvated ion, which could not be observed when a protic solvent (methanol, water) was used for electrospray, could be obtained after dipolar aprotic solvents like DMSO and DMF were added to the protic solvent (methanol). Charge reduction by charge transfer to DMSO to DMF on CID occurs for Cu^{2+} at r = 3 (see Table 2):

$$Cu^{2+}(DMSO)_3 \rightarrow Cu^+(DMSO)_2 + DMSO^+ \tag{32}$$

Due to the large difference, IE (Cu^+) = 20.3 eV and IE (DMSO) = 9.0 eV, reaction 32 must be strongly exothermic. A thermochemical treatment similar to that represented in equations 4–7 for the Ca^{2+} H_2O system and leads to an estimate of the activation energy $\Delta H_{11}^{\ddagger}$ would be possible. Unfortunately the required supporting thermochemical data such as $\Delta H_{3,0}(Cu^{2+})$ and $\Delta H_{2,0}(Cu^+)$ for DMSO are not available, so that a numerical value for $\Delta H_{11}^{\ddagger}$ cannot be obtained. Because Cu^{2+} also undergoes charge reduction when DMSO and DMF are used, the prediction can be made that no clusters of Cu^{2+} can be produced by forward clustering (association) reactions starting with the naked ion and DMSO, DMF molecules. The same holds also for other M^{2+} ions whose IE(M$^+$) is high.

The most powerful prevention of charge reduction is obtained with polydentate polycyclic complexing agents. Thus, the triply charged Co^{3+} could be obtained[71] in the gas phase when the hexadentate agent sepulchrate was added to the methanol solution of a Co^{3+} salt that was electrosprayed. The structure of the Co^{3+} sepulchrate is shown below. The ionization energy IE(Co^{2+}) = 33.5 eV is very high and this makes the stabilization provided by the sepulchrate even more impressive. The Co^{3+} sepulchrate, when produced from methanol solutions and equilibrated in N_2 containing ~ 1 mtorr H_2O at room temperature, was quite heavily hydrated. An envelope from $n = 5$ to $n = 10$ peaking at ~ $n = 7$ water molecules was observed.[71] Also

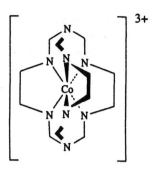

observed, and with higher intensity, were the charge reduced species $(Co\ Sep\text{-}H)^{2+}$ $(H_2O)_n$ with $n = 0$ to 3, and $(Co\ Sep\text{-}2H)^{+}$ $(H_2O)_n$ with $n = 0$ to 2. Evidently, the $CoSep^{3+}$ undergoes a charge reduction reaction in which protic hydrogens of the sepulchrate are used to protonate departing water molecules:

$$(CoSep)^{3+}\ (H_2O)_n = (CoSep\text{-}H)^{2+}\ (H_2O)_k + H_3O^{+}\ (H_2O)_{(n-1-k)} \tag{33}$$

$$(CoSep\text{-}H)^{2+}\ (H_2O)_k = (CoSep\text{-}2H)^{+}\ (H_2O)_m + H_3O^{+}\ (H_2O)_{(k-m-l)} \tag{34}$$

The charge reduction reactions occurred prior to any collisional activation in the second quadrupole Q_2, which indicates that the activation energy for this reaction is very low. These results show that in the presence of a multiple charge even N–H hydrogens can become sufficiently acidic and engage in protonation of water molecules.

Polyatomic Positive Ions

The most important multiply charged polyatomic positive ions are compounds with two or more basic groups which when protonated lead to doubly or poly-charged ions. Typical examples are diamines such as the double protonated α, ω alkyldiamines, $H_3N(CH_2)_pNH_3^{2+}$, and the most important class, the polyprotonated peptides and proteins, which have multiple basic residues. Charge reduction for these systems occurs through proton transfer from one of the protonated basic sites to a solvent molecule. Such a reaction is shown below for the monohydrate of a doubly protonated diamine:

$$^{+}H_3N(CH_2)_p\ NH_3^{+} \cdot OH_2 = {}^{+}H_3N(CH_2)_p\ NH_2 + H_3O^{+} \tag{35}$$

This reaction is analogous to deprotonation reactions of multiply protonated ions by neutral bases B such as:

$$^{+}H_3N(CH_2)_pNH_3^{+} + B \xrightarrow{a} {}^{+}H_3N(CH_2)_pNH_3^{+} - B$$

$$\xrightarrow{b} {}^{+}H_3N(CH_2)_pNH_2 + BH^{+} \tag{36}$$

In reaction 35, activation energy has to be provided to the precursor ion by collisions or other means and charge reduction will occur when the activation energy is lower than that for the desolvation reaction. In reaction 36, the "solvation" of the ion by B, i.e. reaction a, provides the activation energy and proton transfer and charge reduction will occur if the activation energy for reaction b is less than that for the reverse of reaction a.

The presence of a second positive charge can cause deprotonation even though the base B, such as H_2O, has a lower gas-phase basicity than the functional groups $R-NH_2$. Relationships between the gas-phase basicity of B, of $R-NH_2$, and the Coulombic repulsion energy required to lead to the occurrence of reaction 36 have been proposed by Williams and coworkers[76] and Gronert.[77] It appears that theoretical calculations such as performed by Gronert[77] are the best means for predicting the distance between the two charges, i.e. at what value of p, at which deprotonation occurs.

When experiments with electrospray producing $H_3N(CH_2)_pNH_3^{2+}$ ions were performed in this laboratory,[56] it was found that doubly protonated ions could be obtained from methanol-water solutions only when p > 4. For p ≤ 4 only the singly protonated ions were observed, even though the doubly protonated ions are known to be present in the solution. We attribute the failure to observe the doubly protonated ions with p ≤ 4 to the occurrence of charge reduction by deprotonation. Probably methanol, whose gas-phase basicity is greater than that of water, is involved in the deprotonation. The diprotonated diammines, p > 4, could all be dehydrated down to the naked ion either in CID experiments or at higher temperature.

The quaternary methylated diammines $(CH_3)_3N(CH_2)_pN(CH_3)_3^{2+}$ could be obtained from a methanol–water solution also for lower p values, i.e. p ≥ 2. The greater stability of these doubly charged ions can be attributed to the more dispersed charge and to the absence of protic hydrogens.

Polyatomic Negative Ions

While unsolvated multiple charged monatomic positive ions are stable in the gas phase (see subsection on monatomic positive ions) this is not the case for multiply charged negative ions. Doubly charged monoatomic ions like S^{2-}, which are known to be present in the condensed phase, are unstable with respect to electron detachment in the gas phase. A theoretical search for double negatively charged small polyatomic ions by Boldyrev and Simons[78] has shown that all diatomic, triatomic, and tetratomic dianions that were examined, including the well known (in the condensed phase) carbonate, CO_3^{2-} ion, are unstable to loss of one electron in the gas phase. Even the extremely important condensed phase dianions SO_4^{2-} and PO_4^{3-} were found to be unstable.[79]

Table 3. Hydration Number r at which Charge Reduction of Doubly Charged Hydrated Negative Ions Occurs[a]

Ion	r	Ion	r
CO_3^{2-}	$>15^b$	$S_2O_6^{2-}$	0^c
$HOPO_3^-$	$>15^b$	$SO_3C_2H_4SO_3^{2-}$	0^c
SO_4^{2-}	2–3	$O_3SO_2SO_3^-$	0^c
SeO_4^{2-}	2–3	$CO_2(CH_2)_nCO_2^{2-}, p>4$	0^c
		$CO_2(CH_2)_nCO_2^{2-}, p<4$	$>15^b$
$S_2O_3^{2-}$ (thiosulfate)	2–3	$1,4\text{-}CO_2C_6H_4CO_2^{2-}$	0^c

Notes: [a]Charge reduction for all these ion hydrides occurs by intracluster proton transfer (see equation 37). Data from refs. 80–82.
   [b]Doubly charged ion could not be produced from methanol–water mixture as solvent.
   [c]Doubly charged ion could be dehydrated down to the naked ion without charge reduction.

The hydrated sulfate $SO_4(H_2O)_n^{2-}$ could be produced by electrospray in the gas phase,[80] but neither the triply charged orthophosphate PO_4^{3-} nor the doubly charged $HOPO_3^{2-}$ were observed as the naked ion or the hydrate.[81] CID of the hydrated sulfate led to simple desolvation down to $n = r = 4$. The decomposition of the $r = 4$ hydrate led to charge reduction by intracluster proton transfer:

$$SO_4(H_2O)_4^{2-} = SO_3OH^-(H_2O)_k + OH(H_2O)_{3-k}^- \qquad k = 0, 1 \qquad (37)$$

The lowest SO_4^{2-} hydrate observed was $SO_4(H_2O)_2^{2-}$; however some SO_4^- was also detected which could be resulting from very low yields of SO_4^{2-} by dehydration of $SO_4H_2O^{2-}$, followed by electron detachment from SO_4^{2-}. These results[80] are in agreement with the theoretical study of Boldyrev and Simons[79] who predicted that SO_4^{2-} is unstable to loss of one electron. They also showed that the interaction with only two water molecules is sufficient to stabilize the doubly charged ion in the gas phase. Hydrated polysulphates such as $S_2O_6^{2-}$ and $S_2O_8^{2-}$ can be produced by electrospray in the gas phase and can be collisionally dehydrated down to the naked ions. These larger ions can provide increased stabilization of the charge by charge delocalization onto the larger number of electronegative oxygen atoms, especially since these are also spaced farther apart.

Results on the stability of dianions on dehydration obtained in our laboratory are summarized in Table 3. These results provide a clear illustration of how the delocalization of negative charge onto more electronegative ions, and increasing distances between the charge bearing atoms and functional groups, leads to stabilization of the dianions. The doubly charged ions listed in Table 3 should be interesting objects for experimental determinations of the second and first electron affinities as well as for theoretical studies.

B. Thermochemistry of Ion–Molecule Complexes and Specifically Ion–Hydrates Obtained from Equilibrium Determinations

The apparatus and method for ion–molecule equilibrium determinations were described in Section III.A. Here we give a brief summary of results obtained and of their significance and provide, when appropriate, projections for future work.

Doubly Charged Metal Ions M^{2+}

The desirability to obtain thermochemical information on ion–ligand interactions in complexes, $M^{2+}L_n$, involving alkaline earth and transition metal M^{2+} ions was one of the principle aims which led to the development of apparatus for electrospray generated ions.[70] Unfortunately only modest progress has been achieved so far.

Data obtained[56] for the hydration of the isoelectronic ions Na^+ and Mg^{2+} are given in Table 4. The free energy changes $\Delta G^{\circ}_{n-1,n}$ for the successive hydration of the ions are determined near room temperature. The data illustrate the much higher solvating power of the doubly charged ion. Thus, the same free energy, 6.6 kcal/mol, is released for the addition of the fourth molecule to Na^+ and the twelfth molecule to Mg^{2+}, i.e. $\Delta G^{\circ}_{3,4}$ $(Na^+) \approx \Delta G^{\circ}_{11,12}$ $(Mg^{2+}) = -6.6$ kcal/mol. The first solvation shell interaction for Na^+ and a second shell interaction for Mg^{2+} must be involved. The very large solvating power for Mg^{2+} in the second shell must be attributed to the double charge and small size of Mg^{2+}. Undoubtedly, the water molecules in the first shell will be highly polarized such that the hydrogen atoms attain a highly protic character which leads to the formation of strong hydrogen bonds to the water molecules of the second shell. The variable temperature reaction chamber was developed only relatively recently and therefore determinations of $\Delta H^{\circ}_{n-1,n}$ and

Table 4. Comparison of Hydration Energies of Singly Charged Na^+ and Doubly Charged Isoelectronic Mg^{2+} Ions[a]

Ion	Ligand	(n–1,n)	$\Delta G^{\circ}_{n-1,n}$ (kcal/mol)
Na^+	H_2O	(3,4)	6.6
		(4,5)	4.7
Mg^{2+}	H_2O	(10,11)	7.2
		(11,12)	6.6
		(12,13)	6.0
		(13,14)	5.5
		(14,15)	5.2
		(15,16)	4.9

Notes: [a]Data from Klassen et al.[56] Free energy $\Delta G^{\circ}_{n-1,n}$ is for 293 K.

$\Delta G^{\circ}_{n-1,n}$ values extending to lower n, which are accessible only at high temperatures, have not been obtained yet.

It is clear that determinations of thermochemical data at very low n will not be accessible with the equilibrium technique. Theoretical calculations by Klobuk-owsky[74] for ion hydrates lead to $\Delta E_{1,0}(Mg^{2+}) = 80$ kcal/mol, $\Delta E_{2,1}(Mg^{2+}) = 70$ kcal/mol, $\Delta E_{1,0}(Ca^{2+}) = 52$ kcal/mol, and $\Delta E_{2,1}(Ca^{2+}) = 46$ kcal/mol. Previous determinations of ion clustering equilibria[83] have been possible only for enthalpy changes up to ~ 40 kcal/mol. This indicates that ion equilibria determinations will be readily applicable only down to $n = 3$ or 2 for alkaline earth and transition metal ions. Determinations for lower n would be accessible in principle with the CID threshold technique, see Section C. However, as discussed in the preceding Section A, the very low n hydration or complexation energies may not be directly accessible via CID thresholds due to competing charge reduction reactions. The vast amount of important and interesting ion–solvent and ion–ligand energies that could be determined with M^{2+} ions produced by electrospray remain a challenge that we hope will be met in the near future.

Doubly protonated α–ω Diamines and Doubly Deprotonated Acids

In this section, we discuss the change of hydration exothermicity and exoergicity for doubly charged ions, where the charge resides principally on two separate functional groups and the distance between the two charged groups can be changed. The relative magnitudes of the hydration energies reveal to what extent the charge is stabilized and the dependance on the nature of the two charged groups, the distance between them, and the nature of the moiety connecting them.

First, we examine the sequential hydration of a doubly protonated diamine, $H_3N(CH_2)_{12}NH_3^{2+}$. The sequential enthalpies, $\Delta H^{\circ}_{n-1,n}$, free energies $\Delta G^{\circ}_{n-1,n}$, and entropies $\Delta S^{\circ}_{n-1,n}$, obtained from van't Hoff plots of the hydration equilibria,[62] are given in Table 5. The $\Delta H^{\circ}_{n-1,n}$ values follow an interesting pattern: $\Delta H^{\circ}_{0,1} = \Delta H^{\circ}_{1,2} = -15.7$ kcal/mol, then a drop of exothermicity occurs with $\Delta H^{\circ}_{2,3} = \Delta H^{\circ}_{3,4} = -13.5$ kcal/mol. This pattern is very different from the gradual decrease of $\Delta H^{\circ}_{n-1,n}$ with increase of n observed for singly charged ions.[8,15] This difference is easily under-

Table 5. Energetics of the Sequential Hydration of $H_3N(CH_2)_{12}NH_3^{2+}$ [a]

$n-1,n$	$-\Delta G^{\circ}_{298}$ (kcal/mol)	$-\Delta H^{\circ}$ (kcal/mol)	$-\Delta S^{\circ}$ (cal/deg mol)
0,1	9.7	15.7	20.1
1,2	8.8	15.7	23.2
2,3	7.0	13.4	21.5
3,4	6.3	13.6	24.5
4,5	5.1	—	—
5,6	4.4	—	—

Note: [a] Data from Blades et al.[62]

stood. Since the charges in the diammonium ion are separated, the strongest interactions occur when the first two water molecules go to different charged sites. Furthermore, because in the dodecane the two charged sites are widely separated, the interaction of the second water molecule with the second site is essentially identical to the first with the first site.

The hydration of alternate charged sites by sequentially added water molecules is also fully in accord with the observed $\Delta S^\circ_{n-1,n}$ changes (see Table 5). Thus the observed values,

$$\Delta S^\circ_{0,1} - \Delta S^\circ_{1,2} = 3.1 \text{ cal/degree} \cdot \text{mol}$$

$$\Delta S^\circ_{2,3} - \Delta S^\circ_{3,4} = 3.0 \text{ cal/degree} \cdot \text{mol} \qquad (38)$$

can be predicted by considering the expected statistical factor ratios for the forward and reverse rates k_f/k_r, of the given $(n-1,n)$ equilibrium. The statistical factors for the forward and reverse reaction of $(0,1)$ are $S_f/S_r = 6/1$, because there are six sites for the H_2O to occupy when coming in, but only one H_2O leaves. For the $(1,2)$ reactions there are only 3 sites available on the second charged site for the incoming molecule, but two molecules can leave from the dihydrate, thus $S_f/S_r = 3/2$. Since $\Delta H^\circ_{0,1} = \Delta H^\circ_{0,2}$, the corresponding values for ΔS° should differ only due to the statistical factors such that:

$$\Delta S^\circ_{0,1} = \Delta S^\circ_{1,2} = R \ln 4 = 2.75 \text{ cal/degree} \cdot \text{mol} \qquad (39)$$

The same symmetry factor ratio can be derived also for the $(2,3)$ and $(3,4)$ equilibria. The value 2.75 cal/degree·mol is very close to the experimental result given in equation 38, considering that the experimental error expected is more than 1 cal/degree·mol.

The changes of thermochemistry for the $(0,1)$ and $(1,2)$ hydrations, with the number p of CH_2 groups present, is presented in Table 6. The charged $-NH_3^+$ substituent on the one end of the molecule increases the protic character of the hydrogens on the $-NH_3^+$ group on the other end, and when a hydrogen from this group interacts with a water molecule, a stronger H bond is formed. Even though the distances between the two charged groups are very big, this effect is quite noticeable. Thus $\Delta H^\circ_{0,1}$ (p = 7) = -17.8 kcal/mol is 2.6 kcal/mol higher than $\Delta H^\circ_{0,1}$ $(CH_3(CH_2)_6NH_3^+) = -15.2$ kcal/mol (Blades et al.[62]), where a second charge is not present. This H-bond (strengthening) effect is expected to decrease as the distance between the two ammonium groups increases and such changes with increasing values of p are indeed observed (Table 6).

Since the hydration energy changes with p are driven by changes of Coulombic repulsion, a linear relationship between $-\Delta H^\circ_{0,1}$ and the reciprocal distance between the charged centers may be suspected, since the Coulombic energy is proportional to the reciprocal of the distance. The quantity, p + 2, may be assumed to be approximately proportional to the distance between the charged centers. We arrive

Table 6. Hydration of α,ω-Diammonium Alkanes and
Bis-trimethylammonium Alkanes[a]

	$-\Delta G_{298}^{\circ}$ (kcal/mol)		$-\Delta Hsup$ (kcal/mol)		$-\Delta S^{\circ}$ (cal/degree mol)	
p	(0,1)	(1,2)	(0,1)	(1,2)	(0,1)	(1,2)
			$H_3N(CH_2)_pNH_3^{2+}$			
12	9.7	8.8	15.7	15.7	20.1	23.2
10	10.3	9.3	16.8	16.8	22.8	25.2
9	10.3	9.3	16.5	16.3	20.8	23.4
8	10.5	9.7	16.9	16.8	21.5	23.8
7	11.1	9.8	17.8	17.2	22.6	23.7
6	11.5	10.4	17.8	17.3	21.0	23.1
5	$(12.2)^b$	$(11.3)^b$	$(18.6)^b$	$(18.4)^b$	$(21.5)^b$	$(23.9)^b$
			$(CH_3)_3N(CH_2)_pN(CH_3)_3^{2+}$			
6	4.4	~3.8	—		—	
5						
4	5.1	4.5	$(10.0)^c$	—	$(16.5)^c$	—
3	6.1	5.0	11.0	10.7	16.5	19.0
2	7.0	6.4	12.7	12.2	19.3	19.8

Notes: [a]Data from Blades et al.[62]

[b]Due to low yield of doubly protonated, $p = 5$, ion obtained with electrospray, measurements of equilibria at only one temperature were feasible. These led to $\Delta G_{1,2}^{\circ} = 10.1$ kcal/mol and $\Delta G_{1,2}^{\circ} = 8.9$ kcal/mol at $T = 398$ K. The values given in the table were obtained by assuming $\Delta S_{0,1}^{\circ} = 21.5$ cal/deg mol and $\Delta S_{1,2}^{\circ} = 23.9$ cal/deg mol. These values correspond to averages obtained for the longer chain length ($p > 5$) diammonium ion.

[c]ΔH° was obtained assuming $\Delta S_{0,1}^{\circ} = 16.5$ cal/deg mol.

at $p + 2$, by assuming that the positive charge is located largely on the ammonium hydrogens, but is somewhat shifted towards the nitrogen. Since the N–H bonds are somewhat shorter than the C–N and C–C bonds, we allow 0.5 bond equivalent distance between the positive charge and the nitrogen and 1 bond equivalent distance for each C–N and C–C bond. A plot of the present enthalpy data, $-\Delta H_{0,1}^{\circ}$ versus $1/(p + 2)$, shown in Figure 13, does indicate the presence of a linear relationship, although the quality of the plot is somewhat affected by experimental error in the $\Delta H_{0,1}^{\circ}$ values. The $\Delta H_{0,1}^{\circ}$ values for the hexamethyldiammonium ions were also determined (see Table 6). The hydration energies for the same p are much lower than those for the diamines. The major reason for this is the very much weaker hydrogen bonding of H_2O to the $-N(CH_3)_3^+$ groups. This effect is observed also for the singly charged monofunctional group compounds, i.e. $(CH_3)_4N^+$ (see Meot-Ner et al.[89c]).

Hydration energies of doubly charged anions[82] are given in Table 7. The values obtained are also shown in Figure 14 where the $-\Delta G_{n-1,n}^{\circ}$ are plotted versus n. The data are based on those equilibrium constants $K_{n-1,n}$ which could be determined at

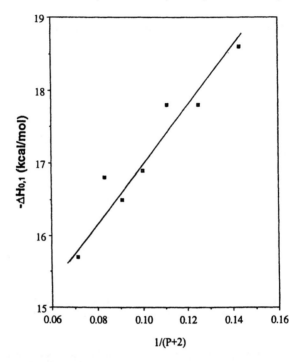

Figure 13. Hydration enthalpy, $\Delta H_{0,1}$, for hydration of $H_3N(CH_2)pNH_3^{2+}$ plotted versus $1/p+2$ which is approximately proportional to the reciprocal of the distance between the two charged centers. This leads to a straight line. Such a relationship can be expected because the Coulombic energy is proportional to the reciprocal of the distance.

room temperature when the H_2O partial pressure was in the range 1–50 mtorr. For very strongly hydrating species only the higher $(n-1,n)$ equilibria were observed while the determinations for a more weakly hydrating species could be obtained at lower $(n-1,n)$ including $(0,1)$ for the weakest hydrating species. Examination of the plots for the individual anions from left to right, i.e. from low to high n reveals at a glance the order of increasing hydration strength: 1,5-napthalenedisulfonate; tetrathionate $O_3SSSSO_3^{2-}$; persulfate $O_3SOOSO_3^{2-}$; 1,2-ethanedisulfonate $SO_3CH_2CH_2SO_3^{2-}$; alkanedicarboxylates $CO_2(CH_2)_kCO_2^{2-}$, where $k = 4-8$; dithionate $O_3SSO_3^{2-}$; thiosulfate SO_3S^{2-}; selenate SeO_4^{2-}; and sulfate SO_4^{2-}. The slope for each plot increases as n is decreased. The values of n at which rapid increases of the slope occur, follow the same compound order as above. Thus the rapid increase occurs at $n \approx 1$ for 1,5-naphthalenedisulfonate and at $n \approx 6$ for the sulfate anion.

The order observed in the plots is easily rationalized. First one can consider the cases where two functional groups are present leading to two distinct charge centers. The doubly charged anions like SeO_4^{2-}, $S_2O_3^{2-}$ (thiosulfate), and SO_4^{2-} could then be

Table 7. Hydration Free Energies of Doubly Charged Anions[a]

Ion	$-\Delta G^\circ_{2,3}$	$-\Delta G^\circ_{3,4}$	$-\Delta G^\circ_{4,5}$	$-\Delta G^\circ_{5,6}$
	Doubly charged ions of dicarboxylic acids			
$CO_2(CH_2)_4CO_2^{2-}$	8.4	7.5	6.4	5.8
$CO_2(CH_2)_5CO_2^{2-}$	8.0	7.2	6.2	5.5
$CO_2(CH_2)_6CO_2^{2-}$	7.8	7.1	6.1	5.5
$CO_2(CH_2)_7CO_2^{2-}$	7.6	6.9	5.9	5.2
$CO_2(CH_2)_8CO_2^{2-}$	7.4	6.7	5.8	5.1
$1{,}4\ CO_2C_6H_4CO_2^{2-}$	8.4	7.3	6.3	5.9

Doubly charged anions of some oxo acids of sulfur and selenium

$(n-1,n)$ and $-\Delta G^\circ_{n-1,n}$ (kcal/mol)

Ion										
sulfate SO_4^{2-}	(5,6)	8.5	(6,7)	7.5	(7,8)	6.7	(8,9)	6.0	(9,10)	5.5
	(10,11)	5.0								
selenate SeO_4^{2-}	(5,6)	8.0	(6,7)	7.0	(7,8)	6.3	(8,9)	5.6		
thiosulfate $S_2O_3^{2-}$	(4,5)	8.3	(5,6)	7.3	(6,7)	6.5	(7,8)	5.9	(8,9)	5.3
	(9,10)	4.7								
dithionate $O_3SSO_3^{2-}$	(7,8)	5.1								
ethane disulfonate $SO_3C_2H_4SO_3^{2-}$	(1,2)	8.7	(2,3)	7.5	(3,4)	6.6	(4,5)	597	(5,6)	5.3
persulfate $O_3SOOSO_3^{2-}$	(1,2)	8.5	(2,3)	7.5	(3,4)	6.7	(4,5)	5.9	(5,6)	5.4
tetrathionate $O_3SSSSO_3^{2-}$	(0,1)	9.1	(1,2)	7.8	(2,3)	6.7	(3,4)	6.1	(4,5)	5.4
1,5-naphthalene-disulfonate	(0,1)	7.8	(1,2)	7.0	(2,3)	6.3	(3,4)	5.5		

Note: [a]All values are in kcal/mol. Standard state 1 atm, 293 K. Data from Blades et al.[82]

considered as special cases where the distance between the charge centers is very small. Two major factors determine the observed order of hydration energies: (a) The distance between the two charged groups. This most important factor is a consequence of the intense Coulombic repulsion between the two charges which is present in the gas phase; and (b) The charge delocalization within the functional group, i.e. $-SO_3^- > -CO_2^-$. The order of stabilization by charge delocalization is examined in the following subsection which deals with singly charged ions. It is established there, that the stabilization increases with the number of equivalent oxygens over which the charge is delocalized and with the size of the central atom. A large central atom leads to stronger stabilization because it increases the distance between the negative oxygen atoms.

The importance of the distance between the charges is most clearly demonstrated by the observed increase of hydration exothermicities of the dicarboxylates $CO_2(CH_2)_kCO_2^{2-}$ with decreasing k. The increase in the order $SO_3S_2SO_3^{2-}$,

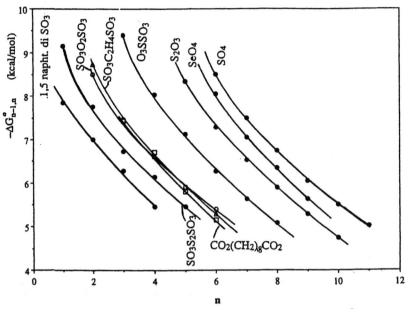

Figure 14. Hydration free energies $-\Delta G^\circ_{n-1,n}$ at 293 K for reaction: $A^{2-}(H_2O)_{n-1} + H_2O = A^{2-}(H_2O)_n$; versus n for several doubly charged acid anions as indicated for each plot. Hydration strength increases in the order: 1,5-naphthalenedisulfonate to SO_4^{2-}. From Blades, A. T.; Klassen, J. S.; Kebarle, P. *J. Am. Chem. Soc.* **1995**, *117*, 10563, with permission.

$SO_3O_2SO_3^{2-}$ $SO_3CH_2CH_2SO_3^{2-}$, $O_3SSSO_3^{2-}$ is also mainly determined by the decreasing distance between the two charged $-SO_3^-$ groups.

The effect of the nature of the charge delocalization within each functional group is well illustrated by the observed substantial increase of hydration energy from $SO_3CH_2CH_2SO_3^{2-}$ to $CO_2CH_2CH_2CO_2^{2-}$. The results in Figure 14 show that the hydration energies of the dicarboxylates become the same as that for ethanedisulfonate only after the carboxylate chain increases to $k = 8$, an impressive difference.

A third factor due to the stabilizing effect of charge delocalization from the functional group to the moiety connecting the two groups might also be expected. Different moieties connecting two charged groups separated by the same distance could be expected to lead to differing hydration energies due to the differential stabilization that might be provided by these moieties. Evidence for such an effect was obtained[82] by comparing the hydration energies of $1,4\text{-}CO_2C_6H_4CO_2^{2-}$ and $CO_2(CH_2)_4CO_2^{2-}$. These two dianions had nearly the same hydration energies (see Tables 8–10) even though the distance between the two carboxy groups in the benzene dicarboxylate is significantly smaller, ~5.6 Å, relative to ~7.5Å in the alkene dicarboxylate. This means that the benzene moiety provides better charge stabilization relative to the alkyl chain.

Table 8. Hydration Free Energies of Some Protonated Alkylamines, Alkyldiamines, and Peptides[a]

M^{2+}	$n-1,n$	$-\Delta G^\circ_{n-1,n}$	M^{2+}	$n-1,n$	$-\Delta G^\circ_{n-1,n}$
$n\text{-}C_3H_7NH_3^+$	1,2	6.4	$GlyH^+$	0,1	9.7
	2,3	4.6		1,2	7.2
$n\text{-}C_8H_{17}NH_3^+$	0,1	8.5		2,3	5.4
	1,2	6.1	Gly_2H^+	0,1	8.8
	2,3	4.6		1,2	6.2
$NH_2(CH_2)_2NH_3^+$	0,1	7.8	Gly_3H^+	0,1	6.7
$NH_2(CH_2)_3NH_3^+$	0,1	6.2		1,2	5.8
$NH_2(CH_2)_4NH_3^+$	0,1	5.3	Gly_4H^+	0,1	5.8
$NH_2(CH_2)_5NH_3^+$	0,1	5.5	$LysH^+$	0,1	5.2
$NH_2(CH_2)_6NH_3^+$	0,1	5.5	$N\text{-}CH_3CO\ LysH^+$	0,1	5.8
$NH_2(CH_2)_7NH_3^+$	0,1	5.6	Gly-Lys H^+	0,1	4.9
$NH_2(CH_2)_8NH_3^+$	0,1	5.4	Lys-Tyr-LysH$_2^{2+}$	0,1	7.1
$NH_2(CH_2)_{10}NH_3^+$	0,1	5.1		1,2	6.2
$NH_2(CH_2)_{12}NH_3^+$	0,1	4.9		2,3	5.6

Note: [a]From Klassen et al.[56]

Table 9. Hydration Free Energies of Singly Charged Anions of Dicarboxylic Acids[a]

Ion	$-\Delta G^\circ_{0,1}$	$-\Delta G^\circ_{1,2}$
$CO_2HCO_2^-$	6.3	
$CO_2HCH_2CO_2^-$	5.3	4.0
$CO_2H(CH_2)_2CO_2^-$	5.4	~4.0
$CH_3OCO(CH_2)_2CO_2^-$	8.4	6.1
$CO_2H(CH_2)_3CO_2^-$	5.5	~4.2
$CO_2H(CH_2)_4CO_2^-$	5.7	4.4
$CO_2H(CH_2)_5CO_2^-$	5.5	—
$CO_2H(CH_2)_6CO_2^-$	5.4	—
$CO_2H(CH_2)_7CO_2^-$	5.3	—
$CO_2H(CH_2)_8CO_2^-$	5.0	—
$1,2\ CO_2HC_6H_4CO_2^-$	4.7	—
$1,4\ CO_2HC_6H_4CO_2^-$	7.4	—
$cis\text{-}CO_2HC_2H_2CO_2^-$	5.0	—
$trans\text{-}CO_2C_2H_2CO_2^-$	7.3	5.2

Note: [a]All values are in kcal/mol. Standard state 1 atm, 293 K. From Blades et al.[58]

Table 10. Free Energies of Hydration of Anions A$^-$
and of Acid Dissociation, $\Delta H = A^- + H^+$

Ion	$-\Delta G^\circ_{n-1,n}$ $(n-1,n)^a$			ΔG°_{acid} ^{b} (kcal/mol)
	0,1	1,2	2,3	
Carboxylates, RCO$_2^-$				
$CH_3CO_2^-$	9.4	6.8	5.2	341.5
$C_2H_5CO_2^-$	9.3	6.6	5.1	340.3
HCO_2^-	9.2	6.8	5.1	338.2
$CH_3OCH_2CO_2^-$	8.6	6.1	4.8	
$HOCO_2^-$	8.50	6.2	4.6	$(334.0)^c$
$CH_3OCO_2^-$	8.3	6.1	4.6	$(331.8)^c$
$CH_2FCO_2^-$	8.3	5.9	4.6	331.0
$C_6H_5CO_2^-$	8.1	5.7	4.5	331.7
$CH_2ClCO_2^-$	7.7			328.8
$CH_3CHOHCO_2^-$	7.7	5.3		$(326.6)^c$
$CHF_2CO_2^-$	7.5	5.3		323.5
$CH_2ICO_2^-$	7.2	5.3	4.3	$(322.6)^c$
$CH_2CNCO_2^-$	7.1	5.2		323.7
$CHCl_2CO_2^-$	6.8	4.9		321.9
$CF_3CO_2^-$	6.8	4.7		316.3
$CCl_3CO_2^-$	5.8			$(309.6)^c$
Anions of Some Oxo Acids of N, P, S, Cl, and I				
NO_2^-	~8.5	6.00	4.5	332.3 ± 4.5
NO_3^-	7.1	5.2	3.9	318.18
$H_2PO_2^-$	8.4	6.4	4.7	
$(HO)HPO_2^-$	7.8	6.6	5.2	
$(HO)_2PO_2^-$	7.6	6.1	4.8	
PO_3^-	6.27	4.90		303.8
D-ribose 5-phosphate	5.8	4.5		
adenosine 5'-phosphate	5.4			
$CH_3SO_3^-$	6.9	5.4	4.4	315.00
$C_7H_{15}SO_3^-$	6.3	5.0		
$C_6H_5SO_3^-$	5.9	4.4		
$C_2H_5OSO_3^-$	5.8	4.6		
$HOSO_3^-$	5.9	4.7		302.6
$CF_3SO_3^-$	4.6	3.8		299
ClO_2^-	9.0	6.1	4.8	
ClO_3^-	6.2	4.7		
ClO_4^-	4.8			
BrO_3^-	6.5	5.0		
IO_3^-	6.4	5.0	4.2	
IO_4^-	4.3			

Notes: [a] Free energy change in kcal/mol, standard state 1 atm. for hydration equilibria $(n-1,n)$: $A^-(H_2O)_{n-1} + H_2O = A^-(H_2O)_n$ at 293 K. For origin of data see Blades et al.[58,59]

[b] Free energy change for reaction: $HA = H^+ + A^-$ in kcal/mol. Standard state 1 atm., 298 K. For origin of data other than those in parentheses, see Blades et al.[58]

[c] Estimated acidity value based on relationship with $\Delta G^\circ_{0,1}$.

Singly Charged Ions: Relationship between Hydration Energies and Stabilization by Intramolecular Hydrogen Bonding

In the preceding section, we established that the gas-phase hydration exothermicities provide a good measure for the degree of stabilization of the ionic charge on the ionic functional group that is to be hydrated. One special form of such stabilization is the presence of an intramolecular hydrogen bond. Ions where the charged group is stabilized by an intramolecular hydrogen bond exhibit low hydration exothermicities. The intramolecular hydrogen bond formation involves cyclization. Typical examples are protonated alkane diamines H_2N-$(CH_2)_pNH_3^+$, see Scheme 1 for the cyclized hydrogen bonded structure of p = 2. The original experimental evidence[84,85] for the occurrence of stabilization and cyclization was based on observations that the protonation of the bifunctional base was much more exothermic (high $-\Delta H°$ values) than that of a single function base like $CH_3(CH_2)_pNH_3$. Furthermore, the protonation of the bifunctional base was accompanied by a large loss of entropy (high $-\Delta S°$ values), which were consistent with H-bond formation and loss of freedom due to cyclization. It was found[84] that the rings became strained for p < 3. Thus, the cyclization exergonicity, $-\Delta G°_{cyc}$ for NH_3-CH_2-$CH_2NH_3^+$ was only ~4 kcal/mol, while that for the higher analogues up to p = 6 was ~12 kcal/mol.[84,85]

The hydration free energies for the singly protonated diamines have been determined[56] and are given in Table 8. The hydration energies $\Delta G°_{0,1}$ are plotted versus p = x in Figure 15. At high p, the $-\Delta G°_{0,1}$ values are essentially constant and very much smaller then the hydration energies observed for alkyl amines where intramolecular H-bonding is absent. Thus $-\Delta G°_{0,1} = 5.4 \pm 0.1$ kcal/mol for p = 4 to 8, while, $-\Delta G°_{0,1} \sim 8.5 \pm 0.3$ kcal/mol for the alkyl monoamines, $C_8H_{17}NH_3^+$ (see Table 8). At low p, where the cyclization is strained and the H bond weaker, larger hydration exergonicities are observed.

Also shown in Figure 15 are the hydration energies for the dicarboxylates $HOCO(CH_2)_pCO_2^-$ from Table 8, which also undergo H-bond-induced cyclization (see Scheme 2). Very similar trends are observed for this system particularly at x > 4, after adjusting the horizontal axis for the dicarboxylates as x = p + 2, so as to

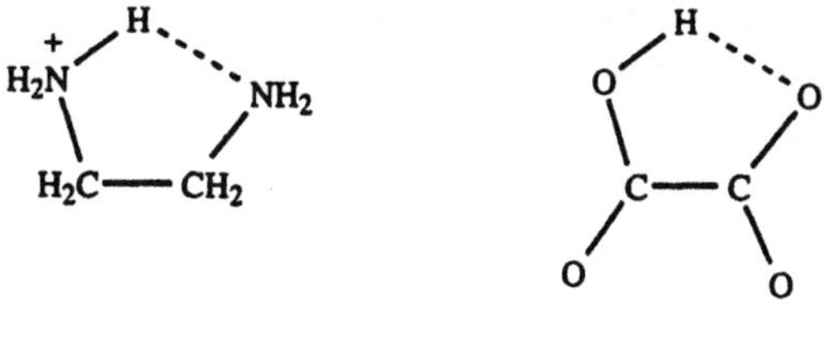

I   II

Scheme 1.

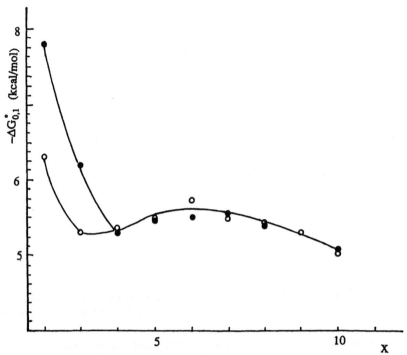

Figure 15. Plot of hydration energies $\Delta G^{\circ}_{0,1}$ for the monohydration of dicarboxylate anions $HOOC(CH_2)_kCO_2^-$ (open circles, o). Also shown are the $\Delta G^{\circ}_{0,1}$ for the protonated diamines $H_2N(CH_2)_kNH_3^+$ (closed circles, ●). Both types of compounds have small $-\Delta G^{\circ}_{0,1}$ values which are due to stabilization of the ions by an intramolecular hydrogen bond which leads to cyclization. The horizontal scale x has been adjusted, x = k (amines); x = k+2 (dicarboxylates) in order to compare cyclic structures with equal number of atoms in ring. Large increases of $-\Delta G^{\circ}_{0,1}$ for x ≤ 3 are due to ring strain which reduces the stabilization of the anion. From Blades, A. T.; Klassen, J. S.; Kebarle, P. *J. Am. Chem. Soc.* **1995**, *117*, 10563, with permission.

compare rings of equal number (see Schemes 1 and 2). The increase at low x observed for the dicarboxylate system is smaller particularly for x = 2, i.e. for the oxalate anion given in Scheme 2. The major cause for this low value should be the charge-stabilizing effect of the electron withdrawing $-CO_2H$ group which is in α position to the CO_2^- group (see Scheme 2, and Blades et al.[82]).

Examples for the presence and absence in intramolecular H-bonds reflected in the hydration energies are shown below for *cis*- and *trans*-dicarboxyethylene.[82] A similar presence and absence of H-bonding occurs for the phthalic acid isomers, $1,2\text{-}CO_2HC_6H_4CO_2^-$, $-\Delta G^{\circ}_{0,1} = 4.7$ kcal/mol, and $1,4\text{-}CO_2HC_6H_4CO_2^-$, $-\Delta G^{\circ}_{0,1} = 7.4$ kcal/mol.

Once it was established that hydration energy differences can be used as indicators for the presence of strong intramolecular hydrogen bonds, the presence of such

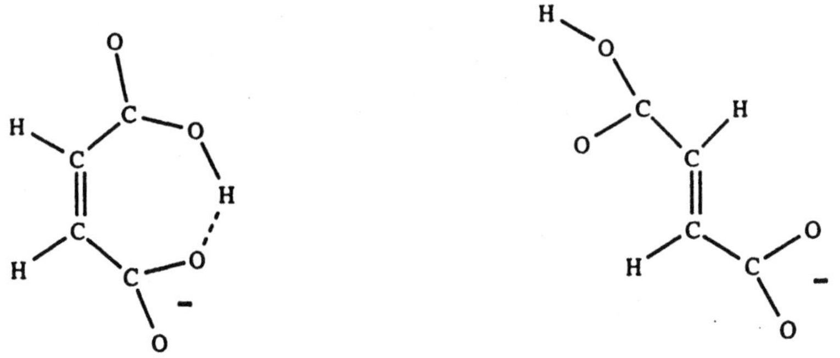

$-\Delta G^{\circ}_{0,1} = 5.0$ kcal/mol

$\Delta G^{\circ}_{0,1} = 7.3$ kcal/mol

Scheme 2.

bonds could be detected also in more complex systems such as protonated peptides[56] and deprotonated nucleotides.[81]

Relationships between Hydration Energies and Gas-Phase Acidities and Basicities

Experimental determinations undertaken prior to the discovery of electrospray as a source of ions have shown[86,87] that the bond strength of H-bonded complexes XH ---A$^-$ increases with the gas-phase acidity of XH and the gas-phase basicity of A$^-$. This relation has been examined[82] for the special case where A$^-$ were a variety of anions produced by electrospray and XH = OH$_2$, on the basis of the hydration energy data (see Table 8) and gas-phase basicities $\Delta G^{\circ}_{base}(A^-) = \Delta G^{\circ}_{acid}$ (AH) corresponding to the free energy change for the gas-phase reaction:

$$AH = A^- + H^+ \qquad (40)$$

A positive correlation between $-\Delta G^{\circ}_{0,1}$ (A$^-$) and ΔG°_{base} (A$^-$) is expected because both quantities should decrease as the electronic stabilization of the charged center of A$^-$ increases. Thus changes in the anion that introduce electron-withdrawing groups, which delocalize the charge and stabilize it, will weaken the hydrogen bond in the hydrate and also weaken the A$^-$– H$^+$ bond. A fairly linear relationship between $-\Delta G^{\circ}_{0,1}$ (A$^-$) and $\Delta G^{\circ}_{base}(A^-)$ was observed (see Figure 5 in Blades et al.[82]). On the basis of this relationship, several gas-phase acidities ΔG°_{acid} (AH) could be estimated for acids AH which cannot be easily generated in the gas phase, because they are very nonvolatile or because they are unstable in the gas phase. This includes (HO)$_2$CO (carbonic acid), CCl$_3$CO$_2$H, (HO)$_3$PO, HClO$_2$, HClO$_3$, HClO$_4$, HBrO$_3$, HIO$_3$, HIO$_4$, and others (see Table 10).

An examination of the data for oxo acids[82] (Table 10) shows that the factors leading to stabilization of the oxo anions and thus to a decrease of $-\Delta G^{\circ}_{0,1}$ (A$^-$) and ΔG°_{acid} (AH) can be expressed by a few simple rules stated in order of decreasing importance:

1. The number of equivalent oxygens over which the negative charge in the oxo anion is delocalized.
2. The presence of the electron-withdrawing power induced by electron-withdrawing substituents.
3. The nature of the central atom.

These rules will be illustrated by a few examples. The oxo anions of phosphorous show the following hydration energies (kcal/mol):

$-\Delta G^{\circ}_{0,1}$:   8.4   7.9   7.6   6.2

The most stabilized anion is PO_3^- which has three equivalent charge delocalizing oxygens. The other three anions have only two such oxygens. The stability differences for these three anions are due to the electron-withdrawing effect of the OH group. The stability increases as the number of OH group increases from zero to two.

The rule concerning the number of charge carrying oxygens is also illustrated by the acidities for the three groups of molecular ions: NO_2^-, NO_3^-; ClO_2^-, ClO_3^-, ClO_4^-; and IO_3^-, IO_4^- (see Table 10).

When comparing anions with equal number of charge bearing oxygens and with same substituents but different central atoms, one finds that the anion stability increases in the order $C \leq N \leq P \leq S \leq Cl \leq I$. This is not an order of changing electronegativity of the central atom, but rather an order of increasing size. Clearcut examples of this effect are the $-\Delta G^{\circ}_{0,1}$ values which have the following order $NO_3^- > PO_3^-$ and $ClO_4^- > IO_4^-$.

Linear relationships between hydration energies of protonated bases BH$^+$, i.e. $-\Delta G^{\circ}_{0,1}$ (BH$^+$) and the gas-phase basicities of B, corresponding to the free energy change for the reaction,

$$BH^+ = B + H^+ \qquad GB(B) \tag{41}$$

are also observed.[86,87] In this case, the hydration exergonicity, $-\Delta G^{\circ}_{0,1}$ (BH$^+$) increases as $-\Delta G^{\circ}_B$ (B) decreases. A decrease of $-\Delta G^{\circ}_B$ (B) indicates a more available proton for the $-H^+$---OH_2 hydrogen bond. Recently, it was shown[62] that such a linear relationship exists also between the hydration energies of the doubly pro-tonated bases, $^+H_3N(CH_2)pNH_3^+$ (see Table 5), and the gas-phase basicities of the singly protonated bases, $^+H_3N(CH_2)pNH_2$. The slope of this relationship was very different from the relationship involving $-\Delta G_{0,1}^\circ$ (BH+) and GB(B). The difference of slope is consistent with the expected large Coulombic repulsion effects in the doubly protonated bases which make a large contribution to the gas-phase basicities of the $^+H_3N(CH_2)_pNH_2$ bases.

Transfer and Exchange Equilibria

Proton Transfer and Electron Transfer Equilibria. The experimental deter-mination used for the data discussed in the above subsections of Section IV.B were obtained from ion–molecule association (clustering) equilibria, for example equa-tion 9. A vast amount of thermochemical data such as gas-phase acidities and basicities have been obtained by conventional gas-phase techniques from proton transfer equilibria,[3,7–12,87d–87g] while electron affinities[88] and ionization energies[89] have been obtained from electron transfer equilibria.

No determinations of proton transfer or electron transfer equilibria have been performed so far with the present apparatus, even though such measurements should be perfectly feasible. The reasons why such measurements were not yet performed are easily summarized. In general, one would use electrospray for the production of ions which are difficult to produce by the conventional gas-phase methods. When proton transfer equilibria and singly charged ions are involved, the inability to produce the singly deprotonated ion A^- or singly protonated ion BH^+ by conven-tional methods is generally due to the fact that the neutral AH or B is nonvolatile, strongly absorbed on the apparatus walls, or unstable. With electrospray, one may be able produce the ions A^- or BH^+ from solution but the fact that the neutral AH or B cannot be put in the gas phase still precludes the determinations of proton transfer equilibria.

Gas-phase acidities and basicities could still be determined with the kinetic bracketing method which relies on the fact that exothermic proton transfer reactions generally proceed at collision rates. The bracketing method could be applied to obtain more accurate gas-phase acidities for some of the anions discussed in the preceding section, such as ClO_2^-, ClO_3^-, ClO_4^-, BrO_3^-, IO_3^-, IO_4^-, etc. Most of these anions, A^-, correspond to rather strong acids AH in the gas phase and there might not be a sufficient range of volatile acids with known and similar acidity to AH to permit the close bracketing which is required for accurate determinations.

Difficulties of a different kind occur for proton transfer equilibria involving doubly charged ions, i.e. doubly deprotonated acids A_2^{2-} or doubly protonated bases $B_2H_2^{2+}$. For example, proton transfer reactions, involving the doubly protonated base B_2H^{2+} and a reference base B_0,

$$B_2H_2^{2+} + B_0 = B_2H^+ + B_0H^+ \tag{42}$$

lead to two singly charged ions and the Coulombic repulsion between these two ions leads to a large activation energy for the reverse reaction.[77] Therefore, equilibria cannot be determined. However, the bracketing method can still be applied.[76] Even then, the experiments lead only to an apparent gas-phase basicity which can be converted to the thermodynamic gas-phase basicity only on the basis of certain assumptions concerning the Coulombic repulsion energy.[76,77]

Equilibria Where a Neutral Molecule Is Exchanged. The difficulties discussed above for proton transfer and electron transfer equilibria involving multiply charged ions are not present when neutral molecules (ligands) which are complexed to a given ion are exchanged. Equation 43 is a typical example:

$$M^{Z+}(H_2O) + CH_3OH = M^{Z+}(CH_3OH) + H_2O \tag{43}$$

The enthalpy change for reaction 43 corresponds to the difference between the M^{Z+} affinities for H_2O and CH_3OH, i.e. $M^{Z+}A(H_2O)$ and $M^{Z+}A(CH_3OH)$, where $M^{Z+}A(L)$ corresponds to the enthalpy charge for the reaction involving the ligand L:

$$(ML)^{Z+} = M^{Z+} + L \tag{44}$$

Exchange equilibria have been determined with an ICR apparatus[91] with $M^+= Li^+$ and calibrated to the known $Li^+A(H_2O)$, obtained from Li^+, H_2O association equilibria. This has led to the Li^+ affinities for over 50 different ligands L. Exchange equilibria determinations are generally measured at a single temperature so that only the ΔG° values are obtained; however since the ΔS° changes are small, approximate ΔH values can also be deduced. Exchange equilibria can also be determined when the ion is multiply charged. Thus, multiply charged ions obtained by electrospray can, in principle, be used for determining exchange equilibria and this procedure could lead to valuable thermochemical data. So far, to our knowledge, such equilibria have not been determined with any ICR or FTICR apparatus.

Recently, we have started exchange equilibria measurements[92a] involving complexes of the Ag^+ ion, such as:

$$Ag^+(H_2O)_2 + CH_3OH = Ag^+(H_2O)(CH_3OH) + H_2O \tag{45}$$

$$Ag^+(H_2O)(CH_3OH) + CH_3OH = Ag^+(CH_3OH)_2 + H_2O \tag{46}$$

$$Ag^+(H_2O)_2 + 2CH_3OH = Ag^+(CH_3OH)_2 + 2H_2O \tag{47}$$

Determination of the equilibrium constants for reactions 45 and 46 leads to $-\Delta G^\circ_{45}$ and $-\Delta G^\circ_{46}$ and summing these two values one obtains the free energy change for the complete exchange, $-\Delta G^\circ_{47}$. Such equilibria were determined for several different ligands: H_2O, CH_3OH, $CH_3CO_2CH_3$, C_6H_6, CH_3COCH_3, $(C_2H_5)_2O$, tetrahydrofuran c-C_4H_8O, CH_3CN, $(CH_3)_2S$, and $(CH_3)_2SO$. The values obtained for the double-exchange equilibria ΔG_{47}° were combined to provide a scale of relative $\Delta G_{0,2}^{\circ}$ values which was calibrated to the absolute $\Delta G_{0,2}^{\circ}$ value determined by Holland and Castleman[92b] for $Ag^+(H_2O)_2$, thus obtaining the $\Delta G_{0,2}^{\circ}$ values for all of the above ligands.[92a] Holland and Castleman[92b] determined their values using sequential association equilibria,

$$Ag^+ + H_2O = Ag^+(H_2O) \qquad (0.1) \qquad\qquad (48)$$

$$Ag^+(H_2O) + H_2O = Ag^+(H_2O)_2 \qquad (1.2)$$

which were determined over the temperature range 500–710 K. van't Hoff plots were then used to obtain: $-\Delta H_{n-1,n}^{\circ}$ and $-\Delta S_{n-1,n}^{\circ}$ data. The Ag^+ ions were produced by thermionic emission.[92a] Such association equilibria determinations would be possible in principle also with the present apparatus with electrospray produced ions. However temperatures above 500 K are not as yet accessible with the present ion source (see Figure 7).

The exchange equilibria (equation 47) and the scale of exchange equilibria involving ligands with progressively higher Ag^+ affinities lead ultimately to very strongly bonded complexes[92a] such as,

$$Ag^+(DMSO)_2 = Ag^+ + 2DMSO \qquad \Delta H = 66 \text{ kcal/mol} \qquad (49)$$

which would not have been accessible with the apparatus used by Holland and Castleman. We expect that exchange equilibria involving electrospray produced singly and multiply charged ions will become a major source of thermochemical data in the near future.

C. Thermochemical Data from CID Threshold Measurements

Treatment of Threshold Curves

Determinations of the energy thresholds, E_0, were obtained by fitting the product ion energy profiles with the CRUNCH program developed by Armentrout and coworkers.[67–69] The fitting procedure is based on the equation,

$$\sigma = \sigma_0 \sum_i g_i (E + E_i + E_{rot} - E_0)^n / E \qquad (50)$$

where σ is the cross section, which is proportional to the observed product ion intensity under single collision conditions. E_i is the internal vibrational energy of the precursor ion whose relative abundance at a given temperature is g_i, where $\Sigma_i g_i = 1$. E_{rot} is external rotational energy of the precursor ion which can contribute to the decomposition, and E_0 is the threshold energy, corresponding to the energy required for the dissociation reaction at 0 K. E is the kinetic energy in the center of mass frame,

$$E = E_{\text{lab}} \, m_{\text{CG}}/(m_{\text{CG}} + m_{\text{ion}}) \qquad (51)$$

where E_{lab} is the kinetic energy of the precursor ion attained in the acceleration to collision cell Q_2 (see Figure 9), and m_{CG}, m_{ion} are the masses of the collision gas and the ion precursor.

The fitting procedure based on equation 50 treats n and E_0 as variable parameters which are determined through the best fit. The vibrational energies E_i are calculated from the frequencies of the normal vibrations of the precursor ion. The frequencies are generally obtained from low level ab initio quantum chemical calculations or semiempirical methods such as AM1 and PM3. An important feature underlying equation 51 is the assumption that the residence time of the excited ions in Q_2 is sufficiently long such that the observed dissociation threshold involves the process where the internal energy of the products is close to 0 K, i.e. essentially all the internal energy of the precursor ion is used up in the dissociation at the observed threshold.

An example of an experimental product ion curve fitted by equation 50 is shown in Figure 16. The precursor ion, Na$^+$ N-methylacetamide, was obtained by electrospray. The CID product is Na$^+$ corresponding to the dissociation:

$$\text{Na}^+ \, N\text{-methylacetamide} = \text{Na}^+ + N\text{-methylacetamide} \qquad (52)$$

The solid curve fitting the experimental points was obtained[64] with the CRUNCH fitting procedure corresponding to equation 50. The E_0 = 38.9 kcal/mol should correspond to the energy required at 0 K for the dissociation reaction 52. It should be noted that the experimental product ion curves generally showed [63-66] small tails toward low energy at the threshold which are due to instrumental limitations associated[63,64] with the sampling from relatively high pressure (10 torr) and a collision cell that is a quadrupole rather than an octapole as used by Armentrout and others.[67] Therefore, in the fitting procedure, we generally omitted the lowest intensity points of the threshold (see Figure 16). The relatively good fit obtained with the theoretical curve indicates that the tailing problem is not serious for this ion; however, ions of higher mass exhibit noticeable tails.

The fits of the experimental profiles with equation 50 are based on the assumption that the energized reactant ions, whose total energy is equal or larger than E_0, will decompose within the residence time t of the precursor ion in Q_2 before entering the mass analyzer Q_3. For the present apparatus, t is estimated to be between 30 and 60 μs and complete dissociation within that time, for ions with energies just above E_0, can be expected only for relatively small polyatomic ions. Ions with many internal degrees of freedom will attain the required dissociation rate only at internal energies which are somewhat higher than E_0. This "kinetic shift",[93] if not taken into account, will lead to E_0 values that are too high.

Armentrout et al.[69] have developed a method with which the kinetic shift can be taken into account in the curve fitting procedure. The fitting procedure, which corrects for the kinetic shift, includes only the fraction of the precursor ions which

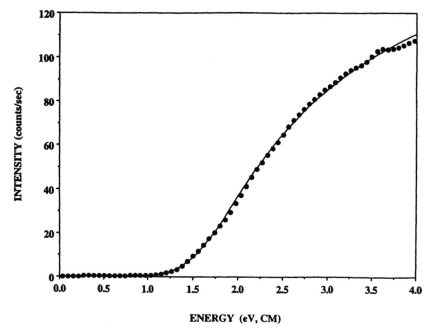

Figure 16. Appearance curve of Na^+ from CID of Na^+N-methylacetamide. The calculated curve (solid line), fitted using experimental points from 1.4–4.0 eV, corresponds to $n = 1.25$ and $E_0 = 1.69$ eV (38.9 kcal/mol). The cross section model in equation 50 is used. From Klassen, J. S.; Anderson, S. G.; Blades, A. T.; Kebarle, P. *J. Phys. Chem.* **1996**, *100*, 14218, with permission.

decompose during the residence time t. This fraction is given by $1 - \exp(-kt)$, where k is the rate constant for the decomposition. k, which can be evaluated with the RRKM formalism, is a function of the internal energy of the ion and the threshold energy E_0. For evaluation of k with RRKM one needs to know the vibrational frequencies of the transition state and this means one also needs to know the structure of that transition state. For dissociation like that occurring in equation 52, loose transition states are expected and procedures for obtaining the required parameters for such transition states have been proposed.[69] Dissociations which proceed by tight transition states require the determination of the vibrational frequencies in the transition state, which can be obtained with semiempirical calculations such as AM1 and PM3.

Once the energy E_0 for the dissociation reaction has been established, the enthalpy change for the reaction at a given temperature T (such as the standard temperature $T = 298$ K) can be obtained from a thermodynamic cycle involving the reaction at 0 K and at T K. Thus, for the dissociation reaction of ion M^+ and ligand L,

$$ML^+ = M^+ + L \tag{53}$$

where M^+ is a monoatomic ion, one obtains,

$$\Delta H_T^o = E_o - E_{vib}(ML^+) + E_{vib}(L) + 5/2\, RT \qquad (54)$$

where $E_{vib}(ML^+)$ and $E_{vib}(L)$ are the internal (vibrational) energies at the temperature T, and the term $5/2\, RT$ is due to the extra translational energy of the products which is $3/2\, RT$ and the volume expansion term $p\Delta V = RT$. The vibrational frequencies for the products have to be obtained from theoretical calculations.

Reaction Enthalpies for $ML^+ = M^+ + L$, where $M^+ = Na^+$ and K^+ and $L =$ Acetamide, N,N-ethylacetamide, N,N-Dimethylacetamide, Glycine, Glycinamide, Glycylglycine and Succinamide

The data obtained[64] from CID threshold measurements are given in Table 11. The ΔH_{298}^b and ΔH_{298}^d values given in the table illustrate the magnitude of the kinetic

Table 11. Energies for Reaction: $M^+L = M^+ + L$ from CID Threshold Determinations[a]

$M^+ = K^+$	$E_{o(OK)}^b$	$\Delta H_{298}^{o\,b}$	$\Delta H_{298}^{o\,c}$	$\Delta H_{298}^{o\,d}$	ΔH_{298}^o lit
Me_2CO	24.2	24.4	24.4	24.4	26.0^e
Me_2SO	31.3	31.6	30.2	31.1	35 ± 3^e
$HCONMe_2$	32.0	32.0	27.1	29.5	31.0^e
$MeCONH_2$	29.7	29.6	29.6	29.7	$(26.2)^f$
$MeCONHMe$	32.3	31.9	28.3	30.4	$(24.8)^f$
$MeCONMe_2$	32.9	32.2	25.5	29.0	31^e
H_2NCH_2COOH	29.2	30.1	29.4	30.0	—
$M = Na^+$					
$MeCONH_2$	34.3	34.7	34.7	34.7	$(36.0)^g$
$MeCONHMe$	38.9	39.0	33.7	35.7	$(38.4)^g$
$MeCONMe_2$	46.9	46.7	35.2	37.5	—
H_2NCH_2COOH	36.3	37.5	35.7	36.6	$(38.5)^h$
$H_2HCH_2CONH_2$	45.1	45.1	39.5	41.4	—
$(H_2NCOCH_2)_2$	62.3	61.8	41.5	44.5	—
$H_2NCH_2CONHCH_2COOH$	64.6	64.7	40.1	42.9	—

Notes: [a] From Klassen et al.,[64] all energies in kcal/mol.

[b] Without correction for kinetic shift, E_o internal energy change, ΔHG^o enthalpy change for reaction: $M^+L = M^+ + L$.

[c] With correction for kinetic shift, the two lowest frequencies are 30 cm^{-1} for both K^+ and Na^+ complexes.

[d] With correction for kinetic shift, the two lowest frequencies are 10 cm^{-1} for Na^+ complexes and 5 cm^{-1} for K^+ complexes. This set is considered to be the best. For a discussion, see Klassen et al.[64]

[e] Experimental determinations based on ion-molecule equilibria, Sunner et al.[94]

[f] Theoretical calculations, HF/6-31G, combined with a semiempirical correction. Roux and Karplus.[35c]

[g] Theoretical calculations, HF/6-31G*, combined with a semiempirical correction. Roux and Karplus.[35c]

[h] Theoretical calculations, 6-31+G(2d) MP2, Jensen.[95]

[i] Succinamide.

shift discussed in the previous section. For the potassium complexes which involve smaller ligands and lower binding energies, only one ligand, Me_2CONMe_2, leads to a noticeable kinetic shift of ~3 kcal/mol. Very substantial kinetic shifts occur for the stronger bonded sodium complexes when the ligands are big. Thus for L = glycylglycine, $H_2NCH_2CONHCH_2CO_2H$, the kinetic shift is close to 22 kcal/mol. It is evident, that for the large compounds, corrections for the kinetic shift are absolutely essential.

The enthalpy changes with the nature of the ligand, as observed in Table 9, are discussed in the original work.[64] Here we will provide only a brief outline. All the ligands used, except Me_2SO, are carbonyl compounds. Me_2CO leads to somewhat weaker bonding than $MeCONH_2$; see ΔH_{298}^{d} values for K^+ of 24.4 and 29.7 kcal/mol. The lowest energy structure for acetone, calculated by Weller et al.,[96]

corresponds to the M^+ being aligned with the molecular dipole. The lowest energy structure for acetamide, calculated by Roux and Karplus,[35c] shows M^+ tilted by a few degrees away from the C–O axis. This tilt actually corresponds to an alignment with the molecular dipole of acetamide which is tilted relative to the C–O axis. This is due to partial π-electron donation from the NH_2 to the carbonyl carbon, resulting in an increase of the partial negative charge on the oxygen and formation of a partial positive charge on the nitrogen.

The bonding of glycine to M^+ is stronger than that for acetamide; see ΔH_{298}^{od} values for Na^+ of 34.7 kcal/mol and 36.6 kcal/mol. The bonding to glycine is stronger because it involves dicoordination to the carbonyl group and the amino group (see structure below) according to Jensen.[95] A somewhat stronger bonding still is

provided by glycinamide, ΔH_{298}^{od} (Na^+) = 41.4 kcal/mol, because electron donation from the NH_2 group enhances the negative charge on the carbonyl oxygen.

Still stronger bond energies are observed for glycylglycine and succinamide; see ΔH^{od}_{298} values for Na$^+$ of 42.9 kcal/mol and 44.5 kcal/mol. The probable structures based on AM1 and MNDO calculations[63] are:

The strong bonding with these compounds is due to dicoordination with two carbonyl oxygens. The stronger bonding in succinamide relative to glycylglycine should be due to the better electron pair donor ability of the NH$_2$ group relative to OH.

The bonding of K$^+$ and Na$^+$ to N-methylacetamide is of interest[64] in studies of the interaction of these ions with peptides and proteins, and particularly studies of the ion transport through transmembrane channels such as the gramicidin channel. Roux and Karplus[35] have used the complexation of the given alkali ion with two N-methylacetamide molecules and two water molecules as a model for interactions occurring in transmembrane channels.

The very strong bonding observed for Na$^+$ and glycylglycine, $\Delta H^{\circ}_{298} = 43$ kcal/mol (Table 9), is also of interest in bioanalytical mass spectrometry. Thus, multiply sodiated peptides and proteins would be expected from electrospray of solutions which were made basic by the addition of sodium hydroxide. Acidic solutions suppress the formation of sodium adducts since protonation of the basic residues of the peptides lead to Coulombic repulsions which reduce the uptake of the more weakly bound sodium. Synthetic polymers and oligomers containing multiple carbonyl and amide groups lead to sodiated ions when exposed to MALDI in suitable matrices.[97] The formation of sodium adducts is not surprising considering the strong bonding of sodium to carbonyl and amide groups demonstrated by the thermochemical measurements presented in Table 9.

Threshold Energies for CID of Protonated Glycine, Glycinamide, N-Methylglycinamide, Glycylglycine, Glycyl Glycinamide, and Gly-Gly-Gly

Tandem mass spectrometry has become an important tool for determining the sequence of amino acids in protonated peptides[98] and the sequence of bases in deprotonated nucleic acids such as DNA.[99] Despite the importance and widespread use of CID-MS to sequence peptides and nucleic acids, the mechanistic details of the dissociation processes are poorly understood. A better understanding of the major dissociation pathways is of interest not only from a fundamental point of view, but also for practical reasons such as refinements of the CID-MS as a sequencing technique.

Postulations of dissociation mechanisms have been largely based on the observed fragment ions, their possible structures, and qualitative considerations as to possible low-energy transition states which would lead to the observed fragment ions. The basic approach in the work from our laboratory is to use the threshold curves to obtain the threshold E_0 values for a given dissociation and then compare this experimental E_0 with activation energies obtained from quantum mechanical calculations where the transition state model corresponds to a given proposed mechanism for the dissociation. Agreement between the experimental E_0 and the theoretical activation energy will strongly support the validity of the proposed mechanism. Obviously, the success of this approach will depend on the accuracy of the experimental and theoretical determinations. Theoretical computations with large basis sets are very costly for large molecules, and therefore the approach can be applied only to mono-, di- and tripeptides and even more stringent limitations on size apply for nucleotides. The evaluation of E_0 values from threshold curves becomes more difficult with increasing size of the precursor ion because of increasing kinetic shifts. Therefore, a limitation to small size precursor ions is imposed also for this reason.

We have applied the above approach to both protonated peptides[65] and deprotonated nucleotides[66] with some success. Here we will give only a brief and partial account of results obtained for protonated glycines.[65] At low collision energies (<100 eV) protonated peptides dissociate primarily along the backbone of the amide bonds producing **b** ions, which contain the N-terminus and are formally acylium ions $H(-HNCHRCO-)_n^+$, and **y** ions, which contain the C-terminus and correspond to smaller protonated peptides, the smallest of these being the protonated terminal amino acid. Dissociation of the **b** ions leads to formation of smaller **b** ions as well as **a** (immonium) ions. Sequence information is obtained from the mass difference of successive fragment ions of the same type (e.g., $\mathbf{b}_n - \mathbf{b}_{n-1}$).

Protonated glycine, glycinamide, and *N*-methylglycinamide weré chosen as starting points for the threshold studies because glycine is the simplest amino acid and *N*-methylglycinamide could be considered the simplest model for a peptide. Furthermore a wealth of experimental[100,101] and theoretical information[102] is available in the literature for these compounds. Protonated glycine leads to only one abundant fragment, the immonium ion $H_2N=CH_2^+$ at low collision energies <10 eV (CM). The mechanism proposed by Tsang and Harrison[100a] is shown below:

$$^+H_3N-CH_2-COOH \xrightarrow{a} H_2N-CH_2-COOH_2^+$$
$$\mathbf{Ia} \qquad\qquad\qquad \mathbf{Ib}$$

$$\xrightarrow{b} H_2N-CH_2-C{\equiv}O^+ + H_2O \xrightarrow{c} H_2N=CH_2^+ + CO + H_2O \qquad (55)$$

It proceeds by protonation of the hydroxyl group, reaction a, which is endothermic by ~43 kcal/mol (Hoppilliard,[102a] MP2/6-31G*//3-21G). Excited I_b species decompose to the acylium ion and H_2O. The acylium ion is unstable and decomposes spontaneously to the immonium ion, $H_2NCH_2^+$. The transition state for the formation of the immonium ion corresponds to the transition state for reaction b. This transition state was determined to be only ~1.5 kcal/mol higher than the energy of I_b, although this evaluation was based on low-level ab initio calculations,[65] (3-21G). The theoretically predicted activation energy for immonium ion formation is thus E_A(theoret) $\approx$ 45 kcal/mol. The structures and relative energies of I_a, I_b, and the transition state $I_b^{\ddagger}$ are given below:

I_a I_b $I_b^{\ddagger}$
ΔE = 0 kcal/mol   ΔE = 43 kcal/mol   ΔE ≈ 45 kcal/mol

The experimentally determined threshold energy E_0 obtained for the immonium ion and corrected for the kinetic shift is E_0 = 44 kcal/mol (see Table 10) and thus in very good agreement with the predicted activation energy. This agreement provides strong support for the proposed mechanism.[100a]

Immonium ions $NH_2=CH_2^+$ are formed as major CID products from protonated glycinamide and N-methylglycinamide. The proposed mechanism is analogous to that shown in reaction 55 with –OH replaced by NH_2 or $NHCH_3$. The proton transfer corresponding to step a is less endothermic; however the energy increase to reach the transition state for reaction b from the amide protonated I_b is larger such that the theoretically calculated activation energy does not change much: Glycinamide E_A (3-21G) $\approx$ 49 kcal/mol while the experimental threshold value for glycinamide corrected for the kinetic shift is E_0 = 44.2 kcal/mol (see Table 12). Theoretical ab initio calculations for the activation energy for the formation of the $NH_2CH_2^+$ ion from N-methylacetamide have not been performed. The threshold energy E_0 = 50 kcal/mol, corrected for the kinetic shift, was obtained for the formation of the immonium ion from the N-methylacetamide precursor ion (see Table 12). This value is close to that for glycinamide and is consistent with both reactions proceeding by the same mechanism.

Two major product ions were observed from the CID of protonated glycylglycine (Gly-Gly)H$^+$ at low kinetic energies 10 eV (CM). These were the immonium ion $NH_2CH_2^+$ a_1 and the protonated glycine (Gly)H$^+$ which is the y_1 ion. The a_1 ion is

Table 12. CID Threshold Energies $(E_0)^a$

Precursor Ion (MH^+)	Fragment Ion		$E_0{}^b$	$E_0{}^c$
$^+H_3NCH_2COOH$	$^+H_2N{=}CH_2$	(a_1)	43.9	44.4
$^+H_3NCH_2CONH_2$	$+H_2N{=}CH_2$	(a_1)	44.2	44.9
$^+H_3NCH_2CONHMe$	$+H_2N{=}CH_2$	(a_1)	49.9	60.0
$^+H_3NCH_2CONHMe$	$+H_3NMe$	(y_1)	<49.9	59.8
$(Gly–Gly)H^+$	$+H_2N{=}CH_2$	(a_1)	43.7	64.9
$(Gly–Gly)H^+$	$^+H_3NCH_2COOH$	(y_1)	<37.3	54.1
$(Gly–Gly–NH_2)H^+$	$[MH^+{-}NH_3]$	(b_2)	20.4	21.4

Notes: ^{a}Threshold energies, E_0, in kcal/mol. Data from Klassen et al.[65] Fragment ion notation a_1, y_1, b_2 as used in peptide sequencing work. For details see Harrison et al.[100c]
^{b}Threshold energies include correction for the kinetic shift.
^{c}Threshold energies obtained using equation 50.

probably formed by a mechanism analogous to that shown for glycine (see equation 55) where OH is replaced by $NHCH_2CO_2H$. Evaluation of the transition state energy for the formation of the a_1 ion with the semiempirical AM1 method led[65] to the value E_A(theoret) $\approx$ 35 kcal/mol, while the experimental threshold result corrected for the kinetic shift was $E_0 = 44$ kcal/mol (see Table 12). The experimental E_0 is very close to the thresholds obtained for the a_1 ion formation from $(GlyNH_2)H^+$ and $(GlyNHCH_3)H^+$, which is consistent with the reactions proceeding by analogous mechanisms. The relatively poor agreement with the AM1 calculated value is probably due to deficiencies in the semiempirical calculation.

The kinetic shifts for the immonium a_1 ion formation from $(Gly)H^+$ (0.4), $(GlyNH_2)H^+$ (0.7), $(GlyNHCH_3)H^+$ (10), and $(Gly-Gly)H^+$ (21), where the values in brackets give the kinetic shift in kcal/mol (see Table 12), illustrate the very rapid increase of the kinetic shift with increasing size of the precursor ion. For $(Gly-Gly)H^+$ the kinetic shift is close to 50% of the true threshold, $E_0 = 44$ kcal/mol. Obviously unless an accurate evaluation of the kinetic shift is possible, reliable threshold values cannot be obtained with precursor ions of this size.

The threshold results obtained[66] for protonated glycylglycinamide, $(Gly-GlyNH_2)H^+$, were very different from those for $(Gly-Gly)H^+$. The most striking feature was the very low onset energy observed for a major fragment ion, $H_2NCH_2CONHCH_2CO^+$ (m/z = 115), which is a b_2 ion formed by loss of NH_3 from the precursor ion. A b_2 ion was not observed from $(Gly-Gly)H^+$. The threshold energy for the b_2 ion from the amide was much lower than that for the unstable b_1 ions leading to a_1 ions observed from $(Gly)H^+$, $(GlyNH_2)H^+$, $(GlyNHCH_3)H^+$, and $(GlyGly)H^+$. This result suggested that the b_2 ion does not have the acylium structure but some other more stable structure. Harrison et al.,[100] who have studied the dissociation of a number of protonated small peptides from their FAB metastable spectra, have proposed a general mechanism for the formation of stable b_2

and b_n ions ($n \geq 2$). These b_n ions are not unstable acyclic acylium ions but cyclic oxazalonolyl cations. The Harrison mechanism adapted for the $(Gly\text{-}GlyNH_2)H^+$ precursor ion is shown below:

$$^+H_3NCH_2CONHCH_2CONH_2 \xrightarrow{\;a\;} H_2NCH_2CONHCH_2CONH_3^+ \; = \quad \mathbf{II_b}$$

$$\mathbf{II_a}$$

$$\downarrow b \qquad\qquad (56)$$

$$H_2N\text{-}CH_2\text{-}C \overset{O}{\underset{NH-CH_2}{\diagdown}} C{=}O \; + \; NH_3 \longleftarrow$$

$$\mathbf{II_c} \qquad\qquad \mathbf{II_b}^{\ddagger}$$

The mechanism indicated in equation 56 shows the transition state $\mathbf{II_b^{\ddagger}}$, obtained[65] with the semiempirical AM1. The intermediate structure preceding the transition state indicates the intramolecular nucleophilic displacement reaction which leads to the cyclization. The energy of the transition state $\mathbf{II_b^{\ddagger}}$ relative to the precursor ion $\mathbf{II_a}$ obtained[65] with AM1 was $E_A(\text{theoret}) \approx 27$ kcal/mol, while the experimental CID threshold energy corrected for the kinetic shift was $E_0 = 20.4$ kcal/mol. Again, the agreement between the E_0 value and the calculation is not very close, but considering the errors that can be expected for the AM1 calculation and the error in the E_0 determination, the values can be considered as close enough to provide good support for the proposed mechanism.[100] It is obviously desirable to obtain theoretical values for the activation energy from ab initio calculations with relatively high-quality basis sets. One can hope that the availability of experimental E_0 values will stimulate such theoretical work.

Stable b_2 ions with low threshold energies were also observed[65] in the CID of protonated Gly-Gly-Gly and Gly-Gly-Gly-Gly, thus providing a confirmation for the generality of the mechanism, equation 56.

The formation of y type ions was observed[65] from several of the precursor ions studied and the threshold energies E_0 were compared with proposed mechanisms. The interested reader can find these results and many additional findings in the original work.[65] The purpose of the present account was to provide an illustration of the contribution that the determinations of threshold curves and threshold energies can provide to the elucidation of the mechanisms involved.

V. CONCLUSIONS

Electrospray (ES) affords the production of several important classes of ions in the gas phase, which could not be obtained by other means. Therefore, it provides an opportunity for a significant extension of Gas-Phase Ion Chemistry studies. The most important examples of the new ions are the singly and multiply charged ions of biological origin such as peptides, proteins, and nucleic acids, and ion-ligand complexes of doubly charged alkaline earth and transition metal complexes.

The thermochemical determinations involving ions obtained via ES, described in this work, provide an illustration of opportunities presented by the ES method. However, these are only the beginning of what we believe should become an important area in gas-phase ion chemistry. The ion equilibria measurements, so far performed in our laboratory, were restricted by the limited upper temperature ($T \leq 500$ K). This restriction was due to the cryopumping used and could be easily overcome and the temperature extended upwards by $\sim$200 K by using conventional pumping.

The ion reaction chamber in the present work was at a relatively high pressure (10 torr), so that conditions were similar to those used with our previous pulsed-electron high-pressure sources.[8,9] Reactors operating at lower pressures such as 1 torr or less should also be suitable. Thus, ES could probably be easily adapted for use with flow tubes such as FA and SIFT.

The CID apparatus for threshold energy determinations used in the present work was not capable of producing "state-of-the-art" accuracy in the determinations. Advances in the combination of ES with energy threshold CID should make this technique of outstanding thermochemical importance.

The extension of analytical mass spectrometry from electron ionization (EI) to chemical ionization (CI) and then to the ion desorption (probably more correctly "ion desolvation") techniques terminating with ES, represents not only an increase of analytical capabilities, but also a broadening of the chemical horizon for the analytical mass spectrometrist. While CI introduced the necessity for understanding ion—molecule reactions, such as proton transfer and acidities and basicities, the desolvation techniques bring the mass spectrometrist in touch with "ions in solution," ion—ligand complexes, and intermediate states of ion solvation in the gas phase. Gas-phase ion chemistry can play a key role in this new interdisciplinary integration.

REFERENCES

1. (a) Rosenstock, K. M.; Wallenstein, M. B.; Wahrhaflig, A. L.; Eyring, H. *Proc. Natl. Acad. Sci. USA* **1952**, *38*, 667. (b) Klots, C. E. *Zeitschrift für Naturforschung* **1972**, *27A*, 1526. (c) Lias, S. G.; Ausloos, P. In *Unimolecular Processes. Research Monographs in Radiation Chemistry*; American Chemical Society: Washington, 1975, Chapter 2.

2. (a) Field, F. H.; Franklin, J. L. *Electron Impact Phenomena*; Academic Press: New York, 1957. (b) Rosenstock, H. M.; Dzaxl, K.; Steiner, B. W.; Herron, J. T. *J. Phys. Chem. Ref. Data* **1977**, *6*, Supplement No. 1.

3. Harrison, A. G.; Kebarle, P.; Lossing, F. P. *J. Am. Chem. Soc.* **1961**, *83*, 777.

4. (a) Stevenson, D. P.; Schissler, D. O. *J. Chem. Phys.* **1958**, *29*, 282. (b) Gioumousis, G.; Stevenson, D. P. *J. Chem. Phys.* **1958**, *29*, 294.

5. Franklin, J. L.; Field, F. H.; Lampe, F. W. *Adv. Mass Spectrom.* **1959**, *1*, 308.

6. Talroze, V. *Pure Appl. Chem.* **1952**, *5*, 455.

7. Bowers, M. T. (Ed.). *Gas Phase Ion Chemistry*; Academic Press: New York, Vols. 1–4.

8. (a) Kebarle, P.; Hogg, A. M. *J. Chem. Phys.* **1965**, *42*, 798. (b) Hogg, A. M.; Kebarle, P. *J. Chem. Phys.* **1965**, *43*, 499. (c) Hogg, A. M.; Haynes, R. N.; Kebarle, P. *J. Am. Chem. Soc.* **1966**, *88*, 28. (d) Kebarle, P. *Ann. Rev. Phys. Chem.* **1977**, *28*, 455.

9. "Techniques for the Study of Ion-Molecule Reactions"; Farrar, J. M.; Saunders, W. H. Jr. (Eds.). Part of series: *Techniques of Chemistry*; Weissberger, Ed.; Wiley-Interscience: New York, 1988.

10. Yamdagni, R.; Kebarle, P. *J. Chem. Soc.* **1976**, *98*, 1320.

11. Szulejko, J. E.; McMahon, T. B. *J. Am. Chem. Soc.* **1993**, *115*, 7839.

12. Bartmess, J. E.; Scott, J. A.; McIver, R. T. Jr. *J. Am. Chem. Soc.* **1979**, *101*, 6046.

13. (a) Cumming, J. B.; Kebarle, P. *Can. J. Chem.* **1978**, *56*, 1. (b) Caldwell, G.; Renneboog, R.; Kebarle, P. *Can. J. Chem.* **1989**, *67*, 611.

14. Lias, S. G.; Bartmess, J. E.; Liebman, J. F.; Holmes, J. L.; Levin, R. D.; Mallard, W. G. *J. Phys. Chem. Ref. Data* **1988**, *17*, Supplement No. 1.

15. Keesee, R. G.; Castelman, A. W. *J. Phys. Chem. Ref. Data* **1986**, *15*, 1011.

16. (a) Arnett, E. M.; Jones, F. M.; Taagepera, M.; Henderson, W. G.; Beauchamp, J. L.; Holtz, D.; Taft, R. W. *J. Am. Chem. Soc.* **1972**, *94*, 4724. (b) Taft, R. W. *Prog. Phys. Org. Chem.* **1983**, *14*, 247.

17. (a) Yamdagni, R.; McMahon, T. B.; Kebarle, P. *J. Am. Chem. Soc.* **1974**, *96*, 4035. (b) Dzidic, I.; Kebarle, P. *J. Phys. Chem.* **1970**, *74*, 1466.

18. (a) Hehre, W. J.; Pople, J. A. *J. Am. Chem. Soc.* **1972**, *94*, 6901. (b) Pross, A.; Radom, L.; Taft, R. W. *J. Org. Chem.* **1980**, *45*, 818.

19. (a) Kraemer, W. P.; Diercksen, G. H. F. *Chem. Phys. Lett.* **1970**, *5*, 463. (b) Diercksen, H. F.; Kraemer, W. P. *Theoret. Chim. Acta (Berlin)* **1972**, *23*, 387.

20. Kistenmacher, H.; Popkie, H.; Clementi, E. *J. Chem. Phys.* **1973**, *59*, 5892.

21. Chandrasekhar, J.; Spellmeyer, D. C.; Jorgensen, W. L. *J. Am. Chem. Soc.* **1984**, *106*, 903.

22. Spears, K. G.; Fehsenfeld, F. C.; McFarland, M.; Ferguson, E. E. *J. Chem. Phys.* **1972**, *56*, 2562.

23. Tonkyn, R.; Weisshaar, J. C. *J. Am. Chem. Soc.* **1986**, *108*, 7128.

24. (a) Buchner, S. W.; Freiser, B. S. *J. Am. Chem. Soc.* **1987**, *109*, 1247. (b) Huang, Y.; Freiser, B. S. *J. Am. Chem. Soc.* **1988**, *110*, 4434.

25. Barker, M.; Bordoly, R. R.; Sedgewick, R. D.; Tyler, A. N. *J. Chem. Soc., Chem. Commun.* **1981**, 325.

26. Pelzer, G.; DePauer, E.; Viet Dung, D.; Marien, J. *J. Phys. Chem.* **1984**, *88*, 5065.

27. Sunner, J.; Morales, A.; Kebarle, P. *Int. J. Mass Spectrosc. Ion Proc.* **1989**, *87*, 287.

28. Vestal, M. *Anal. Chem.* **1984**, *56*, 2590.

29. Schmelzeisen-Redecker, G.; Bütfering, L.; Röllgen, F. W. *Int. J. Mass Spectrosc. Ion Proc.* **1989**, *90*, 139.

30. (a) Yamashita, M.; Fenn, J. B. *J. Phys. Chem.* **1984**, *88*, 4451; **1984**, *88*, 4671. (b) Fenn, J. B.; Mann, M.; Meng, C. K.; Wong, S. F.; Whitehouse, C. M. *Science* **1985**, *246*, 64.

31. (a) Alexander, A. J.; Kebarle, P. *Anal. Chem.* **1986**, *58*, 471. (b) Kebarle, P.; Tang, L. *Anal. Chem.* **1993**, *65*, 972A.

32. (a) Desnoyers, J. E.; Jolicoeur, C. In *Modern Aspects of Electrochemistry*; Bockris, J. O. M.; Conwey, B. E., Eds.; Plenum: New York, 1969, Vol. 5, p. 1. (b) Marcus, Y. *Ion Solvation*; John Wiley: New York, 1985.

33. Karas, M.; Bachmann, D.; Bahr, U.; Hillenkamp, F. *Int. J. Mass Spectrom. Ion Proc.* **1987**, *78*, 53.

34. Gao, J.; Kuczera, K.; Tidor, B.; Karplus, M. *Science* **1989**, *244*, 1069.

35. Roux, B.; Karplus, M. (a) *J. Phys. Chem.* **1991**, *95*, 4856. (b) *J. Am. Chem. Soc.* **1993**, *115*, 3250. (c) *J. Computational Chem.* **1995**, *16*, 690.

36. Kebarle, P.; Ho, Y. In *Electrospray Ionization Mass Spectrometry: Fundamentals, Instrumentation, and Applications*; Cole, R. B., Ed.; John Wiley: New York, 1997.

37. Bailey, A. G. *Electrostatic Spraying of Liquids*; John Wiley: New York, 1988.
38. *Electrosprays: Theory and Applications*; Special issue *J. Aerosol Sci.* **1994**, *25*, Pergamon Press.
39. Pfeifer, R. J.; Hendricks, C. D. *AIAA J.* **1968**, *6*, 496.
40. Taylor, G. I. *Proc. R. Soc. London A* **1964**, *A280*, 383.
41. (a) Cloupeau, M.; Prunet-Foch, B. *J. Aerosol Sci.* **1994**, *25*, 1021. (b) Cloupeau, M. *ibid*, 1143. (c) Horming, D. W.; Henricks, C. D. *J. Appl. Phys.* **1979**, *50*, 2614.
42. Smith, D. P. H. *IEEE Trans. Ind. Appl.* **1986**, *IA-22*, 527.
43. Hayati, I.; Bailey, A. I.; Tadros, T. F. *J. Colloid Interface Sci.* **1987**, *117*, 205.
44. Fernandez de la Mora, J.; Locertales, I. G. *J. Fluid Mech.* **1994**, *243*, 561.
45. Blades, A. T.; Ikomonou, M. G.; Kebarle, P. *Anal. Chem.* **1991**, *63*, 2109.
46. (a) Van Berkel, G. J.; McLuckey, S. A.; Glish, G. L. *Anal. Chem.* **1993**, *64*, 1586. (b) Van Berkel, G. J.; Zon, F. *Anal. Chem.* **1995**, *67*, 2916.
47. Ikonomou, M. G.; Blades, A. T.; Kebarle, P. *J. Am. Soc. Mass Spectrom.* **1991**, *2*, 497.
48. Wampler, F. W.; Blades, A. T.; Kebarle, P. *J. Am. Soc. Mass Spectrom.* **1993**, *4*, 289.
49. (a) Taflin, D. C.; Ward, T. L.; Davis, E. J. *Langmuir* **1989**, *5*, 376. (b) Davis, E. J. *ISA Trans.* **1987**, *26*, 1.
50. Gomez, A.; Tang, K. *Phys. Fluids* **1994**, *6*, 404.
51. Tang, L.; Kebarle, P. *Anal. Chem.* **1993**, *65*, 3654.
52. Davis, C. N. In *Fundamentals of Aerosol Science*; Saw, D. T., Ed.; John Wiley: New York, 1978, p 154.
53. Dole, M.; Mack, L. L.; Hinez, R. L.; Mobley, R. C.; Ferguson, L. D.; Alice, M. B. *J. Chem. Phys.* **1968**, *49*, 2240.
54. (a) Iribarne, J. V.; Thomson, B. A. *J. Chem. Phys.* **1976**, *64*, 2287. (b) Thomson, B. A.; Iribarne, J. V. *J. Chem. Phys.* **1979**, *71*, 4451.
55. (a) Wilm, M. S.; Mann, M. *Int. J. Mass Spectrom. Ion Proc.* **1994**, *136*, 167. (b) Wilm, M. S.; Mann, M. *Analytical Properties of the Nano Electrospray Ion Source.* Submitted.
56. Klassen, J. S.; Blades, A. T.; Kebarle, P. *J. Phys. Chem.* **1995**, *99*, 15509.
57. Klassen, J. S.; Blades, A. T.; Kebarle, P. *J. Am. Chem. Soc.* **1994**, *116*, 12075.
58. Blades, A. T.; Klassen, J. S.; Kebarle, P. *J. Am. Chem. Soc.* **1995**, *117*, 10563.
59. Blades, A. T.; Ho, Y.; Kebarle, P. *J. Am. Chem. Soc.* **1996**, *118*, 196.
60. Reid, N. M.; Buckley, J. A.; French, J. B.; Poon, C. C. *Adv. Mass Spectrom. B* **1979**, *8*, 1843.
61. Kessee, R. G.; Lee, N.; Castleman, A. W. Jr. *J. Am. Chem. Soc.* **1979**, *101*, 2599.
62. Blades, A. T.; Klassen, J. S.; Kebarle, P. *J. Am. Chem. Soc.* **1996**, *118*, 12437.
63. Anderson, S. G.; Blades, A. T.; Klassen, J. S.; Kebarle, P. *Int. J. Mass Spectrom. Ion Proc.* **1995**, *141*, 217.
64. Klassen, J. S.; Anderson, S. G.; Blades, A. T.; Kebarle, P. *J. Phys. Chem.* **1996**, *100*, 14218.
65. Klassen, J. S.; Kebarle, P. *J. Am. Chem. Soc.* In print.
66. Ho, Y.; Kebarle, P. *Int. J. Mass Spectrom. Ion Proc.* Submitted.
67. Armentrout, P. B. In *Advances in Gas Phase Ion Chemistry*; Adams, N.; Babcock, L. M., Eds.; JAI Press: Greenwich, 1992, Vol. 1, p. 83.
68. Dalleska, N. F.; Honma, K.; Armentrout, P. B. *J. Am. Chem. Soc.* **1993**, *115*, 12125.
69. (a) Loh, S. K.; Hales, D. A.; Lian, L.; Armentrout, P. B. *J. Chem. Phys.* **1989**, *90*, 5466. (b) Khan, F. A.; Clemmes, D. E.; Schultz, R. H.; Armentrout, P. B. *J. Phys. Chem.* **1993**, *97*, 7978.
70. Blades, A. T.; Jayaweera, P.; Ikonomou, M. G.; Kebarle, P. *J. Chem. Phys.* **1990**, *92*, 5900.
71. Blades, A. T.; Jayaweera, P.; Ikonomou, M. G.; Kebarle, P. *Int. J. Mass Spectrom. Ion Proc.* **1990**, *101*, 325; **1990**, *102*, 251.
72. Magnera, T. F.; David, D. E.; Stulik, D.; Orth, R. G.; Finkman, H. T.; Michl, J. *J. Am. Chem. Soc.* **1989**, *111*, 4100; 5036.
73. Cassady, C. J.; Freiser, B. S. *J. Am. Chem. Soc.* **1984**, *106*, 6176.
74. Klobukowski, M. *Can. J. Chem.* **1992**, *70*, 589.
75. Spears, K. G.; Fehsenfeld, F. C. *J. Chem. Phys.* **1972**, *56*, 5698.

76. Gross, D. S.; Rodriguez-Cruz, S. E.; Brock, S.; Williams, E. R. *J. Phys. Chem.* **1995**, *99*, 4034.

77. Gronert, S. *J. Am. Chem. Soc.* **1996**, *118*, 3525.

78. Boldyrev, A. I.; Simons, J. *J. Chem. Phys.* **1993**, *98*, 4745.

79. Boldyrev, A. I.; Simons, J. *J. Phys. Chem.* **1994**, *98*, 2298.

80. Blades, A. T.; Kebarle, P. *J. Am. Chem. Soc.* **1994**, *116*, 10761.

81. Blades, A. T.; Ho, Y.; Kebarle, P. *J. Phys. Chem.* **1996**, *100*, 2443.

82. Blades, A. T.; Klassen, J. S.; Kebarle, P. *J. Am. Chem. Soc.* **1995**, *117*, 10563.

83. Kebarle, P.; Searles, S. K.; Zolla, A.; Scarborough, J.; Arshadi, M. *J. Am. Chem. Soc.* **1967**, *89*, 6393.

84. Yamdagni, R.; Kebarle, P. *J. Am. Chem. Soc.* **1973**, *95*, 3504.

85. Meot-Ner (Mautner) M.; Hamlet, P.; Hunter, E. P.; Field, F. H. *J. Am. Chem. Soc.* **1980**, *102*, 6393.

86. (a) Yamdagni, R.; Kebarle, P. *J. Chem. Soc.* **1971**, *93*, 7139. (b) Payzant, J. D.; Yamdagni, R.; Kebarle, P. *Can. J. Chem.* **1971**, *49*, 3308. (c) Cumming, J. B.; French, M.; Kebarle, P. *J. Am. Chem. Soc.* **1977**, *99*, 6999. (d) Caldwell, G.; Kebarle, P. *Can. J. Chem.* **1985**, *63*, 1399. (e) Paul, G. J. C.; Kebarle, P. *Can. J. Chem.* **1990**, *68*, 2070.

87. (a) Larson, J. W.; McMahon, T. B. *J. Phys. Chem.* **1984**, *88*, 1083. (b) Larson, J. W.; McMahon, T. B. *Inorg. Chem.* **1984**, *23*, 2029. (c) Larson, J. W.; McMahon, T. B. *J. Am. Chem. Soc.* **1984**, *106*, 517. (d) Cumming, J. B.; Kebarle, P. *Can. J. Chem.* **1978**, *56*, 1. (e) Caldwell, G.; Renneboog, R.; Kebarle, P. *Can. J. Chem.* **1989**, *67*, 611. (f) Koppel, I. A.; Taft, R. W.; Anvia, F.; Zhu, S. Z.; Hu, L. Q.; Sung, K. S.; DesMarteau, D. D.; Yagupolskii, L. M.; Yagupolskii, Y. L; Vlasor, V. M.; Notario, R.; Maria, P. C. *J. Am. Chem. Soc.* **1994**, *116*, 3047. (g) Taft, R. W.; Topsom, R. D. *Prog. Phys. Org. Chem.* **1987**, *16*, 1.

88. Kebarle, P.; Chowdhury, S. *Chem. Rev.* **1987**, *87*, 513.

89. (a) Meot-Ner (Mautner), M. *J. Phys. Chem.* **1980**, *84*, 2716. (b) Meot-Ner (Mautner), M.; Sieck, L. W. *Int. J. Mass Spectrom. Ion Proc.* **1991**, *109*, 187. (c) Meot-Ner (Mautner), M.; Deakyne, C. A. *J. Am. Chem. Soc.* **1985**, *107*, 469.

90. De Frees, D. J.; McIver, R. T.; Hehre, W. J. *J. Am. Chem. Soc.* **1980**, *102*, 3334.

91. Taft, R. W.; Anvia, F.; Gall, J. F.; Walsh, S.; Capon, M.; Homes, M. C.; Hosn, K.; Oloumi, G.; Vasanwala, R.; Yazdani, S. *Pure and Appl. Chem.* **1990**, *62*, 17.

92. (a) Deng, H. D.; Kebarle, P. *J. Phys. Chem.* Submitted. (b) Holland, P. M.; Castleman, A. W. Jr. *J. Chem. Phys.* **1982**, *76* , 4195.

93. Chupka, W. A. *J. Chem. Phys.* **1959**, *30*, 191.

94. Sunner, J.; Kebarle, P. *J. Am. Chem. Soc.* **1984**, *106*, 6135.

95. Jensen, F. *J. Am. Chem. Soc.* **1992**, *114*, 9533.

96. Weller, T.; Lochman, R.; Meiler, W. *J. Mol. Struct.* **1982**, *90*, 81.

97. Scrivens, J. *Characterization of Oligomeric Systems Using MALDI MS/MS*; presented at 96 Lake Louise Workshop on Tandem Mass Spectrometry, Lake Louise, Canada 1996.

98. Hunt, D. F.; Yates, J. R. III; Shabanovitz, J.; Winston, S.; Hauer, C. R. *Proc. Natl. Acad. Sci. USA* **1986**, *83*, 6233.

99. Phillips, R. D.; McCloskey, J. A. *Int. J. Mass Spectrom. Ion Proc.* **1993**, *128*, 61.

100. (a) Tsang, C. W.; Harrison, A. G. *J. Am. Chem. Soc.* **1976**, *98*, 1301. (b) Dookeran, N. N.; Yalcin, T.; Harrison, A. G. *J. Mass Spectrom.* **1996**, *31*, 500. (c) Yalcin, T.; Khow, C.; Csizmadia, I. G.; Peterson, M. R.; Harrison, A. G. *J. Am. Soc. Mass Spectrom.* **1995**, *6*, 1164.

101. (a) Cordero, M. M.; Houser, J. J.; Wesdemiotis, C. *Anal. Chem.* **1993**, *65*, 1594. (b) Beranova, S.; Cai, J.; Wesdemiotis, C. *J. Am. Chem. Soc.* **1995**, *117*, 9492.

102. (a) Bouchonnet, S.; Hoppilliard, Y. *Org. Mass Spectrom.* **1992**, *27*, 71. (b) Jensen, F. *J. Am. Chem. Soc.* **1992**, *114*, 9533. (c) Bouchoux, G.; Bowries, S.; Hoppilliard, Y.; Mauriac, C. *Org. Mass Spectrom.* **1993**, *28*, 1064.

POSITIVE ION–ELECTRON IMPACT IONIZATION CROSS SECTIONS:
THEORY AND EXPERIMENT

Peter W. Harland and Claire Vallance

Advances in Gas-Phase Ion Chemistry
Volume 3, pages 319–358
Copyright © 1998 by JAI Press Inc.
All rights of reproduction in any form reserved.
ISBN: 0-7623-0204-6

ABSTRACT

Quantum mechanical and selected semiclassical and semiempirical methods for the calculation of electron impact ionization cross sections are described and their successes and limitations noted. Experimental methods for the measurement of absolute and relative ionization cross sections are also described in some detail. Four theoretical methods, one quantum mechanical and three semiclassical, have been used to calculate cross sections for the total ionization of the inert gases and small molecules and the results compared with experimental measurements reported in the literature. Two of the theoretical methods, one quantum mechanical and one semiclassical, have been applied to the calculation of orientation-dependent electron impact ionization cross sections and the results compared with recent experiments.

I. INTRODUCTION

Electron impact ionization was first used in mass spectrometry by Dempster[1] during 1918 in an attempt to generate reproducible mass spectra for use in analytical chemistry, an application of mass spectrometry suggested by J. J. Thomson[2] in 1913. After 80 years of experience in electron impact ionization, a relatively small number of absolute ionization cross sections for molecules have been measured, and fragmentation patterns are not easily predictable using either theoretical models or chemical intuition. Electron impact ionization is used to prepare positive ions for kinetic studies of ion–molecule reactions in the gas phase. Studies of the ionization process and ionization mechanisms, guided by molecular orbital computations, have shown that a mixture of isomeric forms of positive ions can be produced by electron impact.[3] This has been subsequently exploited in selected ion flow tube studies (SIFT) of their ion–molecule chemistry.[4] Measurement of ionization thresholds by electron impact ionization of molecules has provided the major contribution to the thermochemical data bases on ions and free radicals.[5] Enthalpies of formation derived from electron impact studies are used routinely to deduce the products and energetically favorable reaction pathways in ion–molecule studies. Electron impact ionization plays an important role in many other fields of physics and chemistry, including mass spectrometry, plasma processes, and gas discharges, and accurate ionization cross sections are required in order to understand the mechanism of the ionization process as well as for modeling systems such as fusion plasmas and radiation effects in both medical and materials science.

In this chapter we focus attention on the efficiency of ionization, the ionization cross section, and consider some recent experimental measurements and theoretical studies of the ionization process. A sketch of electron impact ionization curves, the variation of the ionization cross section as a function of the electron energy, using CO as an example, are shown in Figure 1. The mass spectrum, collected at the electron energy corresponding to the maximum in the ionization cross section, is also shown, although there will be no further discussion of fragmentation in this

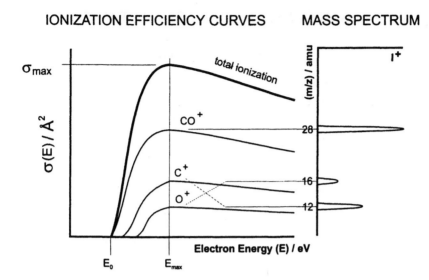

Figure 1. Sketch of the ionization efficiency curves and the "70 eV" mass spectrum for the electron impact ionization of carbon monoxide.

review. Ionization cross sections for the inert gases, measured by several independent groups, lie in a narrow range; this is not always the case for molecules where the range of values reported can be substantial. For example, recent values reported for the maximum total ionization cross section, $\sigma(E_{max})$, for O_2 lie in the range from 2.62 Å^2 [6] to 2.978 Å^2 [7] and for NH_3 from 2 Å^2 to 3.19 Å^2. [8–10] For most molecules only one measurement is available, and for many simple molecules there are no experimental values in the literature. Since accurate measurements of absolute ionization cross sections are difficult, it is desirable to have models of the ionization process available from which reliable cross sections can be calculated. Significant differences between the ionization cross section for collisions at the positive and negative ends of the molecular dipole for a number of symmetric top molecules has been reported recently[11] and a successful model should account for this angular dependence.

Theoretical models of the electron impact ionization process have focused on the calculation of the ionization cross section and its energy dependence; they are divided into quantum, semiclassical and semiempirical. Methods for the calculation of the ionization cross section and experimental techniques developed for the measurement of absolute ionization cross sections will be described in more detail below. Cross sections calculated using the semiempirical additivity method developed by Deutsch and Märk (DM) and their coworkers, [12–14] the binary-encounter-Bethe (BEB) method of Kim and Rudd, [15,16] and the electrostatic model (EM) developed by Vallance, Harland, and Maclagan[17,18] are compared to each other and to experimental data.

II. CALCULATION OF CROSS SECTIONS

A. Quantum Mechanical

It is difficult to treat electron impact ionization rigorously using quantum mechanical methods, one of the main reasons being that the exit channel consists of two free electrons in the field of an ionized atom or molecule, which represents a complex quantum many body problem.[19] This is in addition to the usual considerations of coupling effects, target polarization, and other factors involved in all electron–atom scattering calculations. For these reasons attention has been focused on developing approximate methods which, while physically reasonable, are manageable for use in numerical calculations. This has led to the wide range of semiempirical and semiclassical approximations to the electron impact ionization cross section discussed later, as well as the more rigorous quantum mechanical approaches.

The many-body problem can be readily formulated for bound systems, and can also be extended to the case of a bound system in the presence of a single free electron. However, it has not been yet been possible to apply similar methods to the case of a system containing two free electrons, corresponding to the exit channel case for the electron impact ionization process. Several groups[20] have applied perturbation methods in an attempt to solve this problem. Fadeev[21] used a different approach and reformulated the three-body problem in terms of an infinite number of two-body problems. While these are no longer difficult to solve in theory, in practice they are still computationally unmanageable. So far these kinds of approaches have only been applied to very light atoms such as hydrogen and helium. Calculations on many electron systems generally use some variation on the partial wave approximation.

The Partial Wave Approximation

In the partial wave theory free electrons are treated as waves. An electron with momentum k has a wavefunction $\psi(k,r)$, which is expressed as a linear combination of partial waves, each of which is separable into an angular function $Y_{Lm}(\theta,\phi)$ (a spherical harmonic) and a radial function $R_L(k,r)$,

$$\psi(k, r) = \sum_{Lm} c_{Lm} Y_{Lm}(\theta,\phi) R_L(k,r) \tag{1}$$

where c_{Lm} are expansion coefficients, and L and m are quantum numbers. The radial function is a solution to the associated radial Schrödinger wave equation

$$\left[-\frac{1}{2} \frac{d^2}{dr^2} + \frac{L(L+1)}{2r^2} + V(r) \right] R_L(k,r) = \frac{1}{2} k^2 R_L(k,r) \tag{2}$$

in which $V(r)$ is the potential experienced by the free electron and largely determines the functional form of $R_L(\mathbf{k},r)$. There are several variants of the partial wave approximation depending on the potential function used. In the simplest treatment, known as the plane wave Born approximation, the potential $V(r)$ is set to zero. The target atom has no interaction with the electron, and the resulting radial functions are the spherical Bessel functions. Because of its relative simplicity, this method has been widely applied.[22,23] The first improvement to the plane wave approximation is to assume that the free electron undergoes a Coulomb interaction with an effective nuclear charge Z centered on the ionized target atom, i.e. $V(r) = -Z/r$. In this case, the $R_L(\mathbf{k},r)$ are Coulomb functions and this treatment is known as the Coulomb approximation. Nussbaumer and Moores[24] have applied this to several ionic targets. The distorted wave approximation is a further improvement which includes a term describing the potential due to the electrons of the atomic target, V_{elec}, i.e. $V(r) = -Z/r + V_{elec}$. This type of potential has been used in the study of both atomic and ionic targets.[25–28]

During the electron impact ionization process the incident electron approaches the target with energy E, interacting with the electrons of the target through the Coulomb potential $V_{12}(r_1,r_2)$. An inelastic transition scatters the free electron into a final state of energy E_f at the same time as a bound electron is ejected into a state of energy E_e. Considering only the first-order perturbation, the matrix element describing this direct interaction is,

$$M_d = \langle \psi_b \psi_i | 1/r_{12} | \psi_e \psi_f \rangle \tag{3}$$

where the wavefunctions ψ_b, ψ_i, ψ_e, ψ_f describe the bound, incident, ejected, and final scattered electrons, respectively. The transition probability or probability of ionization is proportional to the square of this matrix element:

$$P(\text{bi,ef}) \sim |M_d|^2 \tag{4}$$

Incorporating the required kinematic factors transforms this transition probability into a cross section,

$$\sigma(E,E_e,E_f) = 16a_o^2/E \sum_L (2L+1)|M_d|^2 \tag{5}$$

where E is the energy of the incident electron and L is the total angular momentum of the electron-target system. To obtain the total cross section we integrate over the final states of the electrons:

$$\sigma(E) = \int_0^{E_{max}} \sigma(E,E_e,E_f)\,dE_e \tag{6}$$

In this so called "half range Born" approximation the upper limit is given by,

$$E_{max} = \frac{1}{2}(E - E_o) \qquad (7)$$

where E_o is the ionization potential of the target. When treating the electron impact process quantum mechanically there are several phenomena which should be taken into account. Resonances in the cross section, electron correlation between the various electrons involved, and also exchange processes due to the indistinguishability of the electrons can all make significant contributions to the cross section and have not been included in the treatment so far.

There are two types of exchange to be considered in the electron impact ionization process. Potential exchange between a free electron and a bound target electron is analogous to the type of exchange encountered in Hartree–Fock calculations on bound state systems and can be easily included in a frozen core approximation by incorporating non-local terms in V_{elec}. Scattering exchange between the two final state free electrons is much more difficult to accommodate. The exchange scattering matrix element is,

$$M_{ex} = <\psi_b \psi_i |1/r_{12}|\psi_f \psi_e> \qquad (8)$$

so that the total ionization probability including exchange is,

$$P(bi,ef) \sim |M_d + M_{ex}|^2 = |M_d|^2 + |M_{ex}|^2 - \lambda|M_d||M_{ex}| \qquad (9)$$

The first two terms are the direct and exchange ionization terms, while the remaining term is an interference term. λ is a phase factor related to the relative phases of the two matrix elements and depends on the interaction between the two electrons. There are several choices for λ. In the Born–Oppenheimer approximation $\lambda = 0$, the interference term is ignored, and since the exchange cross section is often comparable to the direct cross section, the total cross sections obtained are usually far too large. A more realistic approximation is the Born-exchange approximation, of which there are two variants. In the Peterkop formulation[30] $\lambda = 1$. The interference term is maximized, which minimizes the total exchange contribution. In the Rudge and Schwartz formulation[31] λ is chosen to cancel the imaginary part of the scattering matrix element. In both methods the cross section with exchange included is not too different from the direct cross section and both methods give comparable results. In the Born–Oppenheimer approximation the same final state orbitals are used to calculate both matrix elements. In the Born-exchange approximation, overlapping orbitals in a matrix element must be calculated in the same potential so that orbitals on opposite sides of the scattering matrix remain orthogonal. In practice, the fast final-state partial wave is calculated using the potential of the ion (an $N - 1$ electron system) and the slow ejected electron wave in the potential of the initial target (an N electron system). This choice gives total electron impact ionization cross sections in good agreement with experiment for light atoms and ions.

Resonances arise when the energy of the bound target system is the same as that of the ionized state plus a free electron, and can occur in either the entrance or the exit channel of the electron impact ionization process. Resonances involving the inbound electron lead to spikes in the total cross section at specific incident electron energies corresponding to the resonance state, while resonances involving the ejected electron appear as spikes in the differential cross section, $\sigma(E, E_e, E_f)$. Since an exit channel resonance will contribute to the total cross section at all incident electron energies above its threshold it leads to sharp steps in the total cross section at the corresponding energy. These resonances can be considered as excitation into a configuration which is autoionizing and are often called "excitation–autoionization" resonances.

The cross section is first calculated ignoring resonances and the cross sections corresponding to excitation of autoionizing resonances involving the ejected electron are then added. This assumes that all atoms excited to the resonance state will autoionize, which is not true for highly charged ions since the rate of relaxation to a true bound state of the N electron system through photon emission becomes faster than the rate of autoionization from the resonance state with increasing ionic charge. In these cases the contribution from excitations must be modified by multiplying the resonance cross section by an effective branching ratio for autoionization versus stabilization. Accounting for resonances can make quantum mechanical calculations of electron impact ionization cross sections very complicated and several different approaches have been attempted to simplify the problem.[32,33]

Correlation generally has only a minor effect on electron impact ionization cross sections due to the two-body nature of the Coulomb operator $1/r_{12}$ which governs the interaction. The most important correlation corrections to a Hartree–Fock ground-state wavefunction usually involve doubly excited determinants or "pair-correlations." Since $1/r_{12}$ is a two-body operator, the total wavefunctions on each side of an interaction matrix element can only differ by a maximum of two electrons. One of these electrons is the scattered electron undergoing a transition from the incident to the final scattered state, while the other is the orbital corresponding to the bound electron, which becomes the ejected electron wave. If a double excitation is included in the wavefunction for the entrance channel, the initial and final total wavefunctions differ by too many electrons and the matrix element disappears. If the double excitation is included in both the entrance and exit channels, the matrix element no longer disappears, but the contribution to the cross section usually remains small since the "excited" matrix element does not generally add in-phase to the ground-state matrix element. It is much more difficult to treat electron correlation between the two final state free electrons and so far the effects of such correlations have not been included in calculations.

Partial wave methods have been widely used and tested on a variety of systems (see reference 19 and references therein). They are usually accurate to better than 50% for total ionization cross sections of light atoms. Very heavy atoms have been less successfully treated due to the increasing contributions of resonances and electron correlation effects, though cross sections to within a factor of 2 have been obtained in many cases.[23] At present quantum mechanical calculations using the partial wave method are limited to atomic or ionic targets.

An Ab Initio Electrostatic Model (EM)

Reasonable values for the maximum in the ionization efficiency curve for single ionization by electron impact may be obtained for both atomic and molecular species using a method based on a very simple picture of the electron impact event in terms of an electrostatic interaction.[11,17,18] The target experiences a Coulomb potential as the incident electron approaches. When this interaction potential is equal to the ionization potential of the target, it absorbs energy from the field and is ionized. This represents a critical separation and corresponds to the maximum for the cross section. A molecular orbital package such as Gaussian 94[34] can be used to determine the critical separation of the electron and the center-of-mass of the target at which the above condition is met, which leads to a value for the cross section. A useful aspect of these calculations is the fact that because the z-matrix input to the electronic structure calculations specifies a relative orientation of the electron and the molecule this method gives the maximum cross section as a function of the direction of approach of the incident electron, effectively giving the three-dimensional "shape" of the ionization cross section.

The procedure adopted for these calculations begins with a molecular geometry optimization, followed by a calculation of the vertical ionization potential as the difference in the energies of the positive ion at the neutral geometry and the zero-point energy of the neutral molecule. Alternatively, an experimentally determined ionization potential may be substituted. Because the input for the Gaussian calculations requires that the charge distribution be entered in Cartesian coordinates, the "critical energy" E_c (which is the energy of the system at which the model predicts ionization will occur) is then determined for the selected electron–molecule approach geometry. E_c is the neutral energy plus the Coulomb potential due to the electron at the "critical impact parameter" r_c, or, since the Coulomb potential at ionization is equal to the ionization potential E_o of the molecule:

$$E_c = E_{neut} - E_o \tag{10}$$

Next, a series of single-point energy calculations are carried out on the neutral molecule in the presence of a charge distribution consisting of a single electron. The initial trial electron– molecule separation is modified in an iterative procedure to converge to the critical energy E_c. When the critical separation r_c has been determined the maximum electron impact ionization cross section for the orientation under consideration is given by:

$$\sigma = \pi r_c^2 \tag{11}$$

In general, measurements of electron impact ionization cross sections are carried out on randomly oriented molecules so that the total ionization cross section is given by the average over all possible orientations of the molecule with respect to the electron. A convenient approximate total cross section can be estimated using the present method by averaging the cross sections for approach along each of the positive and negative Cartesian axes (Cartesian averaged cross section):

$$\sigma(\text{cart}) = \frac{1}{6}[\sigma_{+x} + \sigma_{-x} + \sigma_{+y} + \sigma_{-y} + \sigma_{+z} + \sigma_{-z}] \tag{12}$$

If the critical separation is determined for a large number of relative geometries of the electron and molecule it is possible to obtain a three-dimensional picture of the probability of ionization as a function of the orientation of the molecule. Effectively, the idea of an ionization cross section, the area the target molecule presents to the electron, is extended to a three-dimensional object defined by the critical distances, with ionization occurring when the electron penetrates the surface enclosing this volume. The volume enclosed by the electron impact ionization surface may be used to obtain an estimate for the cross section (volume averaged cross section):

$$\sigma(\text{volume}) = \pi r^2 = \pi \left(\frac{3V}{4\pi}\right)^{2/3} \tag{13}$$

The EM method has been tested on the inert gases and a range of small molecules and gives good agreement with experimental results in almost all cases.[17] This method will be discussed further in relation to the orientation dependence of the electron impact ionization cross section in a later section. The semiempirical polarizability method described below was developed to calculate E_{max} and to use it with the σ_{max} values obtained from this method in order to calculate the energy dependence of the cross section.

B. Semiempirical and Semiclassical

Due to the complexity of a full quantum mechanical treatment of electron impact ionization, or even a partial wave approximation, for all but relatively simple systems, a large number of semiempirical and semiclassical formulae have been developed. These often make basic assumptions which can limit their range of validity to fairly small classes of atomic or molecular systems. The more successful approaches apply to broad classes of systems and can be very useful for generating cross sections in the absence of good experimental results. The success of such calculations to reproduce experimentally determined cross sections can also give insight into the validity of the approximations and assumptions on which the methods are based.

Semiempirical treatments of the electron impact process attempt to formulate fairly simple equations containing parameters determined experimentally in order to reproduce the measured cross section and possibly determine cross sections for systems which have not been studied experimentally. Semiclassical treatments make classical approximations to the full quantum mechanical scattering problem, providing a large degree of simplification of the problem in the process. They commonly contain empirical adjustable parameters determined from a best fit to known experimental data. Several fairly major approximations must be made in going from the quantum mechanical to the semiclassical treatment,[35] in which the collision must be treated according to classical laws of motion. The initial state of the bound electron must be described classically in terms of some kind of classical orbit. Various approaches have assumed this electron to be either at rest, orbiting at a constant velocity, or to have some predetermined velocity distribution.[36] Because the three-body problem is difficult to handle, the binary-encounter approximation is often used to simplify the electron impact process to a two-body problem. This simplification was first introduced by Thomson[2] and treats the ionization process as a simple two-body collision between the incident electron and an atomic or molecular electron in the target. Ionizing collisions can be classified as fast or slow depending on the relative magnitudes of the incoming electron velocity and the mean orbital velocity of the atomic or molecular electron in the target. Different approximations can be introduced for the two cases. The Born approximation is often applied to fast collisions,[37] while classical scattering theories such as the binary encounter approximation[2] may be used for slow collisions. A point to bear in mind when using these semiempirical and semiclassical approaches is that they can generally only give either partial cross sections σ_i, the cross section for production of an ion of a given charge in a collision, or counting cross sections, $\Sigma_i \sigma_i$, the sum of the partial cross sections. Most experiments measure the total ion current produced by all ionization processes, and the cross sections derived from such measurements will usually always be greater than the counting cross section. For example, a doubly charged ion will contribute twice the current of a singly charged ion, so that unless the number of ionization events can be measured, the quantity measured experimentally is $\Sigma_i i \sigma_i$.

A multitude of semiempirical and semiclassical theories have been developed to calculate electron impact ionization cross sections of atoms and atomic ions, with relatively few for the more complicated case of molecular electron impact ioniza- tion cross sections. One of the earlier treatments of molecular targets was that of Jain and Khare.[38] Two of the more successful recent approaches are the method proposed by Deutsch and Märk and coworkers[12-14] and the binary-encounter Bethe method developed by Kim and Rudd.[15,16] The observation of a strong correlation between the maximum in the ionization efficiency curve and the polarizability of the target resulted in the semiempirical polarizability model which depends only on the polarizability, ionization potential, and maximum electron impact ionization cross section of the target molecule.[39,40] These and other methods will be considered in detail below.

The Deutsch–Märk (DM) Method

This formalism was originally devised for single ionization of ground-state atoms, but has now been successfully applied to the calculation of electron impact ionization cross sections for a range of molecules, radicals, clusters, and excited state atoms. Like many of the semiempirical and semiclassical methods used to describe the electron impact process, the theory has its roots in work carried out by J.J. Thomson, who used classical mechanics to derive an expression for the atomic electron impact ionization cross section,[2]

$$\sigma = \sum_n 4\pi a_o^2 N_n (E_i^H / E_{in})^2 ((u - 1)/u^2) \tag{14}$$

where a_o is the Bohr radius, N_n is the number of electrons in the nth subshell, E_{in} is the ionization potential from the nth subshell, E_i^H is the ionization potential of the hydrogen atom, and $u = E/E_{in}$, where E is the incident electron energy. Grysinski[41] improved Thomson's equation by assuming a continuous velocity distribution for the atomic electrons, leading to the following expression,

$$\sigma = \sum_n 4\pi a_o^2 N_n (E_i^H / E_{in})^2 f(u) \tag{15}$$

where:

$$f(u) = (1/u)((u - 1)/(u + 1))^{3/2} [1 + (2/3)(1 - 1/2u)\ln(2.7 + (u - 1)^{1/2})] \tag{16}$$

Further improvements to the theory require the incorporation of quantum factors such as exchange effects,[42] although, even then, the theory still fails to predict the correct magnitude of the cross section for several fairly simple atoms such as Ne, N, and F.

Deutsch and Märk compared the classical expression with a theory developed by Bethe.[37] Bethe's calculations showed that the ionization cross section for an atomic electron is approximately proportional to the mean square radius $<r_{nl}^2>$ of the appropriate n,l electronic shell. Experiment had also shown a correlation between the maximum in the atomic cross section and the sum of the mean square radii of all outer electrons. This led to the replacement of the Bohr radius with the radius of the corresponding subshell; the ionization cross section is now given by,

$$\sigma = \sum_{n,l} g_{nl} \pi r_{nl}^2 N_{nl} f(u) \tag{17}$$

where r_{nl}^2 is the mean square radius of the n,l shell[43] and g_{nl} are empirically determined weighting factors. This expression gives improved agreement with experimental results except in cases involving strong autoionization channels, where additional corrections are necessary.[44] The weighting factors g_{nl} were initially calculated by Bethe as a function of n and l using hydrogenic wavefunctions.

Deutsch and Märk determined these generalized factors by a fitting procedure using the inert gases and uranium as test cases. Later, when a more sophisticated fitting procedure was used, it was found that the product of the weighting factor and the binding energy for a specific set of n, l, and N_{nl} values is independent of the atomic number.[14] This allowed a weighting factor matrix with elements $g_{nl}E_{nl}$ to be constructed, from which weighting factors can be extracted for any atomic orbital once the binding energy of an electron occupying the orbital is known. Using the weighting factor matrix, inner atomic orbitals tend to receive a much lower weighting than previously used. These orbitals, having a large binding energy, would not be expected to contribute much to the cross section except for very high energy collisions requiring incident electron energies of hundreds or even thousands of electron volts. There are also significant changes in the weighting factors for valence orbitals, so that using the new set of g_{nl} values leads to considerable changes in the calculated cross section.

The DM method may be extended to molecules using a modified form of the additivity rule first proposed by Otvos and Stevenson,[45] which assumes that the molecular ionization cross section is equal to the sum of contributions from the constituent atoms. The sum is now over the molecular energy levels or molecular orbitals. The core orbitals, with large binding energies, contribute very little to the cross section and may be omitted from the sum. In the calculations carried out by the authors using the DM model, and reported later in this chapter, the cross section was only determined up to an energy of 200 eV, so that only molecular orbitals with a binding energy less than this value were included,

$$\sigma = \sum_{i,nl} \pi r_{i,nl}^2 \sum_j (g_{i,nl}^j N_{i,nl}^j) f(u_j) \tag{18}$$

where $u_j = E/E_j$. In order to use the DM method a population analysis must be carried out to determine $N_{i,nl}^j$, the contribution of each atomic orbital to the molecular orbitals considered. For this series of calculations a Mulliken population analysis at HF level was performed. The minimal STO-3G basis set was used since larger basis sets would have made the analysis prohibitively time consuming. According to a Mulliken analysis[46] the orbital population is given by,

$$N_{i,nl}^j = n_i(c_{ji}^2 + \sum_k c_{ji}S_{jk}c_{ki}) \tag{19}$$

where n_i is the occupation number of molecular orbital i, c_{ji} and c_{ki} are the molecular orbital coefficients of atomic orbitals j and k in molecular orbital i, and S_{jk} is the overlap integral between atomic orbitals j and k.

The Binary-Encounter Bethe (BEB) Method

The BEB theory is a simplified version of the binary-encounter dipole (BED) theory developed by the same author.[15] It is based on a combination of two earlier theories known to accurately describe high- and low-energy collisions. Low-impact parameter collisions, occurring at low incident electron energies, are described well by Mott theory,[47] originally developed for collisions between two electrons. High-impact parameter collisions, occurring at high incident electron energies, were shown by Bethe[37] to occur mainly through the dipole interaction between the atom and electron. The Mott theory for low-energy collisions can be augmented by the symmetric form of the binary-encounter theory,[48] which assigns a velocity distribution to a target electron instead of a wavefunction. However, because the dipole contribution is still lacking, the cross section is not well described at high energies, when the majority of collisions are distant and the dipole interaction dominates. The BED model combines this augmented form of the Mott cross section with the Bethe cross section to give an expression which describes the electron impact ionization process over the entire energy range.

The ratio between the binary-encounter and Bethe theories is set by requiring the form of the binary-encounter theory in the high electron energy limit to match that of the Bethe theory, both in the predicted cross section and in the so called "stopping cross section." The stopping cross section describes the efficiency of a collision of a heavy particle and an electron in effecting a loss of kinetic energy of the heavy particle.[49] It is used to evaluate the "stopping power" of the target medium for electrons and is given by the integral of the product of the energy loss cross section and the energy loss of the incident electron. The BED model leads to an expression for the energy distribution of the differential cross section $d\sigma/dW$ for the ejected electron with its energy W for each atomic or molecular orbital. The total cross section is found by integrating $d\sigma/dW$ over the allowed range of W, i.e. from 0 to $1/2(T - B)$, where T is the energy of the incident electron and B is the binding energy of the atomic electron, given by,

$$\sigma_i(t) = (S/(t + u + 1))[D(t)\ln(t) + (2 + N_i/N)((t - 1)/t - \ln(t)/(t + 1))] \quad (20)$$

where $t = T/B$ is the reduced incident electron energy $u = U/B$ is the reduced orbital kinetic energy,

$$S = 4\pi a_0^2 N R^2/B^2 \quad (21)$$

in which $R = 13.6$ eV is the IP of the 1s orbital of hydrogen,

$$D(t) = (1/N) \int_0^{(t-1)/2} (1/(w + 1)) df(w)/dw \, dw \quad \text{with} \quad w = W/B \quad (22)$$

and:

$$N_i = \int_0^\infty df(w)/dw \, dw \quad (23)$$

The quantities required for each orbital in order to carry out a BED calculation are therefore the binding energy B, average orbital kinetic energy U, orbital occupation number N, and dipole oscillator strength df/dw. The parameters U and B are easily calculated using any atomic or molecular wavefunction package. The average orbital kinetic energy U is a theoretical quantity which cannot be directly measured, but experimental values of the binding energy B can be used when available. This is especially useful in the case of the lowest binding energy in order to ensure the correct threshold is obtained, though according to Hwang et al.[16] using theoretical values of B leads to better agreement near the peak in the cross section function. The dipole oscillator strength df/dw is not so straightforward to obtain. The wavefunctions of both the initial and continuum states are required for the calculation of this property, and since the calculation of excited-state wavefunctions is difficult, values for this parameter are only available for a small number of atomic and molecular species.

The BEB model was developed to overcome this problem. The dipole oscillator strength is assumed to have a simple form based on the approximate shape of the function for ionization of ground-state hydrogen:

$$df/dw = N/(w + 1)^2 \tag{24}$$

The integrated cross section per orbital is then given by:

$$\sigma_{BEB} = (S/(t + u + 1))\,[(\ln(t)/2)(1 - 1/t^2) + 1 - 1/t - \ln(t)/(t + 1)] \tag{25}$$

In the expressions for the BED and BEB cross sections for each orbital the first log term represents large impact parameter collisions dominated by the dipole interaction. The remaining terms represent low-impact parameter collisions, described by the augmented Mott cross section. The second log function describes interference between direct and exchange scattering which is included in the Mott cross section.

The Jain–Khare Method

In a similar approach to that used in deriving the BED theory described above, Jain and Khare[38] divided ionizing collisions into two categories: soft or glancing collisions involving a small energy transfer between the incident electron and the target molecule; and hard collisions with a large amount of energy transfer. For high-energy incident electrons and a small energy transfer the Born–Bethe approximation is valid and $\sigma(E,\varepsilon)$, where E and ε are the energies of the incident and ejected electrons, respectively, is proportional to the differential oscillator strength. For hard collisions the cross section is described by Mött scattering theory. The Born–Bethe and Mött cross sections are given by,

$$\sigma_{BB}(E,\varepsilon) = (4\pi a_0^2 R^2/E)(1/W)(df/dW)\ln(CE) \tag{26}$$

$$\sigma_M(E,\varepsilon) = (4\pi a_0^2 R^2/E)\,S\,[1/\varepsilon^2 - 1/(\varepsilon(E - \varepsilon)) + 1/(E - \varepsilon)^2] \tag{27}$$

where W is the energy loss suffered by the incident electron during the collision, R is the Rydberg constant, S is the number of electrons which can participate in hard collisions, df/dW is the differential oscillator strength, and C is a collisional parameter defined by Miller and Platzman.[50] For ionization of a molecule by an incident electron of energy E, with production of a secondary electron of energy ε, Khare[51] proposed the following semiempirical relations,

$$\sigma(E,\varepsilon) = \sigma(E,W) = f'_1(E,\varepsilon)\, \sigma_{BB}(E,\varepsilon) \text{ for } I \le W \le W_o \qquad (28)$$

and,

$$\sigma(E,\varepsilon) = \sigma(E,W) = f'_2(E,\varepsilon)\, \sigma_M(E,\varepsilon) \text{ for } W_o \le W \le \tfrac{1}{2}(E+I) \qquad (29)$$

where I is the ionization potential and W_o is the threshold in the energy loss between the secondary electron being formed by soft and hard collisions. A suggested value was $(I + 60)$ eV. The functions f'_1 and f'_2 are introduced to extrapolate the cross sections to the low energy region where the Born–Bethe and Mött cross sections are expected to give results which are too high:

$$f'_1 = [1 - (2W - I)/E]\, \ln(1 + C(E - I))/\ln(CE) \qquad (30)$$

$$f'_2 = 1 - (2W - I)/E \qquad (31)$$

These formulae for the cross section give a discontinuity at $W = W_o$. To rectify this the two expressions for $W < W_o$ and $W > W_o$ were combined to give a single relation to describe the cross section over the entire energy range,

$$\sigma(E,\varepsilon) = \sigma(E,W) = f_1(E,\varepsilon)\, \sigma_{BB}(E,\varepsilon) + f_2(E,\varepsilon)\, \sigma_M(E,\varepsilon) \qquad (32)$$

where f_1 and f_2 extrapolate to low energies as before and also control the mixing of the Born–Bethe and Mött cross sections:

$$f_1 = 1/(1 + 1/E)\, [1 - (\varepsilon/(E - I))\ln(1 + C(E - I))/\ln(CE)] \qquad (33)$$

$$f_2 = [\varepsilon^3/(\varepsilon^3 + \varepsilon_0^3)]\, (1 - I/E) \qquad (34)$$

The parameter ε_0 was chosen for best agreement with the experimental data of Opal et al.[52] at $E = 500$ eV. Jain and Khare applied this equation to the calculation of ionization cross sections for CO_2, CO, H_2O, CH_4, and NH_3 and achieved fairly good agreement with experiment for all cases except for CO, where the cross section was too low, though the ionization efficiency curve still exhibited the correct shape. The main limitation of this method, which it has in common with the BED theory, is the inclusion of the differential oscillator strengths for the target molecule which restricts the number of systems to which it can be applied.

The Lotz Method

One of the most widely used empirical expressions for the electron impact ionization cross section was that proposed by Lotz.[53] This was designed to reproduce cross sections for the single ionization of atoms and ions from the ground electronic state, and could be used to fit nearly all experimental results from threshold to 10 keV to within 10%. An atomic electron contributes an individual cross section σ_{ei} to the total ionization cross section σ_t whenever the incident electron energy is greater than the binding energy E_i of the electronic shell to which it belongs,

$$\sigma = \sum_{i=1}^{N} \sigma_i \quad \text{with} \quad \sigma_i = n_i \sigma_{ei} \tag{35}$$

where n_i is the number of equivalent electrons in a subshell and N is the number of subshells present. According to Bethe[37] and Rudge and Schwartz[31] we have,

$$\sigma_i = a_i \, n_i \, \ln(E/E_i)/EE_i \quad \text{for} \quad E > E_i \tag{36}$$

where a_i is a constant between 2.6×10^{-14} and 4.5×10^{-14} cm^2 eV^2. Extensive trials for fitting the cross sections yielded a relatively simple equation with three parameters,

$$\sigma_i = a_i \, n_i \, [\ln(E/E_i)/EE_i]\{1 - b_i \exp[-c_i(E/E_i - 1)]\} \quad \text{for} \quad E \geq E_i \tag{37}$$

where a_i, b_i and c_i are constants. The total electron impact ionization cross section is then:

$$\sigma = \sum_{i=1}^{N} a_i \, n_i [\ln(E/E_i)/EE_i]\{1 - b_i \exp[-c_i(E/E_i - 1)]\} \quad \text{for} \quad E \geq E_i \tag{38}$$

In the sum, $i = 1$ is the outermost shell, $i = 2$ is the next inner subshell, and so on. Near threshold, when $E \cong E_1$ (the ionization potential) the above expression reduces to,

$$\sigma = a_1 \, n_1 \, [(E/E_1 - 1)/E_1^2](1 - b_1) \propto E/E_1 - 1 \tag{39}$$

while for large electron energies $E \gg E_i$ we have:

$$\sigma_i = a_i \, n_i \, \ln(E/E_i)/EE_i \propto (\ln E)/E \tag{40}$$

The equation therefore has the correct functional energy dependence for both small and large energies of the impacting electron. Lotz also showed that in most cases the equation could be substituted for the simpler form,

$$\sigma = a \{1 - b\exp[-c(E/E_1 - 1)]\} \sum_{i=1}^{N} n_i \ln(E/E_i)/EE_i \quad \text{for } E \geq E_i \qquad (41)$$

in which $N = 2$ or 3 usually gave good results.

The Bell Method

Bell et al.[54] surveyed a wide range of experimental data on electron impact ionization of light atoms and ions and found that the cross sections could all be fitted to the same simple analytic expression,

$$\sigma(E) = 1/(IE) [A \ln(E/I) + \sum_{i=1}^{N} B_i(1 - I/E)_i] \qquad (42)$$

where E is the incident electron energy, N is the number of subshells, and I is the ionization potential. The coefficients B_i are determined by a least squares fit to the experimental data, with two or three coefficients being sufficient to reproduce the cross section in many cases. The coefficient A is a Bethe coefficient which may be calculated either by fitting the equation,

$$\sigma(E) = 1/(IE)[A \ln(E) + B] \qquad (43)$$

to the high energy form of the Born approximation, or from the equation,

$$A = 1/(\pi\alpha) \int_{1}^{\infty} (\sigma_{ph}/E)dE \qquad (44)$$

where σ_{ph} is the photoionization cross section and α is the fine structure constant.

The Bell equation gives the correct behavior for the ionization cross section at both high- and low-impact energies. In cases where autoionization is important it is not always possible to reproduce the cross section from the single equation above, but if it is used in two separate fits, one from the ionization threshold to the autoionization threshold, and the second above the autoionization threshold, a good fit to the cross section may be obtained over the entire range.

Kistemaker's Modified Additivity Rule

The original additivity rule,[45] according to which a molecular ionization cross section is given by the sum of the cross sections of the constituent atoms, was based on Bethe's observation[37] that the probability of ionization of an atomic electron in the n,l atomic orbital is approximately proportional to the mean square radius of the orbital. They showed that the rule is generally valid for hydrocarbons, but tends to overestimate the cross section in many cases. Bobeldijk et al.[55] suggested that the model could be improved by taking into account the orientation dependence of the cross section. In the case of a long-chain hydrocarbon it seems reasonable to assume that the cross section for ionization for an electron approaching the molecule "head on" will be very different from that for one approaching "side on" due to the vastly different cross-sectional areas which the target presents to the electron in the two orientations. Kistemaker used the additivity rule to determine "parallel" and "perpendicular" cross sections by summing known cross sections of atomic or molecular fragments. In the case of the long-chain hydrocarbon $CH_3(CH_2)_nCH_3$ the parallel cross section $\sigma_\parallel$ would be taken as the cross section for a methyl fragment, while the perpendicular cross section $\sigma_\perp$ is the sum $2\sigma(CH_3) + n\sigma(CH_2)$. The measured cross section, σ, is an average over all orientations and is given approximately by:

$$\sigma = \tfrac{1}{3}\,(\sigma_\parallel + 2\sigma_\perp) \tag{45}$$

Using this approach gives a significant improvement over the original additivity method for large hydrocarbons. This method is limited by the fact that either experimental or theoretical electron impact cross sections of the relevant molecular fragments must be known, but in cases where these quantities are available, it provides a simple method for approximating cross sections for large molecules, which may not be known, from previously existing data for smaller species.

The Polarizability Model

The observed dependence of the atomic electron impact ionization cross section on the mean square radius of the orbital of the ejected electron suggests a possible correlation between the ionization cross section and the polarizablility of the target molecule. This correlation was first noted by Franklin, Field, and Lampe,[56] who explained the correlation in terms of the dependence of both polarizability and electron impact ionization cross section on the dipole matrix. The correlation has more recently been reported for molecular species by Bartmess and Geordiadis,[57] and by Nishimura and Tawara.[58] Vallance and Harland[40] have shown that a strong correlation exists between the maximum in the ionization efficiency curve and the polarizability for a wide range of atomic and molecular species, with the proportionality constant for atomic species depending on the group of the periodic table to which the atom belongs. A stronger correlation was found between the maximum cross section and the quantity $(\alpha/E_0)^{1/2}$, where E_0 is the ionization potential. This second correlation may be explained qualitatively by straightforward electrostatic arguments.[40] The observed shape of the ionization efficiency curve may be rationalized in terms of the interaction of an incoming electron wave with the molecule. While molecular polarizability is strictly described by a second-rank tensor, it is more commonly expressed as an average polarizability α, which can be envisaged as a sphere of charge which is distorted isotropically by the approach of an electron. The polarizability volume therefore relates to the effective size of a molecule as "seen" by an approaching charge. The effective radius of this polarizability sphere is:

$$r_\alpha = (3\alpha/4\pi)^{1/3} \tag{46}$$

The root mean square (rms) radius for the electron distribution is $(3/5)^{1/2}r_\alpha$ and the corresponding rms effective diameter of the molecule is then given by:

$$d_{rms} = 2(3/5)^{1/2}(3\alpha/4\pi)^{1/3} \tag{47}$$

The only size-related quantity for the incident electron is its de Broglie wavelength,

$$\lambda = h/p = (h^2/2mqE)^{1/2} \tag{48}$$

where h is the Planck constant, p is the momentum, m the mass, q the charge, and E the kinetic energy of the incident electron. This suggests that the peak in the ionization efficiency curve may be due to a resonance condition when the incident electron wavelength matches the effective molecular diameter implicit in the polarizability volume. Applying this condition leads to a simple expression for the maximum in the ionization efficiency curve, which correctly predicts the position of the maximum for most atomic and molecular target species:

$$E_{max} = (5h^2/24mq)(4\pi/3)^{2/3}\alpha^{-2/3} \tag{49}$$

The interaction will become weaker as the electron wavelength becomes greater or less than the molecular diameter with a consequential decrease in the cross section. This leads to an expression for the ionization probability as a function of electron energy, giving the shape of the ionization efficiency curve,

$$\sigma \propto 1 - |1 - \lambda_{max}|/\lambda_{max} \tag{50}$$

where λ_{max} is the electron wavelength corresponding to E_{max} and σ_{max}. This function generally falls off too rapidly when $E > E_{max}$ for small molecules compared with the experimental results of Rapp and Englander-Golden[59] and Djuric et al.,[60] though the ionization efficiency curves of the methyl halides[61] are fairly closely matched. Another problem is that the threshold, while reasonably accurate for the methyl halides, is often overestimated for smaller molecules. In order to improve the threshold behavior, the function $(E - E_0)/E$, from the classical binary-encounter treatment of electron impact ionization[62] was incorporated into the expression, improving the predicted threshold to within 1 or 2 eV of the experimental value in most cases. This function was normalized to a peak value of unity. The resulting function is normalized to σ_{max} giving the final expression for the cross section as a function of the incident electron energy,

$$\sigma = \frac{1}{2}\sigma_{max} [A\{(E - E_0)/E\} + \{1 - |1 - \lambda_{max}|/\lambda_{max}\}] \tag{51}$$

where $A = E_{max}/(E_{max} - E_0)$ and $\lambda_{max} = (h^2/2mqE_{max})^{1/2}$

This expression reproduces the experimentally measured ionization efficiency curves surprisingly well, considering the simplicity of the model on which it is based. There is a discontinuity in the function at the maximum (when $\lambda = \lambda_{max}$) but this affects only a small region of the ionization efficiency curve, and satisfactory values of the cross section are still obtained over this region. A great advantage of this method is that it is very simple to apply, depending on only three parameters: the molecular polarizability volume, the ionization potential, and the maximum electron impact ionization cross section. These can be measured or calculated values (from the *ab initio* EM method described above, for example).

III. MEASUREMENT OF CROSS SECTIONS

The experimental determination of a total electron impact ionization cross section requires the measurement of four quantities[19]:

1. the electron current I^- of the ionizing electron beam of known energy;
2. the ion current I^+ produced by collisions of the electron beam with the target gas;
3. the number density n of the gas; and
4. the path length L over which the ions are collected.

In order to measure cross sections, a beam of electrons of known energy is directed through a gas sample of known pressure and the resulting ion and electron currents measured.[63] If mass selective ion detection is used, then partial ionization cross sections σ_z may be determined. These cross sections correspond to the production of z electrons and an ion or ions having total charge $+ze$. Some instruments allow the counting cross section σ_c, also known as the ion production cross section, to be determined:

$$\sigma_c = \sum_z \sigma_z \qquad (52)$$

In general, however, the measured quantity is the total, gross, or electron production cross section σ_t:

$$\sigma_t = \sum_z z\sigma_z \qquad (53)$$

The total cross section is determined from the quantities listed above using the relation:

$$I^+/I^- = nL\sigma_t \qquad (54)$$

Several criteria must be fulfilled by any instrument in order to obtain the ionization cross section accurately.[64] The gas pressure must be low enough to ensure that the electron beam is not scattered. This requires pressures not higher than $\sim 10^{-4}$ torr. In order to preclude space-charge effects in the electron beam the beam current must be maintained below $\sim 10^{-6}$ A. The electron current to all electrostatic elements, apart from the electron collector, should be small enough to ensure that all the electrons are collected and recorded and the electron current collected should be independent of the potential applied to the collector. The same constraints apply to the ion current. If a magnetic field is employed in order to constrain the electron beam, the ratio of ion to electron currents must be independent of the magnetic field strength. The inclusion of a magnetic field is beneficial for two reasons: (1) it constrains the electron beam to the axis which leads to higher signals; and (2) it also minimizes further ionization of the target by high-energy secondary electrons produced in the primary ionizing collisions by constraining these electrons to the axis while the primary ion is accelerated away from the collision region to the collector plate.

A. The Condenser Plate Method

The condenser plate method is the most commonly used method for electron impact ionization cross section measurements on a static gas target. One of the first designs was that of Compton and Van Voorhis.[65] The electrons were produced at a molybdenum filament in the source chamber and focused into the gas-filled scattering chamber through a metal tube which separated the two chambers. The ions produced were collected on a set of five wires in a circular arrangement around the outside of the scattering chamber, while the electrons were detected on a further set of wires at the far end of the chamber. The pressure was measured using a McLeod gauge. Attempts to constrain the electron beam using a wire wrap electromagnet around the chamber in order to produce an axial magnetic field were unsuccessful, and it was necessary to apply a large negative potential to the ion collector, leading to poor resolution of the electron energy. This problem was overcome by setting up an integral equation to represent the observations and solving it by differentiating the experimental curve.

Tate and Smith[66] improved the design of Compton and Van Voorhis, surrounding their instrument by a solenoid to provide a magnetic field which helped to define the electron beam and facilitated the separation of the electrons and ions for ion collection, without introducing a large uncertainty into the electron energy. A schematic diagram is shown in Figure 2. The electrons are emitted from the resistively heated tungsten filament F and pass through the collimating lenses 1, 2, and 3 where they are accelerated before passing into the gas-filled ionization chamber. A field of around 5 Vcm^{-1} between plates C1 and C2 accelerates ions formed towards C1 where they are collected and their current measured on a galvanometer. The same electric field on the guard plates G ensures that the electric

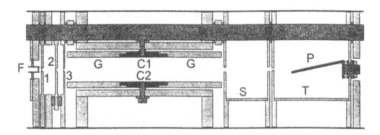

Figure 2. Total ionization source of Rapp et al[59] where: F is the filament; electron lenses are labeled 1, 2 and 3; guard plates G; ion collector C1 and field plate C2; electron collector shield S; electron collector cylinder T; and electron collector plate P.

field in the region of ion collection is homogeneous. The electrons pass through the electron collector shield S into the electron collector cylinder T and collected on the plate P with the electron current measured on a second galvanometer. P should be kept at a high positive potential with respect to T to ensure that all the electrons are collected on the plate and not lost to the chamber walls. Tate and Smith used a variation on this apparatus to measure the cross sections of helium, neon, and argon. Later they measured the ionization cross sections for mercury[67] and molecular targets[68] with much greater accuracy than any previously existing measurements. This was achieved because of a better defined electron beam path length and more accurate knowledge of the electron energy.

Over 30 years later Rapp and Golden[69] made a series of measurements of atomic and molecular electron impact ionization cross sections using an essentially unchanged Tate and Smith type design. They were able to make two kinds of measurements with their ion source. Total ionization cross sections could be measured by maintaining an electric field between C1 and C2 sufficient to collect all the positive ions formed in the fixed path length on the ion collector. The device could also be used to study dissociative ionization by reversing the field so that ions were collected on the positive plate. Because the ions had to move against a retarding field to reach the plate, only dissociated ions, which had gained kinetic energy during the dissociation process, could reach the detector. The electron impact ion source designed by Tate and Smith has a great deal of versatility. It has been used in many subsequent experiments and the basic design is still used today, with recent measurements in good accord with the original values determined by Tate and Smith in their work carried out in the 1930s.

B. The Lozier Tube

The Lozier tube, illustrated in Figure 3, has been used by several groups for a variety of different studies, including the determination of appearance energies and kinetic energies of ion fragments produced in electron impact-induced dissociation

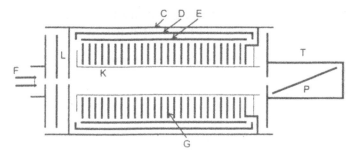

Figure 3. Total ionization source of Tate and Lozier[70] where: F is the filament; L the electron lens; K a cylindrical gauze defining an equipotential electron path; G discs; E ion collection cylinder; D electrometer guard cylinder; C shield; T electron collection cylinder; and P electron collector.

processes[62,63] as well as for total ionization cross section measurements. Electrons are formed at the filament and accelerated through a series of three lenses into the ionization chamber. A magnetic field supplied by an external solenoid constrains the electrons to move along the axis, and they are detected on plate P in the same way as described above for the Tate and Smith source. Inside the ionization chamber the electron beam is surrounded by a cylindrical gauze K. Around the gauze is a direction defining system consisting of a series of thin cylindrical disks G supported on three slotted bars. This in turn is enclosed by the ion collecting cylinder E, the electrometer guard cylinder D, and the shield C. Ions formed in collisions of the target gas with the electron beam pass through the gauze K and a small proportion, whose direction of motion is almost perpendicular to the electron beam, pass through the defining system G towards the ion collector E. In order to measure the kinetic energy of the ions a radial electric field is applied between G and E. The gauze K is present so that an electric field may be applied between G and K to prevent negative ionic products of dissociation processes from reaching the detector.

Tate and Lozier[70] first used the device to determine the kinetic energy distribution of the ions produced at a specified electron energy, the minimum electron energy required to produce ions of a specified energy, and the relative efficiencies of production of ions of a specified energy as a function of incident electron energy. The latter is a measurement of the relative electron impact ionization cross sections. Absolute cross sections cannot be measured using the Lozier tube because the collection efficiency for the ions is less than 100%, and is unknown in practice. Normalization of the data using an accurately determined absolute cross section is required in order to obtain the true cross section. Even after normalization the anisotropy of the angular distribution of the dissociated products leads to significant errors in the determination of total electron impact ionization cross sections using this method. Tozer[71] noted that for polyatomic molecules it is likely that different dissociation products will have different collection efficiencies, so that the method of normalization employed may not be particularly reliable in these cases.

C. The Summation Method

In the summation method a mass spectrometer is used to detect the ions in place of the simple Faraday plate–electrometer arrangements used in the Tate and Smith ion source and the Lozier tube. The measured quantities are partial ionization cross sections rather than gross cross sections since each type of ion fragment resulting from electron impact has a different mass-to-charge ratio and is therefore detected separately. Since the detection efficiency is not unity, only relative cross sections can be measured in this way and the results must be normalized against a known measured value of the absolute cross section in the same way as for the Lozier tube. Care must be taken to ensure that the ion collection efficiency inside the mass spectrometer is independent of the mass-to-charge ratio of the ion, and also of the electron energy in the ion source. If this is the case then the resulting partial ionization cross sections are all on the same scale and differ from the absolute values only by a common constant factor determined by the collection efficiency of the mass spectrometer. The unnormalized total cross section is determined from:

$$\sigma_t = \sum_z z\sigma_z \tag{55}$$

This is then normalized to a known value of the cross section determined by some other method, often the condenser plate method (Section III.A). If a value for the absolute cross section is not available it is often acceptable to assume that the normalization factor is the same as that for some other target molecule for which the partial cross sections have been measured and the total cross section is known. This was done by Märk and Egger,[72] who normalized the cross section for PH_3 using the normalization constant determined from measurements on argon, which were compared to the total cross section for argon measured by Rapp and Englander-Golden[73] using the condenser plate method. One of the first studies using the summation method was that of Schutten et al.[74] who measured the electron impact ionization cross section of H_2O using a special cycloidal mass spectrometer with an extremely high ion extraction efficiency in order to avoid mass discrimination effects. They normalized their data against measurements made in their own laboratory using the condenser plate method. The same study was repeated by Märk and Egger[75] with a few improvements to the method, giving results in fairly good agreement with Schutten et al. The summation method has since been used in the measurement of electron impact ionization cross sections for a wide range of atomic and molecular targets.

Since the summation method allows absolute partial ionization cross sections to be determined, it is straightforward to extract the counting or ion production cross section from the data. Since this is the quantity which is given by many of the current theories the summation method has a distinct advantage over other experimental methods when comparison with theoretical results is required.

D. Other Experimental Methods

There are several other methods which have been used in the experimental determination of electron impact ionization cross sections. Nottingham and Bell[76,77] developed a method specifically for the purpose of accurately determining the absolute electron impact ionization cross section of mercury. A semicircular electron velocity analyzer included in their design ensured that very high energy resolution was possible since only electrons of the required velocity emerged from the analyzer into the ionization chamber. Other aspects of the experiment are similar to the condenser plate method.

Another type of measurement uses gas-filled counters.[19] This method has mainly been used to determine electron impact ionization cross sections at high energies of greater than 100 eV and is therefore well suited for making measurements for comparison with the predictions of the Bethe theory.[37] The method was suggested by Graf[78] and measurements have been carried out by McClure[79] and more recently by Rieke and Prerejchal.[80] The electrons are produced from a beta emitter and velocity selected in a magnetic analyzer. The analysis is based on the use of Poisson statistics applied to the collision processes. The probability q that an electron passes through the ionization chamber without ionizing a target molecule is measured and can then be used to determine the cross section through the relationship,

$$-\ln \theta = nL\sigma \tag{56}$$

where n is the number density of the target gas and L is the path length through the chamber.

The method of crossed beams, originally introduced to enable cross section measurements to be carried out on species that are unstable at ordinary temperatures and pressures,[81] was first used by Boyd and Green[82] and Fite and Brackman.[83] There are both advantages and disadvantages to the method. The lower number densities and smaller collision region compared to the other methods translates into smaller signal levels. It is nearly always necessary to modulate the signal in order to separate signal from background. A general crossed beam experiment is illustrated in Figure 4. A fast neutral particle beam crosses an electron beam at right angles with detection of the electron beam current, the neutral beam intensity, and the mass-selected ion beam current. Because of the directional nature of the neutral beam and the negligible momentum transfer in collisions between the electrons and the neutral species, the positive ions produced by electron impact ionization will have essentially the same direction as the neutral beam. This facilitates separation of the ions from the neutral beam and focuses them into the mass analyzer for detection.

Because the target can no longer be treated as stationary with respect to the electron beam, the cross section is no longer given by the simple relationship,

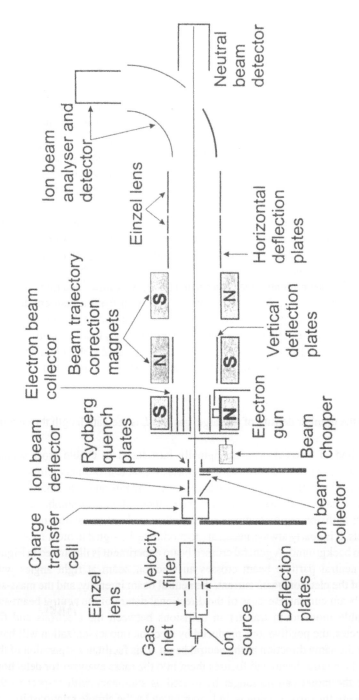

Figure 4. A cross beam machine for the measurement of electron impact ionization cross sections.

$$I^+/I^- = nL\sigma_t \tag{57}$$

but by the expression,[84]

$$I^+/I^- = \sigma(E) \, R \, [(v_e^2 + v_t^2)^{1/2}/v_t v_e] \, F \tag{58}$$

where v_e and v_t are the electron and target beam velocities (assumed to be perpendicular), R is the number of target molecules per second arriving at the target detector, E is the relative energy of the target and electron, and F is a measure of the beam overlap given by,

$$F = \frac{\int j_n(r) j_e(r) \, dr}{\int j_n(r) \, dr \int j_e(r) \, dr} \tag{59}$$

where $j_n(r)$ and $j_e(r)$ are the neutral and electron beam profiles as a function of the distance r from the center of the beams. Because of the difficulty in determining R experimentally, cross sections determined by crossed beam experiments often need to be normalized against a known value (c.f., Lozier tube and summation methods).

The stability of species in a molecular beam makes it possible to measure cross sections which would be impossible to determine by any other method. The development of oriented beam techniques has also allowed an orientation effect on the ionization cross section to be demonstrated for the first time[11] for a series of symmetric top molecules oriented in a combination of inhomogeneous and homogeneous electrostatic fields. Crossed beam measurements also allow the possibility for investigating the dynamics of the ionization process in a way not accessible by other methods since the initial velocities of the collision partners are well-defined and products may be detected as a function of scattering angle if a movable detector is used, and also of velocity using time-of-flight methods.

IV. EXPERIMENT AND THEORY COMPARED

A. The Ionization Cross Section and its Energy Dependence, $\sigma(E)$

Experimentally determined maximum absolute ionization cross sections for the inert gases and a range of small molecules are compared with the predictions of DM, BEB, and EM calculations in Table 1. Atomic orbital coefficients for the DM calculations were determined at the Hartree–Fock level and the EM cross sections are volume averaged for calculations carried out at the HF/6-31G* level. The same data are plotted in Figure 5 with the calculated values on the ordinate and the experimental result on the abscissa. The heavy line represents a direct correspondence between experiment and theory. Although the ab initio EM method performs well for the calculation of σ_{max} and E_{max},[17] the DM and BEB methods allow for the calculation of the cross section as a function of the electron energy, i.e. the ionization efficiency curve. In order to calculate an ionization efficiency curve using the EM method, the value for σ_{max} must be used with the polarizability equation described above. Figure 6 compares experimental and calculated ionization efficiency curves for several of the molecules listed in Table 1.

A critical comparison between experiment and theory is hindered by the range of experimental values reported in the literature for each molecule. This reflects the difficulty in the measurement of absolute ionization cross sections and justifies attempts to develop reliable semiempirical methods, such as the polarizability equation, for estimating the molecular ionization cross sections which have not been measured or for which only single values have been reported. The polarizability model predicts a linear relationship between the ionization cross section and the square root of the ratio of the volume polarizability to the ionization potential. Plots of this function against experimental values for ionization cross sections for atoms are shown in Figure 7 and for molecules in Figure 8. The equations determined

Table 1. Measured and Calculated Maximum Total Ionization Cross Sections in $\mathring{A}^2$

System	σ_{max} (expt)	DM (calc)	BEB (calc)	EM (calc)
He	$0.37^{59,85,86}$, 0.38^{84}, 0.39^{61}	0.37	0.36	0.37
Ne	0.72^{85}, 0.74^{84}, 0.78^{59}	0.69	0.96	1.00
Ar	2.54^{85}, $2.70^{84,87}$, 2.81^{61}, 2.86^{59}	2.72	1.99	2.53
Kr	3.50^{61}, 3.70^{84}, 3.72^{85}, 4.26^{59}	3.96	2.32	3.42
Xe	4.59^{88}, 4.98^{84}, 5.10^{61}, 5.46^{59}	5.05	b	4.13
N_2	2.54^{59}	2.90	2.52	2.82
NO	3.15^{59}	2.64	2.54	3.62
CO	$2.05-2.66^{59,89-91,c,d}$	3.31	2.53	3.13
CO_2	2.05^{89}, 3.27^{90}, $3.55^{59,e}$	4.51	3.57	4.00
NO_2	—	4.02	3.69	4.23
H_2O	2.05^{60}, $4.40^{91,f}$	2.38	2.25	2.09
NH_3	$2.4-3.01^{92-95,c}$	3.31	2.96	3.47
CH_4	3.70^{59}, 4.24^{61}	5.26	3.30	3.90
CH_3F	3.72^{61}	4.99	3.64	4.13
CH_3Cl	6.91^{61}	7.53	5.06	5.57
CH_3Br	8.02^{61}	8.51	5.42	7.71
CH_3I	10.3^{61}	a	a	—
$CHCl_3$	12.25^{61}	b	b	11.31

Notes: [a]Basis set not available for I.

[b]Time consuming and expensive calculation.

[c]Where a range of literature values is available, only the extreme values are shown in the table.

[d]For CO, it is likely that the true cross section lies close to the maximum value in the range. The dipole moment of CO is small and the polarizability of the CO molecule is slightly higher than that for isoelectronic N_2.

[e]The value of 2.05 $\mathring{A}^2$ for CO_2 and the lower end of the range for NH_3 are not consistent with the values for other molecules of similar physical properties.

[f] The value of 4.40 $\mathring{A}^2$ for H_2O is unlikely, the cross section for O_2 is less than 3 $\mathring{A}^2$ and H_2O would be expected to exhibit a lower cross section than O_2.

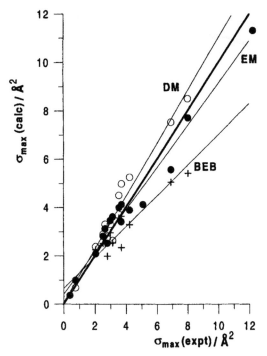

Figure 5. A comparison between measured maximum ionization cross sections and the predictions of BEB (+), DM (o), and EM (●) calculations. The heavy line represents a perfect correspondence between experiment and theory.

from the plots by linear least-square fits are given in Table 2. Figure 9 shows the predicted values of E_{max} from the BEB, DM, and polarizability model plotted against the measured values. The BEB model performs better for small molecules and the DM method performs better for larger molecules with heavy atoms. The semiempirical polarizability model, for which the wavelength of the electron corresponding to E_{max} is equated to the root mean square radius for the electron distribution, performs well for all molecules.

B. The Electron–Molecule Orientation Dependence, $\sigma(\theta)$

In general, measurements of electron impact ionization cross sections are carried out on randomly oriented molecules in the gas phase, so that the total ionization cross sections in Table 1 and Figure 5 are the average over all possible relative orientations, θ, of the molecule with respect to the electron. In the EM approach, the input for the Gaussian[34] calculations requires that the charge distribution be entered in Cartesian coordinates, leading to the critical separation and an approximate total cross section for approach along each of the positive and negative

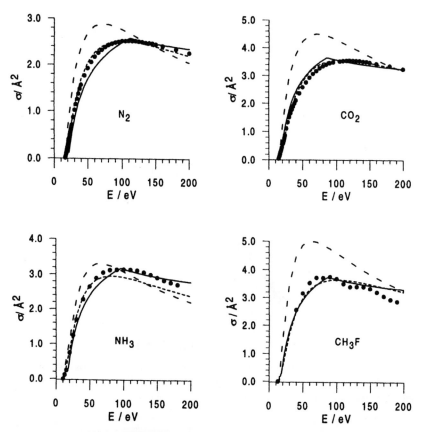

Figure 6. Comparison of electron impact ionization cross sections calculated by the BEB (---), DM (- - -), and the method described here (———) with experimental data (●) for N_2, CO_2, NH_3, and CH_3F.

Table 2. Relationships for the Calculation of Maximum Total Ionization Cross Sections Based on Literature Experimental Data

σ_{max} as a Function of $(\alpha/E_0)^{1/2}$	Linear Fit
Main group elements	$\sigma = 9.635(\alpha/E_0)^{1/2} - 0.718$
Group I metals	$\sigma = 2.637(\alpha/E_0)^{1/2} - 0.166$
Group II metals	$\sigma = 4.505(\alpha/E_0)^{1/2} - 0.591$
Transition metals	$\sigma = 8.041(\alpha/E_0)^{1/2} - 2.960$
Molecules	$\sigma = 17.973(\alpha/E_0)^{1/2} - 4.135$

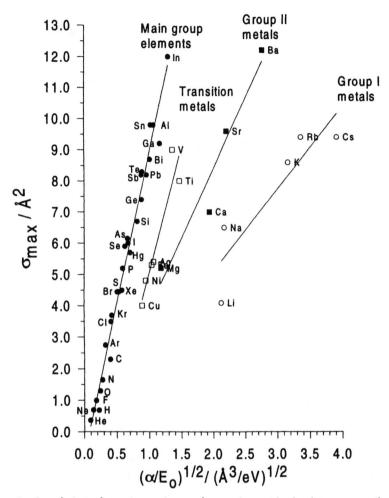

Figure 7. Correlation of atomic maximum electron impact ionization cross sections with $(\alpha/E_0)^{1/2}$, where E_0 is the atomic ionization potential.

Cartesian axes. This Cartesian averaged cross section is in good accord with experimental measurements for molecules such as diatomics, nonlinear triatomics, NH_3, CH_4, and substituted methanes, which can all be regarded as roughly spherical.[18] For rod-shaped molecules, such as CO_2, the cross section averaged in this way is overestimated as too much weight is afforded the larger cross section components for end-on approach. If the critical separation is determined for a large number of relative geometries of the electron and molecule, it is possible to obtain a three-dimensional picture of the probability of ionization as a function of the orientation of the molecule. Effectively, the idea of an ionization cross section, the

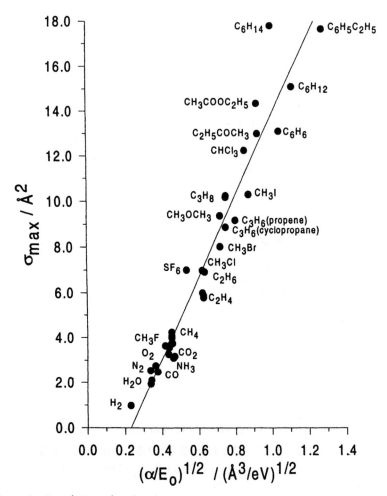

Figure 8. Correlation of molecular maximum electron impact ionization cross sections with $(\alpha/E_0)^{1/2}$, where E_0 is the molecular ionization potential.

area the target molecule presents to the electron, is extended to a three-dimensional object defined by the critical distances, with ionization occurring when the electron penetrates the surface enclosing this volume.

Ionization surfaces calculated for several of the molecules listed in Table 1 are shown in Figures 10 and 11. The volume averaged cross sections determined from the volumes enclosed by these surfaces for nonspherical molecules such as CO_2 are in much better accord with experiment, consistent with the idea that the poor performance of Cartesian averaging for molecules such as CO_2 is due to the large departure from a spherical shape. Improved agreement with experiment is also

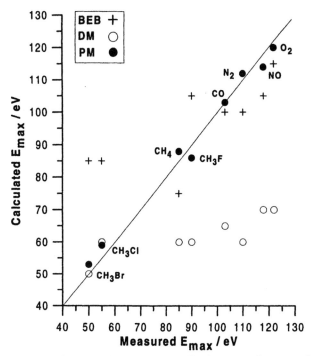

Figure 9. Comparison between the experimentally determined E_{max} and the predictions of the BEB(+), DM(O), and PM(•) models. The line represents a perfect correspondence between experiment and theory.

obtained for NH_3, and for H_2O if the lower experimental value is correct, which seems likely since the cross section of O_2 is less than 3 $Å^2$ and H_2O would be expected to have a smaller cross section than O_2. In most cases the difference between the Cartesian averaged and volume averaged cross sections is small, indicating that averaging the values obtained for approach along each of the positive and negative Cartesian axes does give a reasonable estimate for the cross section.

Recent experiments[11] have shown that for symmetric top molecules, such as methyl chloride, electron impact ionization is more probable at the positive end of the dipole. This follows intuitively from simple electrostatics, since it would be expected that an electron should be more strongly attracted to the positive end of the dipole, leading to a greater ionization probability at this end of the molecule. The effect may be quantified in terms of a steric ratio R.

$$R = \frac{\sigma_+}{\sigma_-} \tag{60}$$

where σ_+ and σ_- are the cross sections for ionization at the positive and negative ends of the dipole, respectively. The experimentally determined steric ratio for the

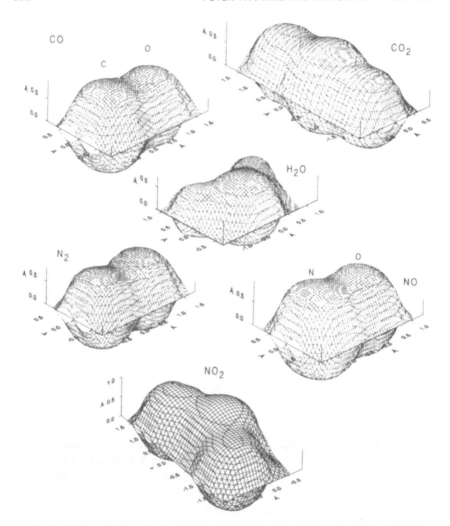

Figure 10. Ionization surfaces for CO, CO_2, H_2O, N_2, NO, and NO_2. Scales are critical separations in Å units calculated from the center-of-mass of the molecule to the projectile electron.

total ionization of CH_3Cl was reported to be 1.6 at 200 eV and the ratio for molecular ion formation was found to be 2.6 at 200 eV.

Using the ab initio EM method, the steric ratio is taken as the calculated cross section for electron approach along the molecular axis towards the positive end of the dipole divided by that for approach towards the negative end of the dipole. The model predicts a steric ratio of 1.41 for the total ionization of CH_3Cl; that is, both the sign and magnitude are in accord with the experimental measurements. The calculated 'volume of ionization' for methyl chloride, as viewed from each end of the molecule, is shown in Figure 11.

Orientation effects can also be investigated using the DM model since the cross section is calculated from a summation over contributions from all the atomic orbitals in a molecule. We assume that an electron incident on the negative end of the molecular dipole results in electron loss from an atomic orbital on the halogen atom, while one approaching the positive end effects electron loss from a carbon or hydrogen orbital. A cross section may be calculated for the chlorine end by summing over Cl atomic orbitals only, and similarly for the methyl end by summing over only C and H orbitals. The ratio of these two cross sections gives a steric ratio for the ionization probabilities at each end of the dipole. This approach gives the

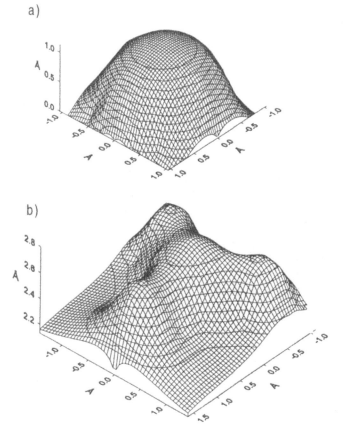

Figure 11. Ionization surfaces for CH₃Cl: **(a)** for an electron approaching the Cl-end of the molecule; **(b)** for an electron approaching the CH₃-end of the molecule.

orientation dependence of the total cross section, which includes contributions from fragmentation processes. The major fragmentation product in the electron impact ionization of CH_3Cl is the CH_3^+ ion. The experimental observations[11] were consistent with the formation of the molecular ion, CH_3Cl^+, through end-on collisions between the electron and the molecule (electron approach along the dipole axis) by loss of an electron from an orbital centered on the chlorine atom or on the methyl group. Formation of the methyl ion appears to follow broadside collisions with the molecule (electron approach perpendicular to the dipole axis), that is, the loss of an electron from the C–Cl bond. The simple scheme of dividing the molecule into a chlorine atom and a methyl group to calculate a steric ratio for ionization incorporates broadside collisions, and so allows for fragmentation processes to occur. The steric ratio calculated by this method for CH_3Cl is 1.37 for an electron energy of 200 eV, that is, the DM model reproduces the experimentally determined preferred direction of ionization and, despite the rudimentary assumptions made, the magnitude is in reasonable agreement with the measured value of 1.6 at 200 eV.

Figure 12 shows steric ratios calculated using the DM method for CH_3Cl from threshold to 200 eV. Preliminary experimental measurements[96] of the steric ratio for CH_3Cl as a function of electron energy show a decrease of ~14% over the range

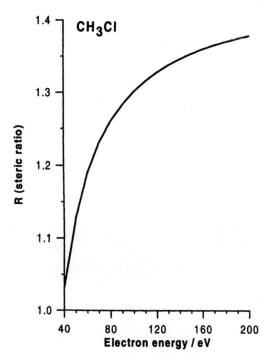

Figure 12. Steric ratio for the total ionization of CH3Cl as a function of electron energy calculated using the DM theory.

from 240 to 100 eV. The calculations support this trend, with a decrease in the steric ratio of 8% over the same range. The value of 1.41 calculated by the ab initio EM approach which corresponds to E_{max}, which is close to 60 eV, is in good accord with the ratio anticipated by an extrapolation of the experimental ratio measured over the range 240 to 100 eV. Qualitatively we can reconcile this decrease in steric ratio with decreasing electron energy in terms of the interaction of the electron wave with the molecule. The de Broglie wavelength of the electron increases from 87 pm at 200 eV to 366 pm at 11.22 eV, the ionization threshold for CH_3Cl. Using tabulated bond lengths and orbital radii, the diameter of the CH_3Cl molecule is around 340 pm. We might anticipate, then, that as the electron energy decreases and the electron wavelength approaches molecular dimensions the steric effect would diminish. Similarly, the steric effect might be expected to increase with decreasing electron wavelength, leveling off as the wavelength approaches orbital dimensions. This would correspond to electron energies around 300 eV, outside the range available in the experiments.

V. CONCLUSIONS

None of the three theories used to calculate electron impact ionization cross sections could be considered to render the others obsolete. The BEB method gives the best fit to the functional form of the ionization efficiency curve for small molecules, it provides a better fit to the experimental data closer to the ionization threshold than the other methods, but it underestimates the maximum ionization cross sections for heavier molecules. The DM method provides a better fit to the ionization efficiency curves for the heavier molecules, especially for electron energies greater than E_{max}, but it tends to overestimate the cross sections for heavier molecules and it underestimates E_{max} for lighter molecules. The EM method performs as well as the other methods for the value of σ_{max} for the light molecules but underestimates the cross sections for heavy molecules by a factor similar to the overestimation of the DM method. The polarizability method outperforms the BEB and the DM methods for the calculation of E_{max} and when combined with the σ_{max} value from the EM calculation reproduces the ionization efficiency curve as well as the BEB method.

Steric ratios for ionization at either end of dipolar molecules may be calculated using the DM and EM ab initio methods. Both methods correctly predict the correct sign for the steric ratio and they are in reasonable agreement with each other and with experiment for the magnitude of the steric ratio for CH_3Cl.

ACKNOWLEDGMENTS

The authors would like to acknowledge financial support from the Foundation for Research, Science and Technology (NZ), the Marsden Fund and the University of Canterbury. We should also like to thank Sean A. Harris for the preparation of Figures 2 and 4.

REFERENCES

1. Dempster, A. J. *Phys. Rev.* **1918**, *11*, 316.
2. Thomson, J. J. *Philos Mag.* **1912**, *23*, 449; Thomson, J. J. *Rays of Positive Electricity and Their Application to Chemical Analyses;* Longmans, Green: London, 1913.
3. Harland, P. W.; McIntosh, B. J. *Int. J. Mass Spectrom. Ion Procs.* **1985**, *67*, 29; Harland, P. W. *Int. J. Mass Spectrom. Ion Procs.* **1986**, *70*, 231; Harland, P. W.; Maclagan, R. G. A. R. *J. Chem. Soc. Faraday Trans.* **1987**, *83*, 2133.
4. McEwan, M. J. *Advances in Gas Phase Ion Chemistry*; Adams, N. G.; Babcock, L. M.; Eds.; JAI Press: Greenwich, CT, London., 1995, Vol. 1.
5. (a) Franklin, J. L.; Dillard, J. G.; Rosenstock, H. M.; Heron, J. T.; Draxl, K.; Field, F. H. *Nat. Stand. Ref. Data Ser., NBS, US*, 1969. (b) Levin, R. D.; Lias, S. G. *Nat. Stand. Ref. Data Ser., NBS, US*, 1982. (c) Lias, S. G.; Bartmess, J. E.; Liebman, J. F.; Holmes, J. L.; Levin, R. D.; Mallard, W. G. *Nat. Stand. Ref. Data Ser., NBS, US*, 1988.
6. Schram, B. L.; De Heer, F. J.; van der Wiel, M. J.; Kistemaker, J. *Physica* **1965**, *31*, 94.
7. Srivastava, S. K.; Krishnakumar, E. *Int. J. Mass Spectrom. Ion Procs.* **1992**, *113*, 1.
8. Märk, T. D.; Egger, F.; Cheret, M. *J. Chem. Phys.* **1977**, *67*, 3795.
9. Djuric, N.; Belic, D.; Kurepa, M. V.; Mack, J. U.; Rothleitner, J.; Märk, T. D. Proc. 12th ICPEAC, Gatlinburg, 1981, p 384.
10. Syage, J. A. *J. Chem. Phys.* **1992**, *97*, 6085.
11. Aitken, C. G.; Blunt, D. A.; Harland, P. W. *J. Chem. Phys.* **1994**, *101*, 11074; *Int. J. Mass Spectrom. Ion Procs.* **1995**, *149/150*, 279.
12. Deutsch, H.; Märk, T. D. *Int. J. Mass Spectrom. Ion Procs.* **1987**, *79*, R1.
13. Margreiter, D.; Deutsch, H.; Schmidt, M.; Märk, T. D. *Int. J. Mass Spectrom. Ion Procs.* **1990**, *100*, 157.
14. Margreiter, D.; Deutsch, H.; Schmidt, M.; Märk, T. D. *Int. J. Mass Spectrom. Ion Procs.* **1994**, *139*, 127.
15. Kim, Y. K.; Rudd, M. E. *Phys. Rev. A* **1994**, *50*, 3954.
16. Hwang, W.; Kim, Y. K.; Rudd, M. E. *J. Chem. Phys.* **1996**, *104*, 2956.
17. Vallance, C.; Harland, P. W.; Maclagan, R. G. A. R. *J. Phys. Chem.* **1996**, *100*, 15021.
18. Vallance, C.; Maclagan, R. G. A. R.; Harland, P. W. *J. Phys. Chem. A*, **1997**, *101*, 3505.
19. Märk, T. D.; Dunn, G. H. (Eds.). *Electron Impact Ionization*; Springer-Verlag: Wien, New York, 1985.
20. McCarthy, I. E.; Sobovics, A. J. In *Theoretical Methods in Controlled Thermonuclear Fusion*; McDowell, M. R. C.; Ferendeci, A. M., Eds.; Plenum Press: New York, 1980.
21. Fadeev, L. D. *Sov. Phys. J.E.T.P.* **1961**, *12*, 1014.
22. Peach, G. *Proc. Phys. Soc.* **1966**, *87*, 375, 381; *J. Phys. B* **1968**, *2*, 1088; *ibid* **1970**, *3*, 328; *ibid* **1971**, *4*, 1670.
23. McGuire, E. J. *Phys. Rev. A* **1971**, *3*, 267; *ibid* **1977**, *16*, 62, 72; *ibid* **1979**, *20*, 445.
24. Moores, D. L. *J. Phys. B* **1972**, *5*, 286; *ibid* **1978**, *11*, L403.
25. Moores, D. L.; Nussbaumer, H. *J. Phys. B* **1970**, *3*, 161.
26. Blaha, M.; Davis, J. *Electron Ionization Cross Sections in the Distorted Wave Approximation*. N.R.L. Memorandum Report 4245, Naval Research Laboratory, Washington, DC, 1980.
27. Trefftz, E. *Proc. Roy. Soc. London* **1962**, *A271*, 379.
28. Stingl, E. *J. Phys. B* **1972**, *5*, 1160.
29. Younger, S. M. *Phys. Rev. A* **1980**, *22*, 111, 1425; *ibid* **1981**, *23*, 1138; *ibid* **1981**, *24*, 1272, 1278; *ibid* **1982**, *25*, 3396; *ibid* **1982**, *26*, 3177.
30. Peterkop, R. K. *Theory of Ionization of Atoms by Electron Impact*, Boulder, Colorado, Colorado Associated University Press: Boulder, CO, 1977.
31. Rudge, M. R. H.; Schwartz, S. B. *Proc. Phys. Soc.(London)* **1966**, *88*, 563.

32. Butler, K.; Moores, D. L. *XII Intl. Conf. on Phys. Electronic and Atomic Collisions*; Datz, S., Eds.; 1981.
33. Laggatuta, K. J.; Hahn, Y. *Phys. Rev. A* **1981**, *24*, 2273.
34. Frisch, M. J.; Trucks, G. W.; Schlegel, H. B.; Gill, P. M. W.; Johnson, B. G.; Robb, M. A.; Cheeseman, J. R.; Keith, T. A.; Petersson, G. A.; Montgomery, J. A.; Raghavachari, K.; Al-Laham, M. A.; Zakrzewski, V. G.; Ortiz, J. V.; Foresman, J. B.; Cioslowski, J.; Stefanov, B. B.; Nanayakkara, A.; Challacombe, M.; Peng, C. Y.; Ayala, P. Y.; Chen, W.; Wong, M. W.; Andres, J. L.; Replogle, E. S.; Gomberts, R.; Martin, R. L.; Fox, D. J.; Binkley, J. S.; Defrees, D. J.; Baker, J.; Stewart, J. P.; Head-Gordon, M.; Gonzales, C.; Pople, J. A. GAUSSIAN 94 (Revision A.1), Gaussian Inc., Pittsburgh, PA, 1995.
35. Rudge, M. R. H. *Revs. Mod. Phys.* **1968**, *40*, 564.
36. Tripathi, D. N.; Rai, D. K. *Ind. J. Pure Appl. Phys.* **1972**, *10*, 185.
37. Bethe, H. *Ann. Physik* **1930**, *5*, 325.; *Z. Physik* **1932**, *76*, 293.
38. Jain, D. K.; Khare, S. P. *Proc. 9th ICPEAC*, Seattle, 1975, p 484; *J. Phys. B.* **1976**, *9*, 1429.
39. Brooks, P. R.; Harland, P. W. *Advances in Gas Phase Ion Chemistry*; Adams, N. G.; Babcock, L. M., Eds.; JAI Press: Greenwich, CT, London, 1996, Vol. 2.
40. Vallance, C.; Harland, P. W. *Int. J. Mass Spectrom. Ion Procs.* in press, 1997.
41. Grysinski, M. *Phys. Rev. A* **1965**, *138*, 305.
42. Vriens, L. *Phys. Rev.* **1966**, *141*, 88.
43. Fraga, S.; Karwovski, J.; Saxena, K. M. S. *Handbook of Atomic Data*; Elsevier: Amsterdam, 1976.
44. Margreiter, D.; Deutsch, H.; Märk, T. D. *Contrib. Plasma Phys.* **1990**, *30*, 487.
45. Otvos, J. W.; Stevenson, D. P. *J. Am. Chem. Soc.* **1956**, *78*, 546.
46. For example: Szabo, A.; Ostlund, N. S. *Modern Quantum Chemistry—Introduction to Advanced Electronic Structure Theory*; Macmillan Publishing: New York, 1982, p 151.
47. Mott, N. F. *Proc. R. Soc. London Ser. A* **1930**, *126*, 259.
48. Vriens, L. In *Case Studies in Atomic Physics*; McDaniel, E. W.; McDowell, M. R. C., Eds.; North-Holland: Amsterdam, 1969, Volume 1, p 335.
49. Dalgarno, A., Eds.; In *Atomic and Molecular Processes*; Bates, R. D., Ed.; Academic Press: London, 1962, p 124.
50. Miller, W. F.; Platzman, R. L. *Proc. Roy. Soc. (London)* **1957**, *A70*, 299.
51. Khare, S. P. *Planet. Space Sci.* **1969**, *17*, 1257.
52. Opal, C. B.; Beaty, E. C.; Peterson, W. K. *Atomic Data* **1972**, *4*, 209.
53. Lotz, W. *Z. Physik.* **1967**, *206*, 205; *ibid* **1968**, *216*, 241, *ibid* **1970**, *232*, 101.
54. Bell, K. L.; Gilbody, H. B.; Hughes, J. G.; Kingston, A. E.; Smith, F. J. *J. Phys. Chem. Ref. Data.* **1983**, *12*, 891.
55. Bobeldijk, M.; Van der Zande, W. J.; Kistemaker, P. G. *Chem. Phys.* **1994** *179*, 125.
56. Lampe, F. W.; Franklin, J. L.; Field, F. H. *J. Am. Chem. Soc.* **1957**, *79*, 6129.
57. Bartmess, J. E.; Geordiadis, J. E. *Vacuum* **1983**, *33*, 149.
58. Nishimura, H.; Tawara, H. *J. Phys. B* **1994**, *27*, 2063.
59. Rapp, D.; Englander-Golden, P. *J. Chem. Phys.* **1965**, *43*, 1464.
60. Djuric, N. L.; Kadez, I. M.; Kurepa, M. V. *Int. J. Mass Spectrom. Ion Procs.* **1988**, *83*, R7.
61. Vallance, C.; Harris, S. A.; Hudson, J. E.; Harland, P. W. *J. Phys. B*, **1997**, *30*, 2465.
62. Massey, H. S. W.; Burhop, E. H. S. *Electronic and Ionic Impact Phenomena*, 2nd edn.; Oxford: Clarendon, 1969, p 187.
63. Field, F. H.; Franklin, J. L. *Electron Impact Phenomena and the Properties of Gaseous Ions*; Academic Press: New York, 1957.
64. Briglia, D. D.; Rapp, D. *J. Chem. Phys.* **1965**, *42*, 3201.
65. Compton, K. T.; Van Voorhis, C. C. *Phys. Rev.* **1925**, *26*, 436.
66. Tate, J. T.; Smith, P. T. *Phys. Rev.* **1930**, *36*, 1293.
67. Smith, P. T. *Phys. Rev.* **1931**, *37*, 808.
68. Tate, J. T.; Smith, P. T. *Phys. Rev.* **1932**, *39*, 270.

69. Rapp, D.; Englander-Golden, P.; Briglia, D. D. *J. Chem. Phys.* **1965**, *42*, 4081.
70. Tate, J. T.; Lozier, W. W. *Phys. Rev.* **1930**, *35*, 1285.
71. Tozer, B. A. *J. Electron. Control* **1958**, *4*, 149.
72. Märk, T. D.; Egger, F. *J. Chem. Phys.* **1977**, *67*, 2629.
73. Rapp, D.; Englander-Golden, P. *J. Chem. Phys.* **1965**, *43*, 1464.
74. Schutten, J.; de Heer, F. J.; Moustafa, H. R.; Boerboom, A. J. H.; Kistemaker, J. *J. Chem. Phys.* **1966**, *44*, 3924.
75. Märk, T. D.; Egger, F. *Int. J. Mass Spectrom. Ion. Phys.* **1976**, *20*, 89.
76. Bell, M. E. *Phys. Rev.* **1939**, *55*, 201.
77. Nottingham, W. B. *Phys. Rev.* **1939**, *55*, 203.
78. Graf, T. *J. Phys. Radium.* **1939**, *10*, 513.
79. McClure, G. W. *Phys. Rev.* **1953**, *90*, 796.
80. Rieke, F. F.; Prepejchal, W. *Phys. Rev.* **1972**, *A6*, 1507.
81. Kieffer, J. K.; Dunn, G. H. *Rev. Mod. Phys.* **1966**, *38*, 1.
82. Boyd, R. L. F.; Green, G. W. *Proc. Phys. Soc. (London)* **1958**, *71*, 351.
83. Fite, W. L. F.; Brackman, R. T. *Phys. Rev.* **1948**, *112*, 1141.
84. Wetzel, R. C.; Baiocchi, F. A.; Hayes, T. R.; Freund, R. S. *Phys. Rev.* **1987**, *A35*, 559.
85. Stephan, K.; Helm, H.; Märk, T. D. *J. Chem. Phys.* **1980**, *73*, 3763.
86. Montague, R. G.; Harrison, M. F. A.; Smith, A. C. H. *J. Phys. B* **1984**, *17*, 3295.
87. Straub, H. C.; Renault, B. G.; Lindsay, B. G.; Smith, K. A.; Stebbings, R. F. *Phys. Rev. A* **1995**, *52*, 1115, and references therein.
88. Stephan, K.; Märk, T. D. *J. Chem. Phys.* **1984**, *81*, 3116.
89. Freund, R. S.; Wetzel, R. C. *Phys. Rev. A* **1990**, *41*, 5861.
90. Hille, E.; Märk, T. D. *J. Chem. Phys.* **1978**, *69*, 4600.
91. Orient, O. J.; Srivastava, S. K. *J. Phys. B* **1987**, *20*, 3923.
92. Srivastava, S. K.; Rao, M. V. V. S. *J. Geophys. Res.* **1991**, *96* (E2), 17563.
93. Khare, S. P.; Meath, W. J. *J. Phys. B* **1987**, *20*, 2101.
94. Bedereski, K.; Wójcik, L.; Adamczyk, B. *Int. J. Mass Spectrom. Ion Phys.* **1980**, *35*, 171.
95. Crowe, A.; McConkey, J. W. *Int. J. Mass Spectrom. Ion Phys.* **1977**, *24*, 181.

INDEX

Advances in Gas-Phase Ion Chemistry

Edited by **Nigel G. Adams**, and
Lucia M. Babcock, *Department of
Chemistry, University of Georgia*

Volume 2, 1996, 266 pp. $109.50/£69.50
ISBN 1-55938-703-3

REVIEW: "Aside from providing state-of-the-art accounts of
forefront areas of research in ion chemistry written by experts
in the field, this volume also affords motivation for further
studies in new research directions. The latter is most stimu-
lating and adds to the already excellent value this volume has
for researchers and graduate students working in gas phase
ion chemistry and related fields."

— *Journal of American Chemical Society*

CONTENTS: Preface. Effect of Molecular Orientation on
Electron Transfer and Electron Impact Ionization, *Phil R.
Brooks and Peter W. Harland.* Experimental Approaches to
the Unimolecular Dissociation of Gaseous Cluster Ions, *Terry
B. McMahon.* New Approaches to Ion Thermochemistry via
Dissociation and Association, *Robert C. Dunbar.* Alkyl Cation-
Dihydrogen Complexes, Silonium and Germonium Cations:
Theoretical Considerations, *Peter R. Schreiner, H. Fritz and
Paul v. R. Schleyer.* Symmetry Induced Kinetic Isotope Effects
in Ion-Molecule Reactions, *Greg I. Gellene.* Ion-Molecule
Chemistry: The Roles of Intrinsic Structure, Solvation and
Counterions, *John E. Bartmess.* Gas Phase Ion Chemistry un-
der Conditions of Very High Pressure, *W. Burk Knighton and
Eric P. Grimsrud.* Index.

Also Available:
Volume 1 (1992) $109.50/£69.50

FACULTY/PROFESSIONAL discounts are available in
the U.S. and Canada at a rate of 40% off the list price
when prepaid by personal check or credit card and
ordered directly from the publisher.

JAI PRESS INC.
55 Old Post Road No. 2 - P.O. Box 1678
Greenwich, Connecticut 06836-1678
Tel: (203) 661- 7602   Fax: (203) 661-0792

Printed and bound by CPI Group (UK) Ltd, Croydon, CR0 4YY

08/05/2025

01864830-0004